CHARGED AND REACTIVE POLYMERS

VOLUME 4

CHARGED GELS AND MEMBRANES – PART II

CHARGED AND REACTIVE POLYMERS

A SERIES EDITED BY ERIC SÉLÉGNY

Vol. 1: POLYELECTROLYTES. *Papers initiated by a NATO Advanced Study Institute on 'Charged and Reactive Polymers', held in France, June 1972.* Edited by Eric Sélégny and co-edited by Michel Mandel and Ulrich P. Strauss

2: POLYELECTROLYTES AND THEIR APPLICATIONS Edited by Alan Rembaum and Eric Sélégny

3: CHARGED GELS AND MEMBRANES – Part I Edited by Eric Sélégny and co-edited by George Boyd and Harry P. Gregor

VOLUME 4

CHARGED GELS AND MEMBRANES

PART II

Edited by

ERIC SÉLÉGNY

Université de Rouen, France

D. REIDEL PUBLISHING COMPANY

DORDRECHT-HOLLAND / BOSTON-U.S.A.

Library of Congress Cataloging in Publication Data

Nato Advanced Study Institute on Charged and Reactive Polymers, 2d,
    Forges-les-Eaux, France, 1973.
    Charged Gels and Membranes.

    (Charged and reactive polymers; v. 3–4)
    Includes bibliographical references and indexes.
        1.   Polymers and polymerization—Electric properties—Congresses.
2.   Membranes (Technology)—Congresses. 3.   Membranes
(Biology)—Congresses. I.   Sélégny, Eric. II.   Title. III.
Series.
QD381.9.E38N2   1973   668   76–6086
ISBN–13:978–94–010–1469–4   e–ISBN–13:978–94–010–1467–0
DOI: 10.1007/978–94–010–1467–0

Published by D. Reidel Publishing Company,
P.O. Box 17, Dordrecht, Holland

Sold and distributed in the U.S.A., Canada, and Mexico
by D. Reidel Publishing Company, Inc.
Lincoln Building, 160 Old Derby Street, Hingham,
Mass. 02043, U.S.A.

This book is dedicated to the
celebration of the 70th anniversary of
Dr. F. Körösy and the 60th anniversary
of Professor G. Manecke who have both had
a long and significant activity in the field.

PROFESSOR GEORG MANECKE

TABLE OF CONTENTS

INTRODUCTION   IX

I. SPECTROSCOPY

CARLA HEITNER-WIRGUIN / Electronic Spectroscopy as a Tool for the
Elucidation of Complex Species in Polyelectrolyte Gels   3
GEORG ZUNDEL / The Influence of Cations on the Conformation of Biological
Membranes and Macromolecules – Infrared Investigations   21

II. NON ISOTHERMAL TRANSPORTS

J. W. LORIMER / Non-Isothermal Transport Phenomena in Charged Gels and
Membranes   45
ANITRA THORHAUG / The Effect of Temperature and Ouabain on the Mem-
brane Properties of a Giant Algal Cell   63

III. CHARGED MOSAIC AND ARRAYS

S. ROY CAPLAN / Charged Membrane Arrays: Mosaics and Stacks   89

IV. CARRIERS AND CHARGES

SALLY KRASNE and GEORGE EISENMAN / The Ion Selectivity of Carrier
Molecules, Membranes and Enzymes   107
O. KEDEM, M. PERRY, and R. BLOCH / Valinomycin-Induced Potassium Speci-
ficity in Charged Membranes   125
R. VAROQUI and E. PEFFERKORN / Ion Exchange and Structural Properties
of Alkali Ion Macromolecular Carriers in Liquid Membranes   137

V. OXIDATION – REDUCTION

GEORG MANECKE / Synthesis, Electrochemistry and Application of some Oxidation-Reduction Polymers   173

VI. FLUCTUATION, OSCILLATION, EXCITATION

FUMIO OOSAWA / Field Fluctuation in Ionic Solutions and Membranes   193
TORSTEN TEORELL / Non-Linear Transport and Oscillations in Fixed Charge
Membranes: Some Possible Biological Implications   205
I. TASAKI / Physico-Chemical Properties of the Nerve Membrane   213

INDEX   239

INTRODUCTION

The introduction to the first of these two volumes on Charged Gels and Membranes has recalled already that both were issued from the second Advanced Study Institute of Forges les Eaux, of which the co-directors were Professors G. E. Boyd and K. S. Spiegler.

However, it seems necessary to add some further remarks for the eventual readers of this one volume only or for those of all four which now constitute the series.*

One discovers that each volume is precisely linked to the next; and the total contains a large number of the very fundamental steps by which *macromolecular physical chemistry finds itself simultaneously at the frontiers of application and of biology.* One often wonders how this is possible.

Research has been the best means of understanding the microscopic elements of life. *Biomimetic phenomena or bioanalogue compounds* in their turn have led to innumerable practical realisations. On one hand, the notion of *'vital force'* receded and is disappearing due to repetitive total and *asymmetric synthesis** of always larger, and more complex, biological molecules. On the other hand discoveries of *inter-relations in physical chemistry disengage the analogies between living and non-living systems:* the interrelations between phenomena, between phenomena and structures or the appearance of these structures under the influence of intermolecular forces or of gradients of more statistical forces.

The biophysical ideas have traditionally been presented throughout the history of physical chemistry. Undoubtedly, one of its first major results was the reproduction of the osmotic pressure of plants by a synthetic semi-permeable barrier followed by interpretation (Pfeffer-van 't Hoff). Related to molecular agitation and to diffusion (Fick-Einstein), osmotic pressure is independent of membrane matter as long as it is inert: it depends only upon the relative dimensions of the pores and of the permeant and non-permeant molecules, from which we have a first inter-relation. But without pores, passive permeation can be selective by the difference in solubility (Henry) or in charge (Donnan) if the membrane possesses itself non-diffusible fixed charges (Teorel-Siever-Mayers).

With the charges and potentials, we have the laws of Nernst, Nernst-Einstein, Nernst-Planck, and Nernst-Hartley.

Before the development of synthetic macromolecular chemistry the technology of membranes had to go a long way to realize basic experiments, keeping one eye on physiology and one on physical laws. One finds a fine description of this progress by K. Sollner in Part I [1].

* The next (fifth) volume of the series will deal with 'Optically Active Polymers'.

The biomimetic phenomena produced are so well in accord with biological ones that the existence of cellular membranes was accepted by a majority of biologists long before they became visible by electron microscopy. Even before the necessary and sufficient constituting molecules and macromolecules had been clearly specified and located: this research itself has been aided by forecasts based on models.

However, biological membranes are as complex as their behaviour, and the construction of models continues and develops on the basis of new results (See the article of T. Teorell in Part I) [1].

Even if some general predictions can be made by using such mathematical languages which are little sensitive in their form to heterogeneities, as irreversible thermodynamics or other mathematics of the engineer, it is still necessary to clarify and experiment upon the particulars and identify the structures.

This means spectroscopic investigations, the two-by-two combination of charges or solubility effects, temperature gradients, oscillations, chemical reactions and asymmetry.

One discovers these approaches in this volume. It would have been possible to separate models from biological studies but the preceding leads to the opposite choice. There are four ways to regroup: methods of study, language of description, nature of observed phenomena and of structures.

Spectroscopy reveals the conformational changes due to metal-complexation of synthetic or natural macromolecules. Formulations and experiments are similar in the non-isothermal studies on synthetic membranes, or on the algae Valonia.

One special mention must be made of the long and methodic preparative work of Dr Körösy, which was necessary to meet the requirement of quality and homogeneity, which are now expected of synthetic membranes. There is a great variety of anionic, cationic, bilayer or mosaic membranes. The experimental studies of several laboratories proceed with such polyethylene-based membranes, the preparation of which is still at work at a time when Dr Körösy is now retired. We feel justified about the dedication of this volume on the occasion of his 70th birthday.

The study of mosaic membranes, of another origin, is described; one can recall that such structures have been advanced as constituting part of biological membranes on interpreting results of electron microscopy.

The reaction of ions with antibiotics, macrocyclic or polyelectrolytic transporters, or their combinations permit these ions to penetrate and permeate lipid or low dielectric liquid membranes. Under the action of external forces this leads to facilitated transport and to strong selectivities.

For active transport pumping against the electrochemical gradient of the transported species, exchange or energetic reactions must be associated with membrane asymmetry. This subject has already been studied with enzymatic reactions (Sélégny) in the first volume of this series [2]; it seemed unnecessary to deal with it again here.

But redox reactions and coupled proton and electron transfers are of primary importance in biological chains of reactions and energy transfers.

We were thus happy to receive Prof. Manecke at the Institute and his contribution in this book with a review on his numerous original redox polymers. It is in agreement with his friends and collaborators that we have made him the surprise of the dedication of this volume for the celebration of his 60th year.

The deformations and transconformations of membranes under constraint or oscillations and fluctuations remind us that no rigid model is completely satisfactory as regards biological motivity: rapid, daily, annual, permanent or accidental oscillations are known. But there are two particular types of major importance.

Cardiac rhythm has much occupied Prof. Teorell; I recall having seen him in the process of deciphering its meanderings by means of equations and numerical calculations during one of his long visits at NIH. He now delivers to us a veritable membrane machine which reproduces the pace.

Dr Tasaki, a rapid author and passionate seeker, is the living justification for our discipline. Since from the model to the biological reality he puts in application on the axone all the means of research that it offers.

We must again thank all those co-editors, organizers, authors, and critics of the Institutes who contributed to those volumes.

To make the interconnections evident a general index for the first four volumes of the series is added to the end of this one.

We hope that this will help our readers, students, researchers, professionals, and authors.

The intention of the editor was to clarify better the discipline of charged macromolecules and to make more evident its impact; to clarify that which the molecular biologist or physiologist can await in this domain of physical chemistry, and that which synthetic practice owes to natural phenomena.

Should the collaborations which alreay exist between participants in the Instituts and Symposia continue to develop, should new ideas be born under the influence of these pages, the objective will be realized.

References

1. *Charged Gels and Membranes*, Part I, E. Sélégny (ed.), G. Boyd and H. P. Gregor (co-editors), D. Reidel Publ. Co. Dordrecht/Boston, 1976.
2. *Polyelectrolytes*, E. Sélégny (ed.), M. Mandel and U. P. Straus (co-editors), D. Reidel Publ. Co., Dordrecht/Boston, 1974.

I

SPECTROSCOPY

ELECTRONIC SPECTROSCOPY AS A TOOL FOR THE ELUCIDATION OF COMPLEX SPECIES IN POLYELECTROLYTE GELS

CARLA HEITNER-WIRGUIN

Dept. of Inorganic and Analytical Chemistry, Hebrew University, Jerusalem, Israel

Abstract. The complex species formed in a stepwise reaction starting from a solvated species to a complex MA_n may be characterised by their selective sorbtion on an ion exchanger and by measuring the absorption spectrum of the sorbed species.

The method is illustrated by a study of complex species of copper, cobalt and uranyl ions. From the spectra and the spectroscopic parameters evaluated, the symmetry and coordination number of the species sorbed were determined.

The properties of the exchangers, which favour the sorbtion of species unstable in solution, are discussed.

1. Introduction

With the increasing interest in the study of ion exchange equilibria, more information was required about the identity and structure of the ions sorbed. Thus from this data it should be possible to formulate the chemical reaction as well as the quantitative mass equilibria. As more and more complex ions were studied some of them coloured such as transition metal ions, attempts were made to define these sorbed ions by measurement of their visible and near ultraviolet spectrum. Among the first workers to realise the importance of the identity of the species sorbed were Lindenbaum and Boyd [1] who studied the spectra of chlorocomplexes in liquid anion exchangers. The techniques of measuring the spectra of solid particles of the ion exchangers varied from author to author as difficulties were encountered. It was therefore quite natural that some of the initial studies were made by measurements of the reflectance spectra which are much easier to carry out on solids. Such measurements were made by Rutner [2] on chlorocomplexes of cobalt and iron sorbed on Dowex 1 and Klier and Ralek [3] on nickel ions, sorbed on synthetic zeolites. Coleman [4] used a particular technique for the study of the chlorocomplexes of cobalt. He prepared flattened disks from the resin spheres, 0.2-0.6 mm thick, placed them in a hole in a platinum foil and then taped the foil into a regular 1 cm cell. Ryan [5, 6] studied the nitrato and chlorocomplexes of uranyl ions sorbed on Dowex 1 by putting the resin particles together with the equilibrium solutions in 1 cm cells. The excess solution was drawn off with a fine tipped pipet in order to ensure close packing and a minimum floating of the resin particles. Other authors [7–9] have measured the spectra of dry particles as such, in mulls or organic solvents. From all these studies no quantitative data was evaluated.

Measurements were also made in other regions of the spectrum and added valuable information for the determination of the properties of sorbed species. Among them were infrared and Raman spectrum [10, 11] and electron spin resonance [12].

Eric Sélégny (ed.), Charged Gels and Membranes II, 3–20. All rights reserved.

The purpose of the present work was to study the stepwise formation of complexes by the use of ion exchangers. In the formation of a complex MA_n generally the only data available about the intermediate species are the individual stability constants. As n increases in these complexes, the techniques for the evaluation of the stability constants become more complicated and less reliable. Therefore an attempt was made to obtain a selective sorption of the various species formed in the stepwise reaction and then to define each species by the measurement and interpretation of its spectrum. The spectra obtained were compared to those known in solution (aqueous and nonaqueous) or to the solid – in most cases the initial species (solvated) and the final one (MA_n) have been defined in the literature. Assignment of the spectra and the evaluation of quantitative parameters enable the determination of the structure of the species sorbed. In solution generally a mixture of the complex species is present and therefore the sum of all their spectra is measured. The selective sorbtion of one or another species on the exchanger may be controlled by (a) changes in the equilibrating solution (b) use of the different ion exchangers.

2. Experimental

A wide range of ion exchangers were used: a strong and a weak cation exchanger (Dowex 50 and IRC-50) a strong anion exchanger (Dowex 1) and a chelating resin of the aminodiacetate type (Dowex A-1). The latter resin Dowex A-1 acts differently according to the acidity of the solution and is therefore very suitable for this type of work [10]. In order to obtain as much information from the spectral measurements as possible, it was necessary to adapt a measuring technique which would enable the evaluation of quantitative intensity parameters such as the molar extinction coefficient (Σ_{mol}) and oscillator strength (f). The technique adopted was the measurement of a nujol suspension of the resin particles (50–100 mesh) in 1 mm cells, centrifuging the samples for 5 min in order to obtain a uniform settling. Σ_{mol} was evaluated from the equation

$$\log I/I_0 = \Sigma lc$$

The c value (concentration of the cation in the exchanger) was obtained by transforming the loading of the resin from meq g^{-1} to moles per liter of exchanger.

Oscillator strength was determined by the formula

$$f = 4.6 \times 10^{-9} \, \Sigma_{max} \Delta \gamma_{1/2}$$

where Σ_{max} is the molar extinction coefficient and $\Delta \gamma_{1/2}$ is the half intensity band width i.e. the width at $1/2 \, \Sigma_{max}$.

Other parameters evaluated were 10 Dq-crystal field splitting, B-nephelauxetic parameter and g values (from esr measurement).

3. **Results and Discussion**

3.1. COPPER COMPLEXES

These complexes were studied principally as a test case since a great amount of data is available on this subject and the results obtained may be compared to this data. Copper ions were sorbed from dilute nitrate and chloride solutions as well as from concentrated chloride solutions (Figure 1) [13]. The spectra of the species sorbed from the dilute solutions are identical, each showing an absorption band at $12\,200\,\mathrm{cm}^{-1}$ with $\Sigma_{mol} \approx 10$. This band is characteristic for the $Cu(H_2O)_6^{2+}$ species. When the equilibrating solution has a high chloride concentration the absorption maximum of the sorbed species is displaced to lower energies $11\,400\,\mathrm{cm}^{-1}$ and Σ_{mol} increases to about 60. The shift to lower energy is in agreement with the relative positions of water and

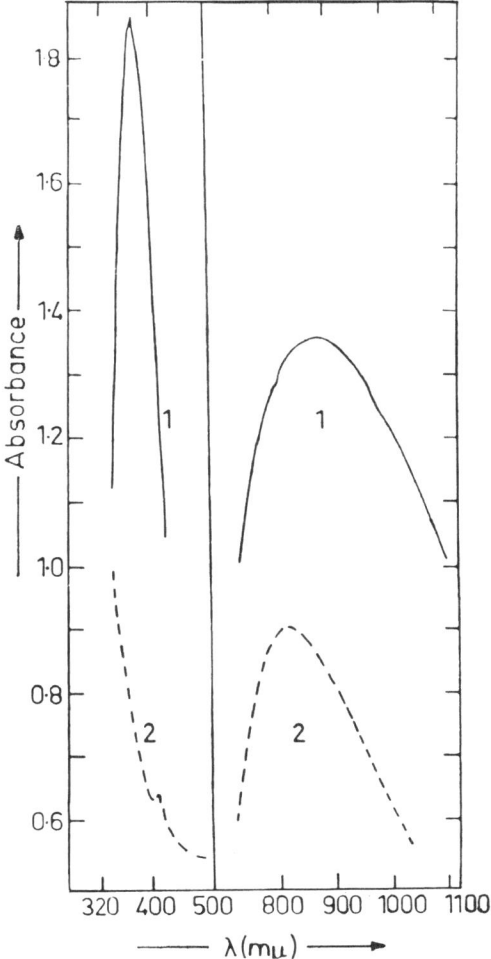

Fig. 1. (1) CuCl₂ sorbed on Dowex 50; (2) Cu(NO₃)₂ sorbed on Dowex 50. [13]

chloride ions in the spectrochemical series while the increase in the molar extinction coefficient may be explained by a decrease in symmetry of the sorbed species which makes the d-d transition less forbidden. The analysis of the resin for copper and chloride showed a ratio $Cu:Cl = 1:1$. The spectra together with the analysis may be assigned to the somewhat distorted octahedral species $[CuCl(H_2O)_5]^+$. The copper species sorbed on the anion exchanger depend on the copper concentration of the equilibrating solution [Figures 2 and 3]. Very broad absorption bands between 7000–10000 cm^{-1} are obtained. The energies of these bands and their Σ_{mol} values correspond roughly to a tetrahedral configuration. From these bands and the position of the charge transfer bands it may be concluded that from low Cl:Cu ratios in solution $CuCl_3H_2O^-$ was sorbed, while at high Cl:Cu ratios $CuCl_4^{2-}$ was sorbed. The spectrum obtained for the sorbed $CuCl_4^{2-}$ ion is identical to that found for $CuCl_4^{2-}$ in nonaqueous solvents [14], single crystals [15] and molten salts [16, 17].

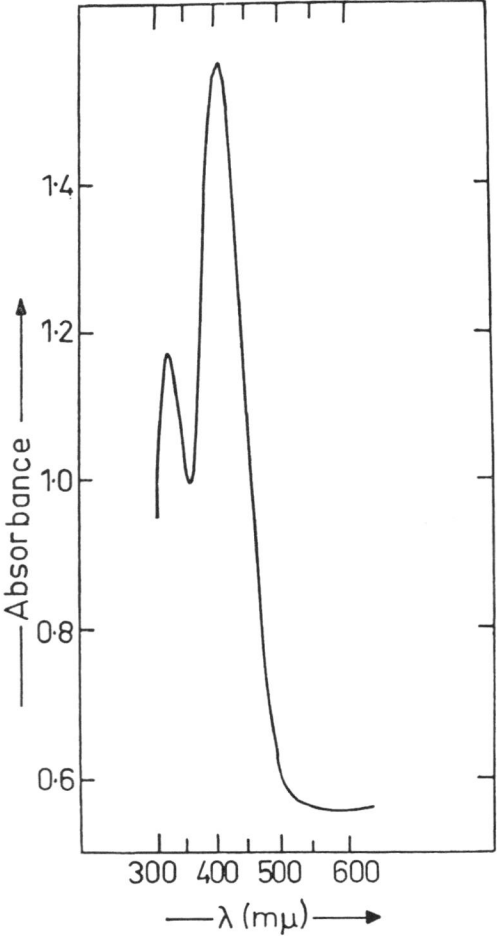

Fig. 2.   Solution of 0.05 M CuCl₂-KCl sorbed on Dowex 1 [13].

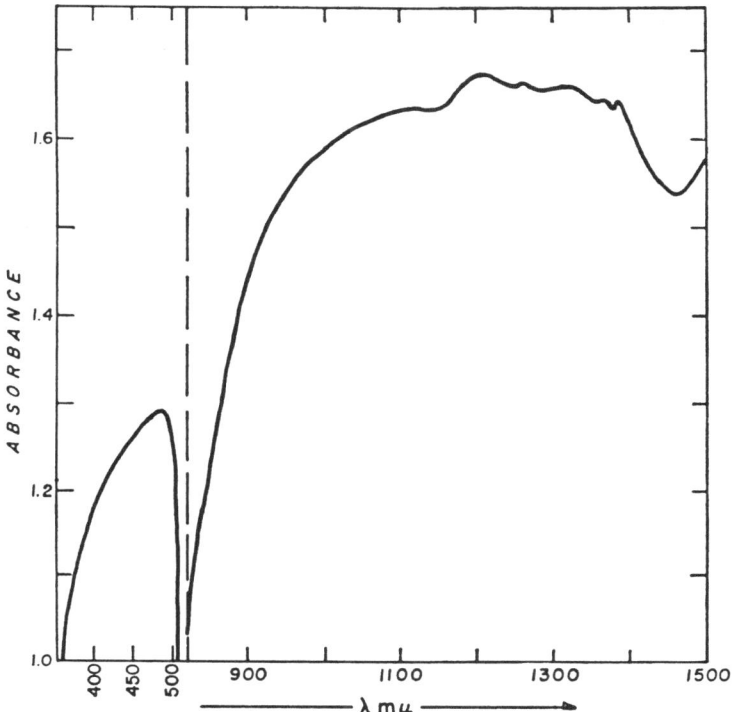

Fig. 3.   Solution of 0.5 M CuCl₂ sorbed on Dowex 1 [13].

The results obtained for copper species sorbed on all types of exchangers are summarised in Table I. These results show clearly that the chelating resin, acts as a strong anionic exchanger in concentrated acid solution and $CuCl_4^{2-}$ is sorbed as on the regular strong anion exchanger Dowex 1. Esr measurements on the sorbed copper species were carried out and are presented in Figure 4. From the g values and hyperfine splitting constants, α^2 was calculated which gives an idea of the covalency of the species sorbed.

3.2. Cobalt complexes

Among the cobalt complexes studied [18, 19, 20, 21, 22], striking results were obtained for the cobalt nitrate system. The nitrate ion is a rather weak complexing agent and no nitrate complexes of cobalt in aqueous solutions are known. Recently Addison and Gatehouse [23] prepared an anhydrous tetranitrato cobaltous complex starting from the metal and dinitrogen tetraoxide in organic solvent mixtures. This compound was later defined by spectral measurements [24] and by X-ray diffraction [25, 26] and found to be one of the rare compounds of cobalt with a coordination number of eight, each nitrate acting as a bidentate ligand. The absorption spectrum of cobalt ions in aqueous solution containing even a large excess of nitrate ions shows no indication whatsoever that any nitrate complex is formed. The same is true for

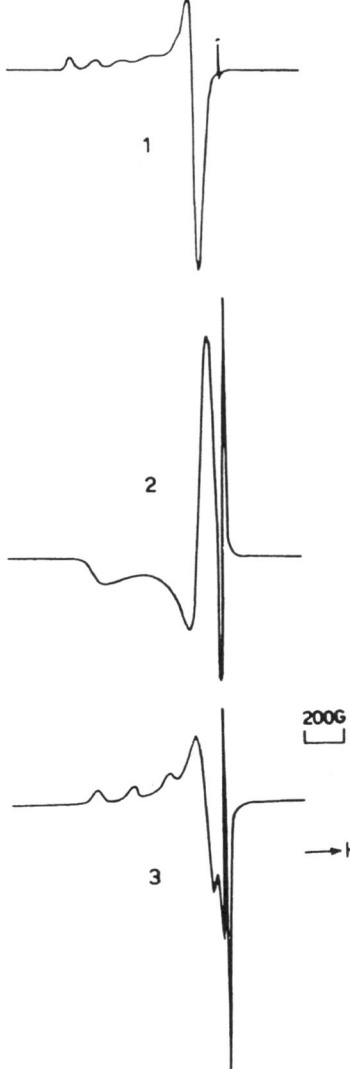

Fig. 4.   ESR spectra of copper ions sorbed on: 1. Dowex 50 W and IRC-50; 2. Dowex 1 and Dowex
A-1 in conc. HCl; 3. Dowex A-1 in neutral solution. *DPPH peak. [10]

cobalt nitrate in ethanol, acetone or acetonitrile solutions. A comparison of the
species sorbed on the cation exchanger in the absence and presence of an excess of
nitrate ions in the equilibrating solution shows a very small redshift of the near
infrared band and an increase in Σ_{mol} from 12 to 33 for the species sorbed from
solutions containing an excess of nitrate ions (Table II). This may be taken as an
indication for the formation of a cationic octahedral mononitrate species. Figures 5
and 6 show the absorption spectra of cobalt species sorbed on cationic and anionic
exchangers. When the anion exchanger is equilibrated with the same type of solution

TABLE I

Spectral and ESR data for various copper species sorbed on ion exchangers [10, 13]

Resin	Species sorbed assumed	Absorption max cm⁻¹	$g_\parallel$	$g_\perp$	A cm⁻¹	B cm⁻¹	α^2
Dowex 50	$Cu(H_2O)_6^{2+}$	12200	2.405	2.082	0.151		0.90
Dowex 50[a]	$CuCl(H_2O)_5^+$	11400, 26600		2.178			
Dowex 1, yellow	$CuCl_4^{2-}$	(7450–10000)24400, 31200	2.365 g_{ave}	2.084	–	–	
Dowex 1, brown[a]	$CuCl_3H_2O^-$	(6100–10000) 20840	g_{ave}	2.123			
Dowex A-1 H⁺ (conc acid)	$CuCl_4^{2-}$	(7450–10000) 24600	2.365	2.084	–	–	
Dowex A-1 H⁺	Cu-chelate	14300, 29300	2.278	2.088	0.183	0.032	0.86
Dowex A-1 Na⁺	Cu-chelate	(11700–15300) 28600	2.258	2.064	0.194	0.031	0.864
IRC-50	$(H_2O)_4Cu\!\!<^{OOC-R}_{OOC-R}$	14500 (broad)	2.32	2.077	0.142	–	0.79
Frozen soln CuCl₂	$CuCl_4^{2-} + CuCl_3^-$		2.34	2.081			
Frozen soln CuIDA	$Cu(IDA)_2$		2.278	2.067	0.170	0.032	0.812

[a] No splitting into $g_\parallel$ and $g_\perp$ is obtained apparently because of the high copper concentration.

TABLE II

Absorption bands of cobalt ions in cm^{-1} sorbed on cation and anion exchangers from aqueous and nonaqueous solutions [22].

Resin	Solvent	Equilibrating media	Near i.r. 1	Vis. 2	UV 3	$f_1^a \times 10^3$	$f_2 \times 10^3$
Dowex 50-Li$^+$	Water	0.1 M Co(NO$_3$)$_2$	6990, 7695 (3.7)^b	19300 (12)	37750, 31200	0.04	0.19
Dowex 50-Li$^+$	Water	0.1 M Co(NO$_3$)$_2$, 2 M NaNO$_3$	8000 (11)	19300 (33)	37750, 31200	0.11	0.50
Dowex 1-NO$_3^-$	Water	1 M Co(NO$_3$)$_2$	8475 (26.4)	18570 (247)	30300	0.41	4.30
Dowex 1-NO$_3^-$	Water	1 M Co(NO$_3$)$_2$, 9 M NaNO$_3$	8475 (53.2)	18520 (340)	29950	0.79	5.34
Dowex 1-NO$_3^-$	Ethanol	1 M Co(NO$_3$)$_2$	8475 (97.6)	18520 (406)	30330	1.59	8.2
Dowex 1-NO$_3^-$	Ethanol	1 M Co(NO$_3$)$_2$, sat. NaNO$_3$	8475 (87.4)	18520 (334)	29950	1.29	7.90
Dowex 1-NO$_3^-$	Acetonitrile	0.05 M Co(NO$_3$)$_2$	8475 (63.4)	18520 (343)	30120	0.94	7.00
Dowex 1-NO$_3^-$	Acetonitrile	0.05 M Co(NO$_3$)$_2$, sat. NaNO$_3$	8475 (13.7)	18650 (162)	29950	0.16	2.86
Dowex 1-NO$_3^-$	Acetone	1 M Co(NO$_3$)$_2$	8400	18720	29950		
Dowex 1-NO$_3^-$	Acetone	1 M Co(NO$_3$)$_2$, sat. NaNO$_3$	8340	18870	29950		

^{a} f = oscillator strength.
^{b} Molar extinction coefficients given in parentheses.

as the cation exchanger a striking change in colour and therefore in absorption spectrum was observed. The anion exchanger showed a deep magenta colour with absorption maxima between 18870 and 18450 cm^{-1} (asymmetric broad peak) and between 8340 and 8475 cm^{-1}. Molar extinction coefficients reach values up to 400 for the visible absorption band. The infrared spectrum of the sorbed species was measured and compared to that of the sorbed nitrate in the absence of cobalt (Figure 7). The appearance of peaks in the spectrum of cobalt nitrate species which are not found in the nitrate form of the exchanger at 1295, 1025, 813 and 1460–1480 cm^{-1} indicate a decrease in the symmetry of the nitrate ion (D_{3h}) to a nitrato ion $O-NO_2^-$ (C_{2v}) in the cobalt complex.

The absorption bands, Σ_{mol}, oscillator strength as well as the infrared spectrum

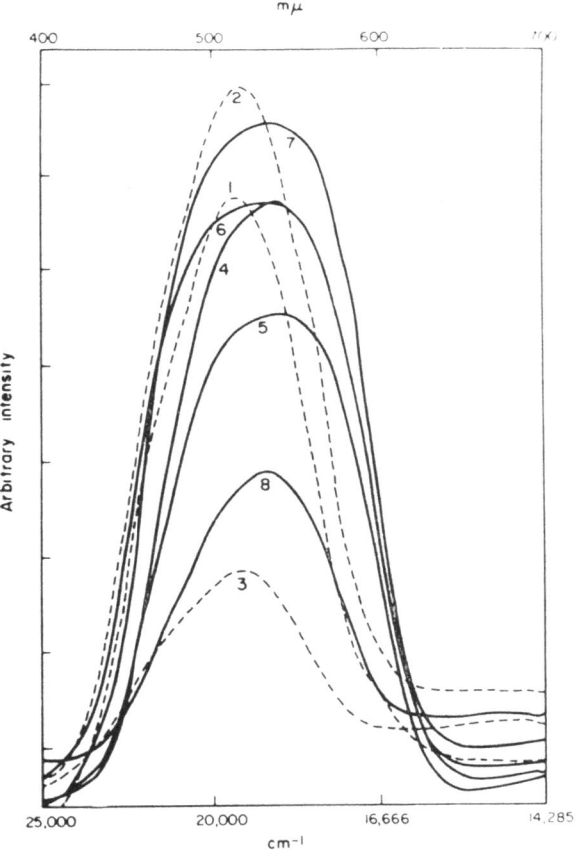

Fig. 5. Visible absorption spectra of cobalt (II) species sorbed on ion exchangers from aqueous and nonaqueous solutions. 1. Dowex 50 W × 8 Li$^+$, 1.0 M Co(NO$_3$)$_2$ in water; 2. Dowex 50 W × 8 Li$^+$, 2.0 M Co(NO$_3$)$_2$ in water; 3. Dowex 50 W × 8 Li$^+$, 1.0 M Co(NO$_3$)$_2$, 2 M NaNO$_3$ in water; 4. Dowex 1 × 8 NO$_3^-$, 1.0 M Co(NO$_3$)$_2$, 9 M NaNO$_3$ in water; 5. Dowex 1 × 8 NO$_3^-$, 0.5 M Co(NO$_3$)$_2$ in ethanol; 6. Dowex 1 × 8 NO$_3^-$, 1.0 M Co(NO$_3$)$_2$ in acetone; 7. Dowex 1 × 8 No$_3^-$, 0.1 M Co(NO$_3$)$_2$ in acetonitrile; 8. Dowex 1 × 8 NO$_3^-$, 0.05 M Co(NO$_3$)$_2$, sat. NaNO$_3$ in acetonitrile. [22]

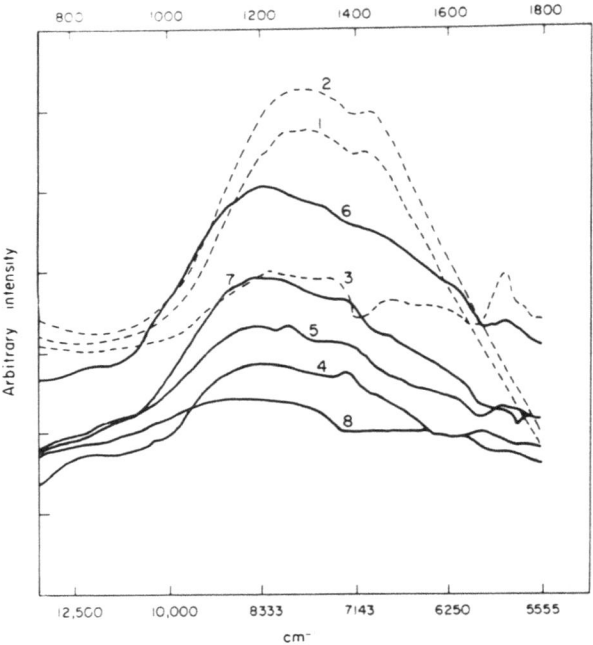

Fig. 6.   Near i.r. absorption spectra of cobalt (II) species sorbed on ion exchangers from aqueous and nonaqueous solutions. 1. Dowex 50 W × 8 Li$^+$, 1.0 M Co(NO$_3$)$_2$ in water; 2. Dowex 50 W × × 8 Li$^+$, 2.0 M Co(NO$_3$)$_2$ in water; 3. Dowex 50 W × 8 Li$^+$, 1.0 M Co(NO$_3$)$_2$, 2 M NaNO$_3$ in water; 4. Dowex 1 × 8 NO$_3^-$, 1.0 M Co(NO$_3$)$_2$, 9 M NaNO$_3$ in water; 5. Dowex 1 × 8 NO$_3^-$, 1.0 M Co(NO$_3$)$_2$ in acetone; 7 Dowex 1 × 8 NO$_3^-$, 0.1 M Co(NO$_3$)$_2$ in acetonitrile; 8. Dowex 1 × 8 NO$_3^-$, 0.05 M Co(NO$_3$)$_2$, sat. NaNO$_3$ in acetonitrile. [22]

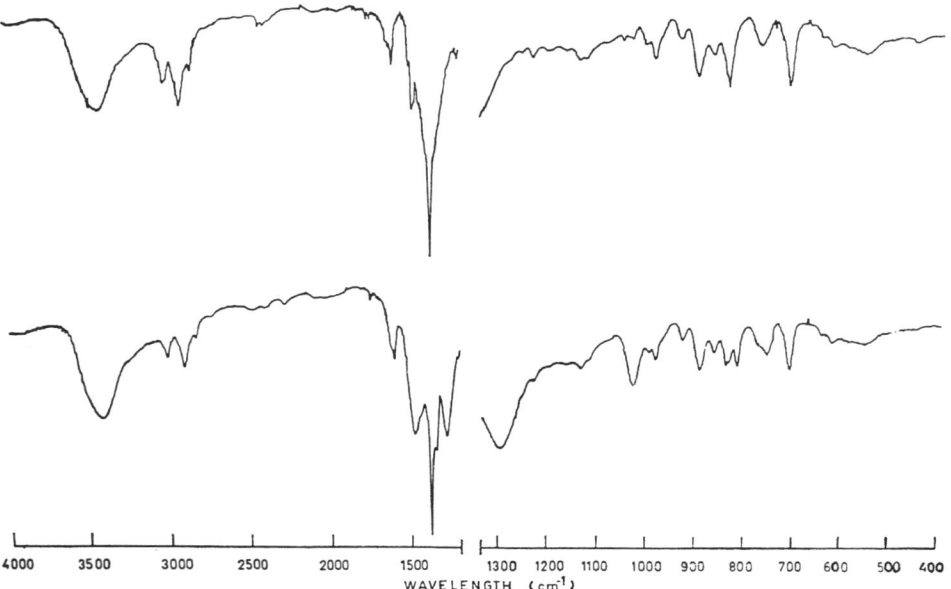

Fig. 7.   Infrared spectra of Dowex 1-NO$_3^-$ (upper curve) and Dowex 1-Co(NO$_3$)$_4{}^{2-}$ (lower curve).

indicate the sorbtion of a tetrahedral $Co(NO_3)_4^{2-}$ species similar to the compound previously described [23–26]. Crystal field splittings $(10\,Dq)$ and nephelauxetic parameters (B) were evaluated [27] and found to agree well with those evaluated for the solid compound $[10\,Dq = 4775\,\mathrm{cm}^{-1}$ (vs $4660\,\mathrm{cm}^{-1})$ and $B = 818\,\mathrm{cm}^{-1}$ (vs $855\,\mathrm{cm}^{-1})]$.

3.3. URANYL COMPLEXES

The uranyl group is a linear cation OUO^{2+} (symmetry $D_{\infty h}$) not much affected by complexing agents. The most usual coordination number of the uranyl group is six while compounds with coordination numbers five and four have also been found. The complexes with coordination number six may be represented as a hexagonal bipyramid where six monodentate or three bidentate ligands are in the same plane as the uranium ion and the two oxygens above and below this plane. The uranyl ion presents a rather complicated absorption spectrum between 330–550 mμ (Figure 8) which arises from one or more vibrationally perturbed electronic transitions [28, 29]. Upon complexation relatively small shifts, changes in the vibrational pattern and in the molar absorption values occur. A variety of complexes of uranyl ions are known and the vibrational pattern of their spectra differ markedly. It is therefore rather difficult to correlate the spectra with the structure of the compounds. From the literature, it appears that the series of tricomplexes (sulfate [30, 31], perchlorate [32], nitrate [33, 5] and acetate- each of these ions acting as a bidentate ligand) have very similar spectra (Figure 9). In all these spectra, the first region starting from the low energy side is very strongly vibrationally perturbed. Siddal et al. [34] have assigned this type of spectra to complexes which have no center of symmetry.

In this work, a series of inorganic uranyl complexes have been studied with various ligands SO_4^{2-} [31], F^-, Cl^-, Br^- [35] SCN^- and NO_3^- by their sorbtion on various ion exchangers. From the spectra measured it may be concluded that species with the same symmetry independent of the type of ligand, present very similar spectra i.e.

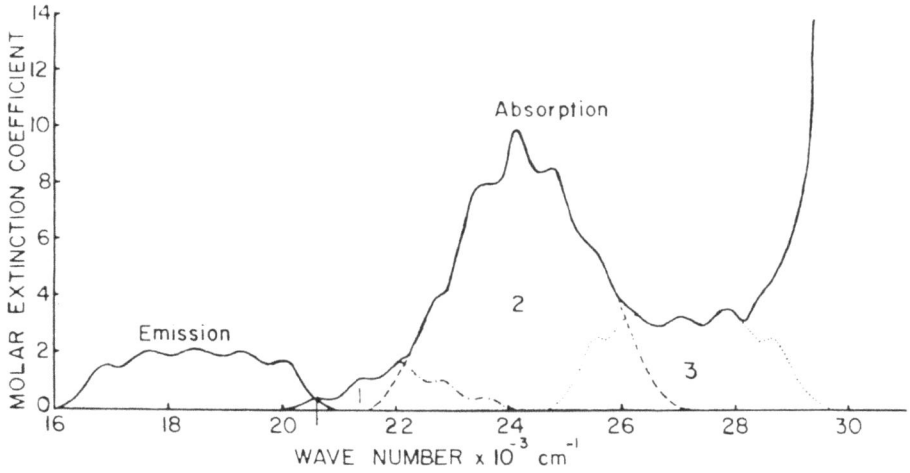

Fig. 8.   The emission spectrum and low energy absorption spectrum of UO_2^{2+}. [28]

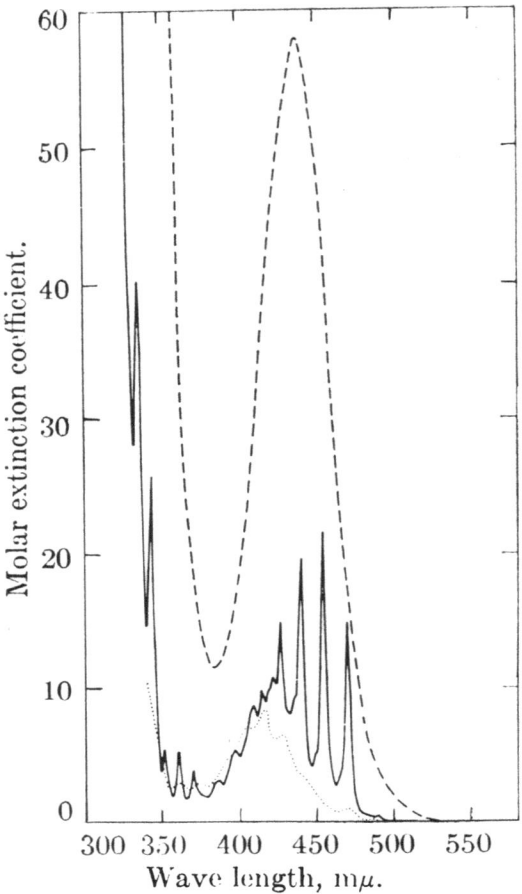

Fig. 9.   Absorption spectra of solid complex U(VI) nitrate salts compared to that of the aqueous uranyl ion: ---[(C₂H₅)₄N] UO₂(NO₃)₄; ———(C₂H₅)₄NUO₂(NO₃)₃; ····· uranyl nitrate in 0.514 M HNO₃.[5]

vibrational pattern. The nature of the ligand causes only a small shift of the spectral curve, according to its place in the spectrochemical series. As may be seen from Figure 10, all the cationic species of the type UO_2X^+, where X is a monodentate ligand show a very similar spectra. These spectra show a similarity to that of the uranyl hexahydrate. This is not surprising as the species $[UO_2X(H_2O)_5]^+$ results from the replacement of one molecule of water by the ligand X. The distortion from the bipyramidal form of the hexahydrate species is only slight since the field strength of the ligands are quite similar to that of water.

Figure 11 shows the spectra of tetracoordinated species of symmetry D_{4h}. In these spectra the first and the second region of the spectrum are strongly vibrationally perturbed, while the molar extinction coefficient of the second region is much higher than the first region. Figure 12 shows the spectra of some tricomplexes, where the

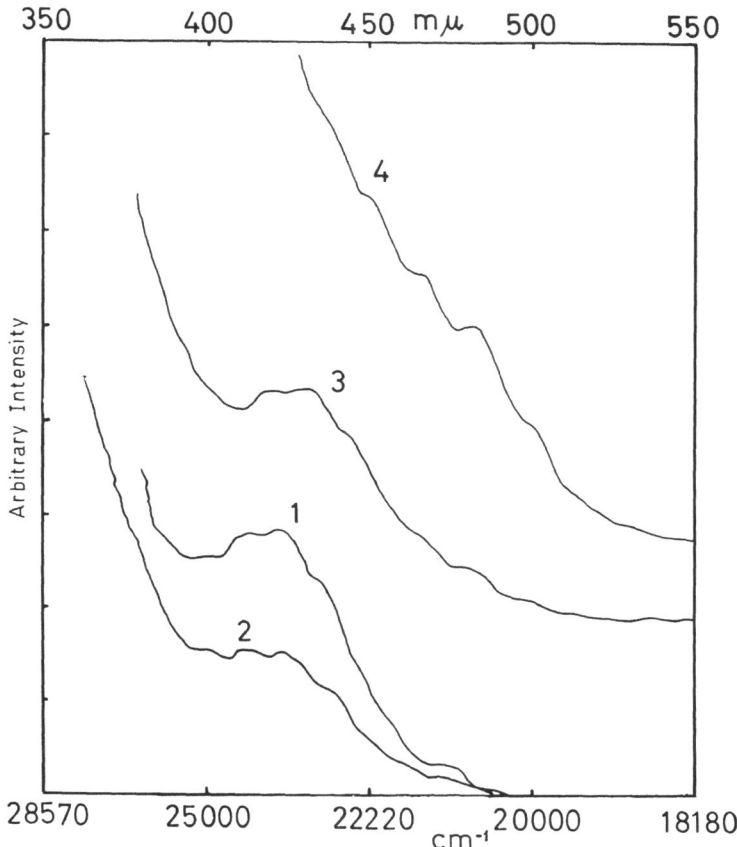

Fig. 10.   Visible spectra of cationic uranyl species: 1. $UO_2NO_3^+$; 2. $UO_2HSO_4^+$;
3. UO_2Cl^+; 4. UO_2Br^+. [36]

ligands all act as bidentate (symmetry D_{3h}). In these spectra, only the first region is very strongly vibrationally perturbed.

A complete vibrationless spectrum is obtained for the thiocyanate (Figure 13) and some nitrate complex (Figure 9) whose structures are not very well understood.

From the work described above it may be seen that a correlation of the spectra obtained for each known species to its structure and coordination number can be made. Using these data, conclusions may be drawn for the structure and coordination number of unknown species from their spectra.

In this study additional information was obtained from the measurement of infrared spectrum for two purposes:

(a) The determination of changes in the symmetry of ligands, e.g. SO_4^{2-} or NO_3^- upon complexation or the type of bonding in the thiocyanate species.

(b) The determination of changes in the asymmetric stretching frequency of the U–O bond caused by various ligands. From these measurements, an estimation of

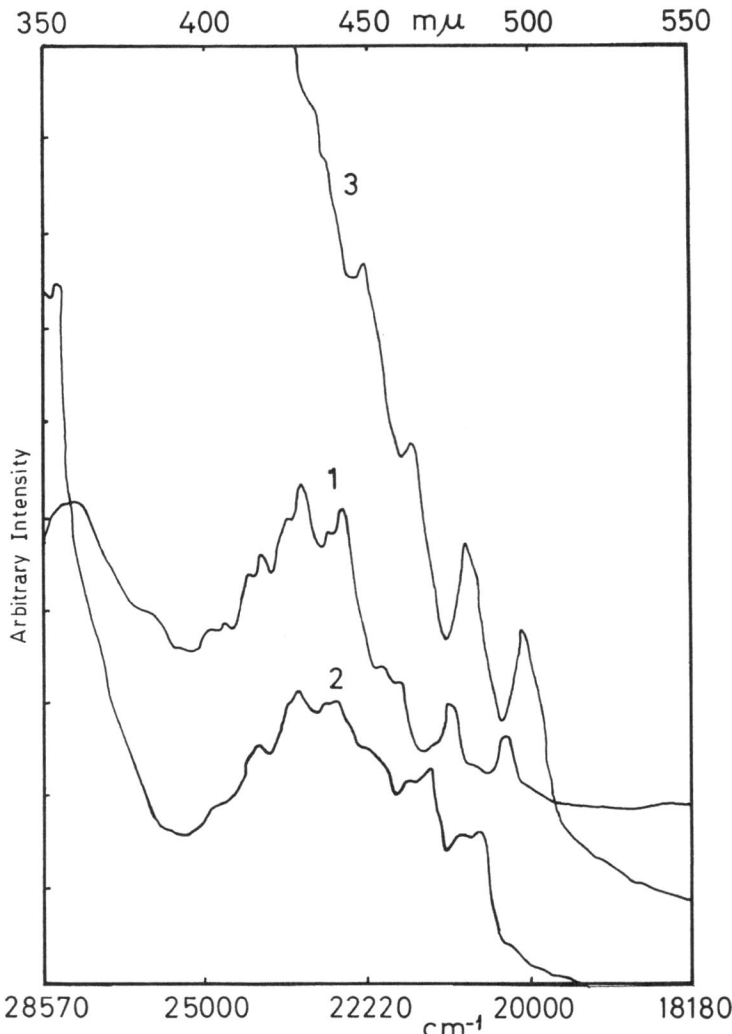

Fig. 11.   Visible spectra of anionic uranyl species symmetry D_{4h}; 1. $UO_2Cl_4^{2-}$; 2. $UO_2(HSO_4)_3H_2O^-$; 3. $UO_2Br_4^{2-}$. [36]

the strength of complexation (ligand field strength) was obtained. These results are in good agreement with those obtained from the visible spectrum.

4. Concluding Remarks

The techniques described in this work may be used to solve problems in two directions:

(a) The determination of the mechanism of sorbtion through the identification of the species sorbed.

(b) The determination of the changes occurring in the geometry and coordination

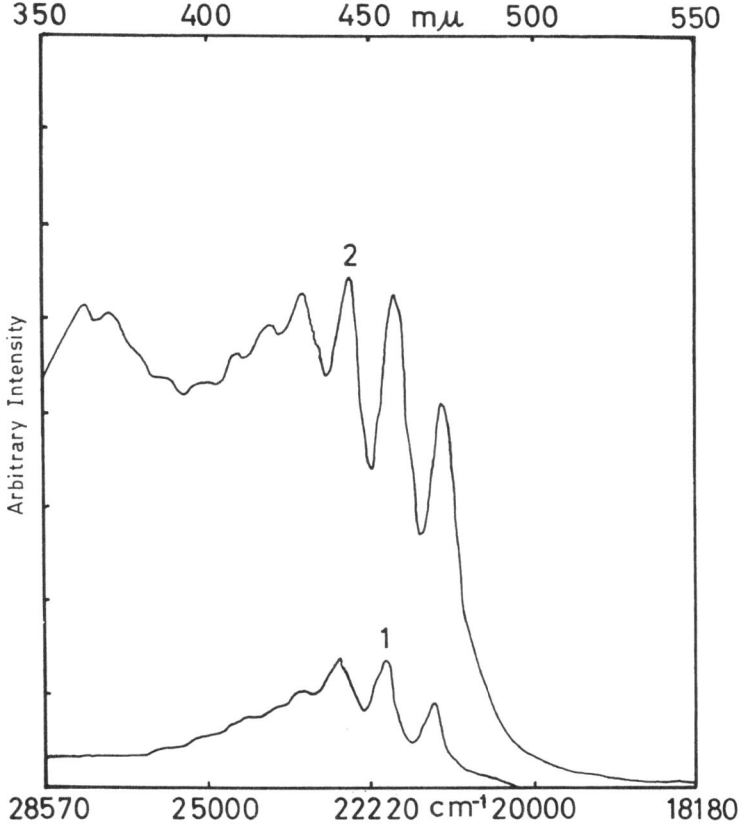

Fig. 12.   Visible spectra of anionic uranyl species symmetry D_{3h}; 1. $UO_2(NO_3)_3^-$; 2. $UO_2(SO_4)_3^{4-}$. [36]

number of the species formed in a stepwise reaction. This is obtained by the quantitative evaluation of the spectroscopic parameters for each species.

Additional valuable data may be obtained by infrared spectroscopy and electron spin resonance, both techniques being easily adapted to ion exchange beads.

In most of the cases studied the final species MA_n was sorbed. As previously mentioned, in the case of cobalt ions, the tetranitrato species was sorbed on the anion exchanger although it does not exist in aqueous solution. On the other hand, Ryan [5] has shown that although the $UO_2(NO_3)_3^-$ and $UO_2(NO_3)_4^{2-}$ are formed as solids (Figure 9), only a mixture of both are sorbed on the exchanger (Figure 14). This problem, however, does not seem to be completely solved. Conditions have been found for selectively sorbing the uranyl trinitrate [36] and its spectrum has been found to be identical to the solid compound. The preparation of the uranyl tetranitrate has been mentioned in the literature [37] but it was not found possible to repeat its preparation and analysis. In order to obtain the tetranitrate from the trinitrate at least one bond of the bidentate nitratogroup has to be opened, to bond a fourth nitrate group, since a higher than six coordinated uranyl species cannot be obtained. Thus the fact that the highest possible species is not sorbed is more a problem of its existence and not of the properties of the exchanger and should be looked upon as an exceptional case.

The last question which one is tempted to ask is why completely unstable species are known to be sorbed on the exchanger. Apparently, two properties of the resin (the anion exchanger) favour this: (a) the large functional group which in the case of Dowex 1 is a large cation i.e. $CH_2N(CH_3)_3^+$. The size of the cation seems to be critical, as in all the nitratocobalt complexes prepared, the cations used were large

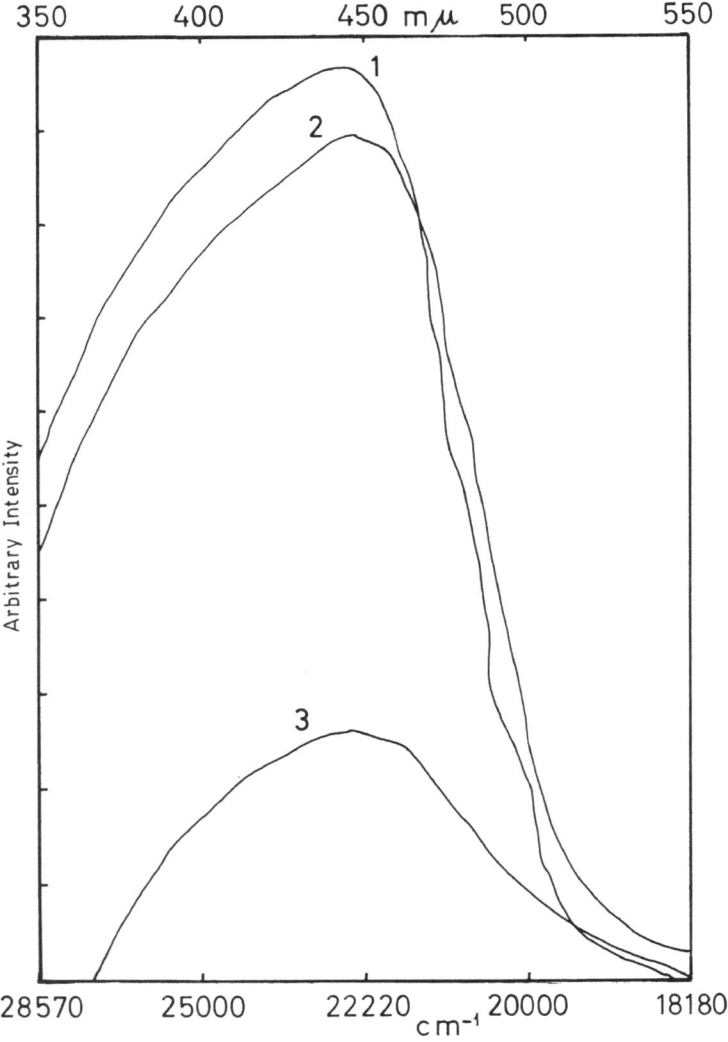

Fig. 13.   Visible spectra of uranyl thiocyanate species sorbed on Dowex 1; 1. from 2 N KCNS; 2. from 6 N KCNS; 3. from 10 N KCNS. [36]

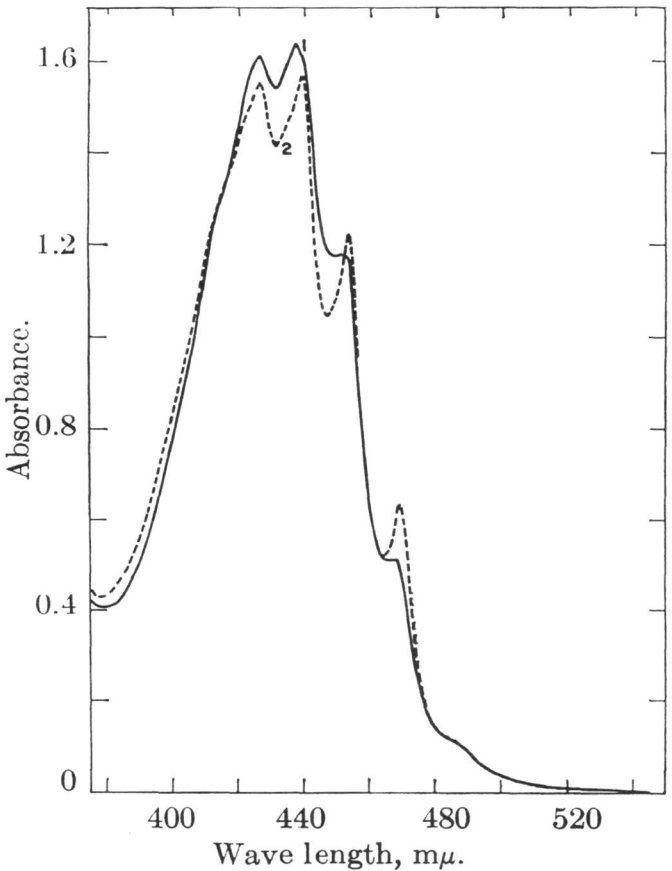

Fig. 14.   Absorption spectra of U(VI) loaded on; 1. high capacity Dowex 1;
2. low capacity Dowex 1. [5]

e.g. $N(CH_3)_4^+$ and $(C_6H_5)_3AsCH_3^+$ [24]. (b) The much smaller solvation shell (i.e. hydration shell), in the resin, which is even more diminished by the presence of a large excess of a neutral salt.

An important conclusion of this work is that the species sorbed are not distorted by the functional group of the ion exchanger. This may be seen clearly from the comparison of the spectra with those of the solid compounds and from the good agreement found in the calculated parameters.

References

1. Lindenbaum, S. and Boyd, G. E.: *J. Phys. Chem.* **67**, 1238 (1963).
2. Rutner, E.: *J. Phys. Chem.* **65**, 1027 (1961).
3. Klier, K. and Ralek, M.: *J. Phys. Chem. Solids* **29**, 951 (1968).
4. Coleman, J. S.: *J. Inorg. Nucl. Chem.* **28**, 2371 (1961).
5. Ryan, J. L.: *J. Phys. Chem.* **65**, 1099 (1961).

6. Ryan, J. L.: *Inorg. Chem.* **2**, 348 (1963).
7. Waki, H., Takahasi, S. and Ohashi, S.: *J. Inorg. Nucl. Chem.* **35**, 1259 (1973).
8. Nortia, T.: *Suomen Kemistilehti* **B41**, 136 (1968).
9. Choppin, G. R., Henrye, D. E. and Buys, K.: *Inorg. Chem.* **5**, 1743 (1966).
10. Cohen, R. and Heitner-Wirguin, C.: *Inorg. Chim. Acta* **3**, 647 (1969).
11. Boyd, G. E.: Lecture presented at Gordon Conference (1971).
12. Faber, R. J. and Rogers, M. T.: *J. Am. Chem. Soc.* **81**, 1849 (1959).
13. Heitner-Wirguin, C. and Cohen, R.: *J. Phys. Chem.* **71**, 2556 (1967).
14. Furlani, C. and Morpurgo, G.: *Theoret. Chim. Acta* **1**, 102 (1963).
15. Ferguson, J.: *J. Chem. Phys.* **40**, 3406 (1964).
16. Gruen, D. M. and McBeth, R. L.: *J. Pure Appl. Chem.* **6**, 23 (1963).
17. Smith, G. P. and Griffith, T. R.: *J. Am. Chem. Soc.* **85**, 4051 (1963).
18. Heitner-Wirguin, C. and Ben-Zwi, N.: *Inorg. Chim. Acta* **4**, 517 (1970).
19. Heitner-Wirguin, C. and Ben-Zwi, N.: *Inorg. Chim. Acta* **4**, 554 (1970).
20. Heitner-Wirguin, C. and Ben-Zwi, N.: *Inorg. Chim. Acta* **6**, 93 (1972).
21. Heitner-Wirguin, C. and Ben-Zwi N.: *Israel J. Chem.* **8**, 913 (1970).
22. Heitner-Wirguin, C. and Ben-Zwi, N.: *J. Inorg. Nucl. Chem.* **33**, 1493 (1971).
23. Addison, C. C. and Gatehouse, B. M.: *J. Chem. Soc.* **613** (1960).
24. Cotton, F. A. and Dunne, T. G.: *J. Am. Chem. Soc.* **84**, 2013 (1962).
25. Cotton, F. A. and Bergman, J. G.: *J. Am. Chem. Soc.* **86**, 2941 (1964).
26. Bergman, J. G. and Cotton, F. A.: *Inorg. Chem.* **5**, 1208 (1966).
27. Ben-Zwi, N. and Heitner-Wirguin, C.: *Israel J. Chem.* **10**, 885 (1972).
28. McGlynn, S. P. and Smith, J. K.: *J. Molec. Spectrosc.* **6**, 164 (1961).
29. Bell, T. and Biggers, R. E.: *J. Molec. Spectrosc.* **18**, 247 (1965).
30. Deptula, C. and Minc, S.: *J. Inorg. Nucl. Chem.* **29**, 221 (1967).
31. Heitner-Wirguin, C. and Gantz, M.: *J. Inorg. Nucl. Chem.* **35**, 3341 (1973).
32. Vdovenko, V. M. V., Skoblo, A. J., and Suglobov, D. N.: *Russ. J. Radiochem.* **6**, 677 (1964).
33. Kaplan, L., Hildebrandt, R. A., and Ader, M.: *J. Inorg. Nucl. Chem.* **2**, 153 (1956).
34. Siddal, T. H., McDonald, R. L., and Stewart, W. E.: *J. Molec. Spectrosc.* **28**, 243 (1968).
35. Heitner-Wirguin, C. and Gantz, M.: *Israel J. Chem.* **12**, 723 (1974).
36. Gantz, M.: Ph. D. Thesis, Hebrew University, Jerusalem (1973).
37. Chernyaev, I. I.: *Complex Compounds of Uranium*, Israel Program for Scientific Translations, Jerusalem, 1966.

THE INFLUENCE OF CATIONS ON THE
CONFORMATION OF BIOLOGICAL MEMBRANES AND
MACROMOLECULES – INFRARED INVESTIGATIONS

GEORG ZUNDEL

Physikalisch-Chemisches Institut, Universität München, Theresienstr. 41, D-8 München 2, W. Germany

Abstract. The IR spectra may provide information on the secondary structure of macromolecules. Splitting and shifting of bands in opposite directions caused by coupling of vibrations of neighboring groups is discussed. Double helix formation is demonstrated by bands in the region 1800–1500 cm^{-1} and the structure of the backbone by bands in the region 1300–1000 cm^{-1}. It is shown that with RNA the 2'OH groups form hydrogen bonds with the O-atom of neighboring base residues. Hence, the difference between DNA and RNA, as far as this is due to the 2'OH group is caused by the structure promotion of these hydrogen bonds. Further, the influence of cations with strong fields on the backbone and especially the structure of Ca^{2+} poly (A) and those formed by Mg^{2+} poly (U) at low temperatures is discussed. With proteins, bands caused by vibrations of groups in the backbone, for instance the Amide I band, are sensitive to changes to the secondary structure. With vesicles prepared from axon membranes it is shown that large parts of the proteins of excitable membranes have antiparallel β-structure when K$^+$ ions are present. In the presence of Na$^+$ and Ca^{2+} these proteins are α-helical. This suggests that the permeability change during excitation of nerve membranes is strongly connected with such cation-induced conformation changes. Finally, the interaction of ions and especially the crosslinking of fixed ions with macromolecules by polyvalent counterions is demonstrated by band splitting caused by removal of degeneracy of vibrations. The relevance of this effect for biological membranes, especially the excitable membranes, is discussed.

1. Introduction

The IR spectra may provide information on the secondary structure of macro-molecules. With regard to these conclusions it is important that the substances must not be – as for the X-ray structure investigations – in crystalline state. Hence these IR structure investigations can be carried out with solutions of macromolecules or amorphous films. Three effects will be discussed: firstly, the coupling of vibrations of neighboring groups which provide information on the mutual orientation of these groups. Secondly, the secondary structure sensitivity of vibrations of the backbone of proteins which is caused by the fact that the same vibrations of all peptide groups are not independent of one another, when these groups are arranged regularly in secondary structures. Thirdly, the removal of the degeneracy of groups with degenerate vibrations caused by interactions with their environment.

2. Secondary Structure of DNA and RNA

Vibrations of molecules may couple. Such couplings, called Fermi resonance, occur when two conditions are fulfilled: Firstly, the wave number values of the vibrations must be similar. Secondly, the vibrations, that is to say their transition moments, must be largely parallel. When these conditions are fulfilled one can observe splitting and shifting in opposite directions of the coupled vibration bands. Conversely, when one observes band splitting or such band shifting one can conclude that the transition moments of these two vibrations are largely parallel. These couplings are not only observed when two vibrations within one group couple but also when two vibrations of neighboring groupings couple. Thus, when one observes such band splittings and shifts, information on the mutual orientation of these two groupings and hence on the structure of the molecules may be deduced.

Only in-plane stretching vibrations of the base residues, which have more or less C=C, C=N and C=O character, and the $-NH_2$ scissor vibration occur in the range 1800–1550 cm^{-1} [1]. The H_2O scissor vibration, which is expected in the spectra of the hydrated films at 1640 cm^{-1}, merges with the overall complex. In the range 1300–1000 cm^{-1} the stretching vibrations of the $>PO_2^-$ groups and vibrations of the pentose residues are observed [1]–[3].

When the double helical structures are formed, band shifts which are largely independent of deuteration occur due to coupling between the stretching vibrations mentioned, in the region 1800–1550 cm^{-1}. They were studied in detail with the aid of the (G+C) pair [4]. On the other hand, a secondary structure-sensitive coupling between the in-plane stretching vibrations and the $-NH_2$ scissor vibration may also occur. This may happen when the $-NH_2$ groups are fixed in the ring planes in connection with the double helix formation. This was studied, for example, with the semiprotonated poly(C) [5]. This effect, however, disappears on deuteration.

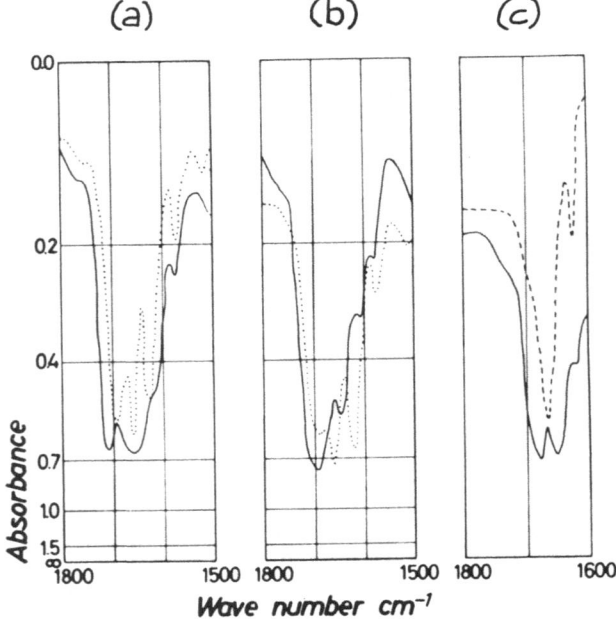

Fig. 1.   IR spectra of calf thymus DNA, (a) films hydrated at 90 % relative humidity of the air ———— H_2O hydrated ····· D_2O hydrated; (b) the same films thoroughly dried; (c) solutions (6 mg/ml) in D_2O ———— at 25 °C, ----- at 92 °C.

A typical 'structure band' occurs at about 1700 cm^{-1} with DNA [6]–[8], double-helical RNA [1] and RNA-DNA hybrids [9]. Its absorbance is proportional to the number of paired bases. This is illustrated in Figure 1. Band splitting (1680, 1645 cm^{-1}) is observed in the D$_2$O-hydrated samples (Figures 1a and c) which disappears on melting (Figure 1c) and decreases on drying (Figure 1b), that is, when the structure breaks down. With respect to the foregoing considerations, this splitting is caused by a coupling of in-plane stretching vibrations [10]. The peak at the larger wave number (1680 cm^{-1}) is the 'structure band'. This 'structure band' appears with H$_2$O-hydrated samples at somewhat larger wave numbers than with D$_2$O-hydrated ones (Figure 1a). Due to coupling with the $-$NH$_2$ scissor vibration, it is shifted even further toward larger wave numbers. Thus band splitting and shifting in opposite directions in the region 1800–1550 cm^{-1} demonstrate double helix formation with polynucleotides.

2.1. SECONDARY STRUCTURE OF RNA BACKBONES IN NEUTRAL AQUEOUS SOLUTION

Couplings of vibrations of the ribose residues and the phosphate groups, observed in the range 1300–1000 cm^{-1}, supply information as to the secondary structure of the backbone [10]. The melting behavior of the backbone of DNA and RNA was studied with these bands [11]. The assignments of the bands in this region are given in [1]–[3] [10].

With regard to the secondary structure of the backbone of RNA in neutral aqueous solutions, two questions are of special interest. Firstly, in contrast to DNA, the ribose residue of RNA contains an OH group. If this group forms a hydrogen bond, which group is the acceptor, and how is the secondary structure of the backbone of the RNA changed by this hydrogen bond in comparison to DNA? The second question is, how do cations change the structure of the RNA backbone? We studied these questions with homo poly (ribonucleotides) with poly (A), poly (C) and poly (U).

2.1.1. The Structure-Promoting Effect of the 2'OH Group

Information concerning the structure-promoting effect of this group is given by shifts which occur due to coupling of a skeleton vibration of the ribose residue and the bending vibration of the 2'OD group in the range 1100–1000 cm^{-1} [10].

The antisymmetric stretching vibration of the ether group of the ribose is observed at about 1060 cm^{-1}. The transition moment of this vibration is in the C(1')–C(4') direction, as illustrated by Figure 2. After deuteration the 2'OD bending vibration is observed at about 1030 cm^{-1}. The transition moment of this vibration is perpendicular to the OD bond. If the 2'OD group is fixed in such a way that the transition moments of these two vibrations are largely parallel, the bands of these vibrations should shift in opposite directions. This effect would provide information on the mutual orientation of these groups.

The dashed spectrum in Figure 2a is the H$_2$O hydrated K$^+$ salt of poly (C), the continuous line is that of the D$_2$O hydrated salt. With the H$_2$O hydrated sample the vibration of the ether group of the ribose residue is observed at 1060 cm^{-1}. With H-D exchange the 2'OD bending vibration emerges at about 1030 cm^{-1}. Further one

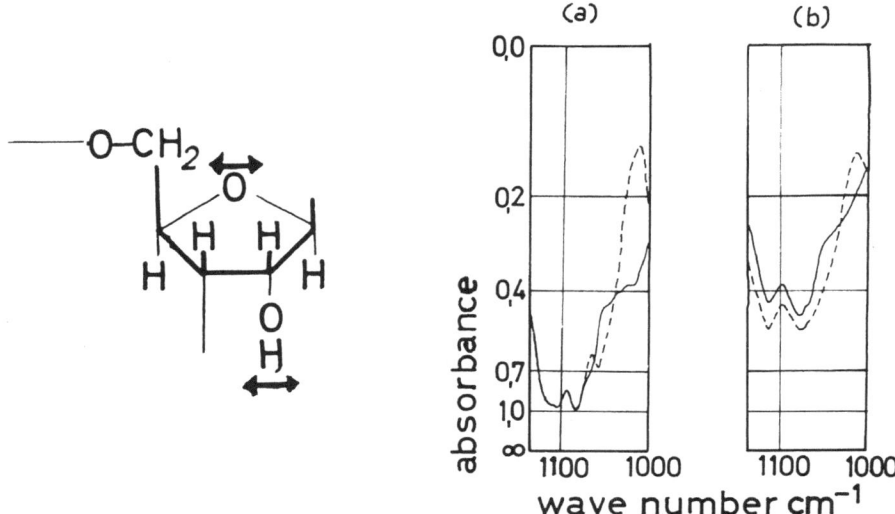

Fig. 2.   Schematic representation illustrating the direction of the transition moments of vibrations and IR spectra of films of poly (C) at 75% relative humidity of the air ——— H_2O hydrated, D_2O hydrated. (a) K^+ salt; (b) Mg^{++} salt.

can see – and this is particularly important – that the ether vibration is shifted toward larger wave numbers by coupling with the 2'OD bending vibration. Hence the vibration of the ether group merges with this band complex. Due to coupling the bands shift in opposite directions. Hence the 2'OD bending vibration is shifted toward smaller wave numbers. Frequently the band of the ether vibration is not pronounced. For instance, with the Mg^{2+} salt of poly (C) this band is only observed as a broad shoulder on the side of this band complex (Figure 2b). One can nevertheless recognize clearly that this shoulder shifts with H-D exchange toward larger wave numbers and merges in the band complex.

These results prove that the transition moments of the vibration of the ether group and those of the 2'OD bending vibration are approximately parallel. Molecular models show that these transition moments are only parallel when the 2'OD group is cross-linked by a hydrogen bond with the ether O-atom of the neighboring ribose residue. In Figure 3a the transition moments are indicated by the lower sticks. These only become parallel when this hydrogen bond is formed. Hence these band shifts caused by coupling prove that the 2'OD groups are cross-linked with the ether O-atoms of the neighboring ribose residues by a hydrogen bond. The formation of such hydrogen bonds was already suggested by Rabczenko and Shugar [12] in the case of poly (U).

The fact that these hydrogen bonds formed by the 2'OH group is of importance for the secondary structure formation is not in contradiction to the observation by Zmudzka and Shugar [13], who found that the 2'-O-methylated Mg^{2+} poly (U) also forms a secondary structure. The melting behavior of this structure is, however, completely different from that of the non-2'-O-methylated

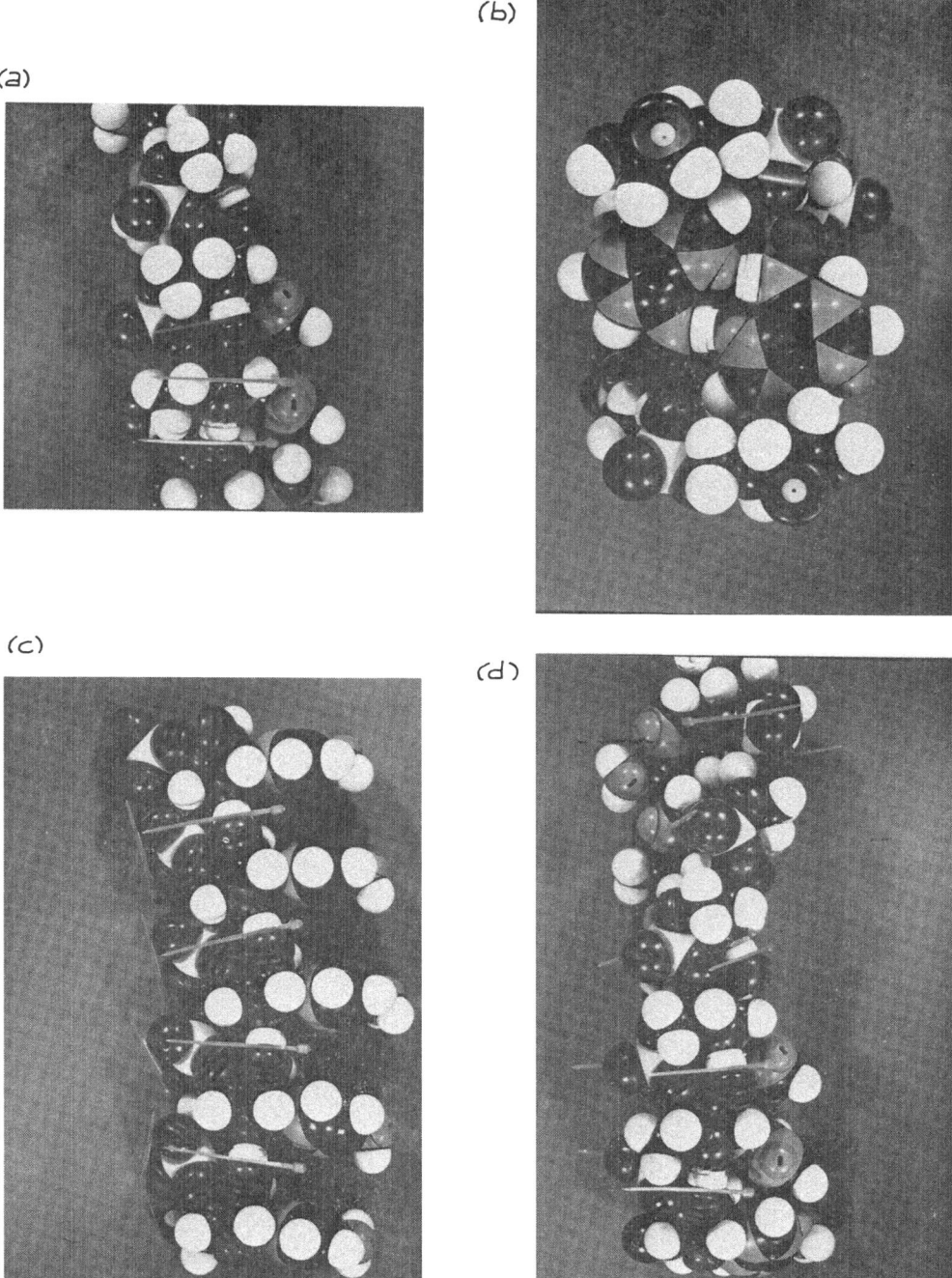

Fig. 3.   Models of RNA molecules. (a) to demonstrate the coupling of the antisymmetric stretching vibration of the C−O−C group and the bending vibration of the 2′OD group (lower sticks). This indicates the formation of the hydrogen bond by the 2′OD group in the backbone. (b) the double helix suggested for the Ca^{2+} salt of poly (A); (c) $>$PO$_2^-$ groups directed outward in the backbone; (d) $>$PO$_2^-$ groups turned to the base residues.

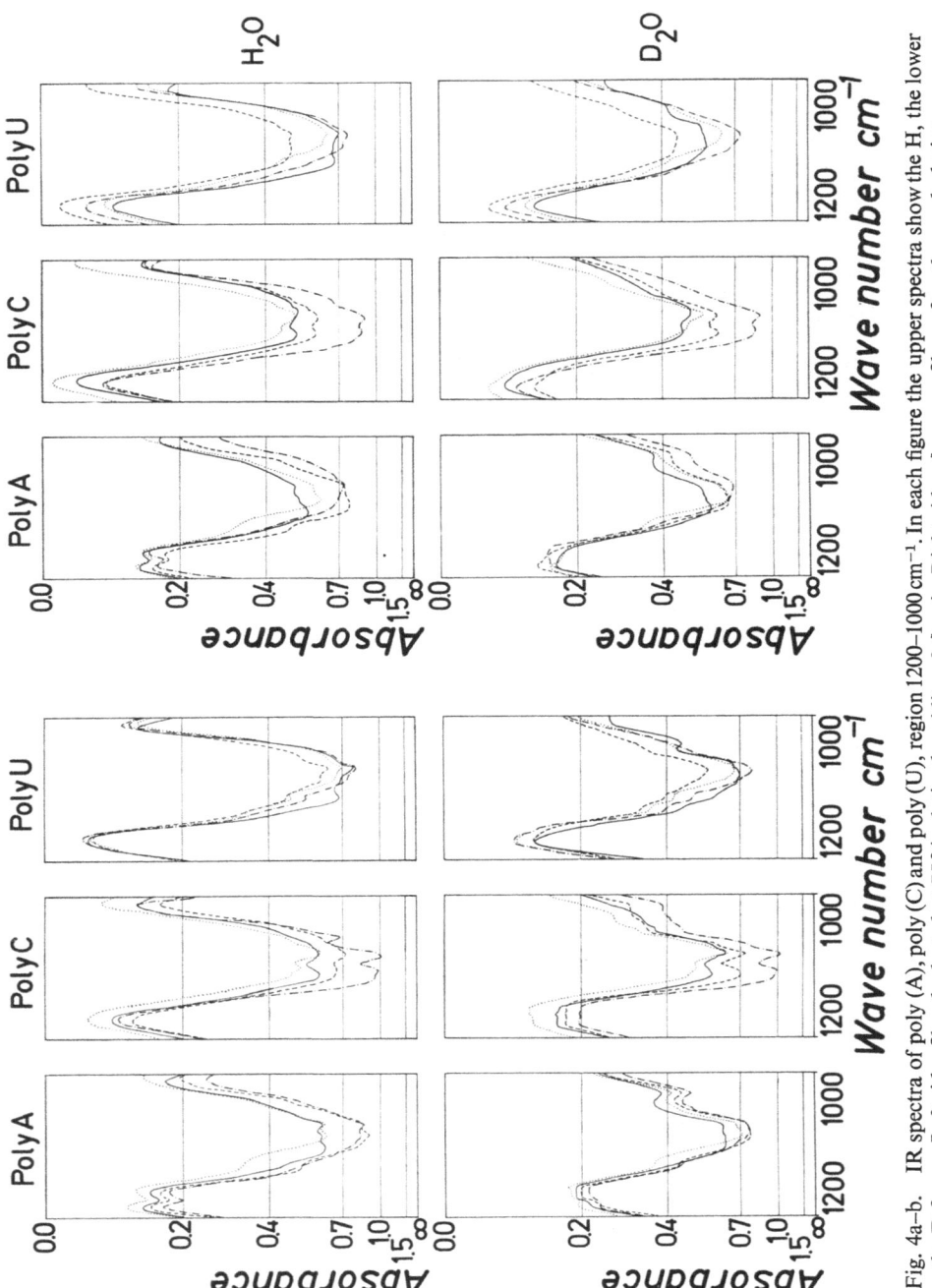

Fig. 4a–b. IR spectra of poly (A), poly (C) and poly (U), region 1200–1000 cm⁻¹. In each figure the upper spectra show the H, the lower the D forms. Left side: films hydrated at 75% relative humidity of the air; Right side: the same films after thorough drying.
(a) ——— Li⁺ ----- Na⁺ ·---·--- K⁺ ····· Cs⁺ salts; (b) ——— Mg⁺⁺ ·---·--- Ba⁺⁺ ····· Cs⁺ salts.

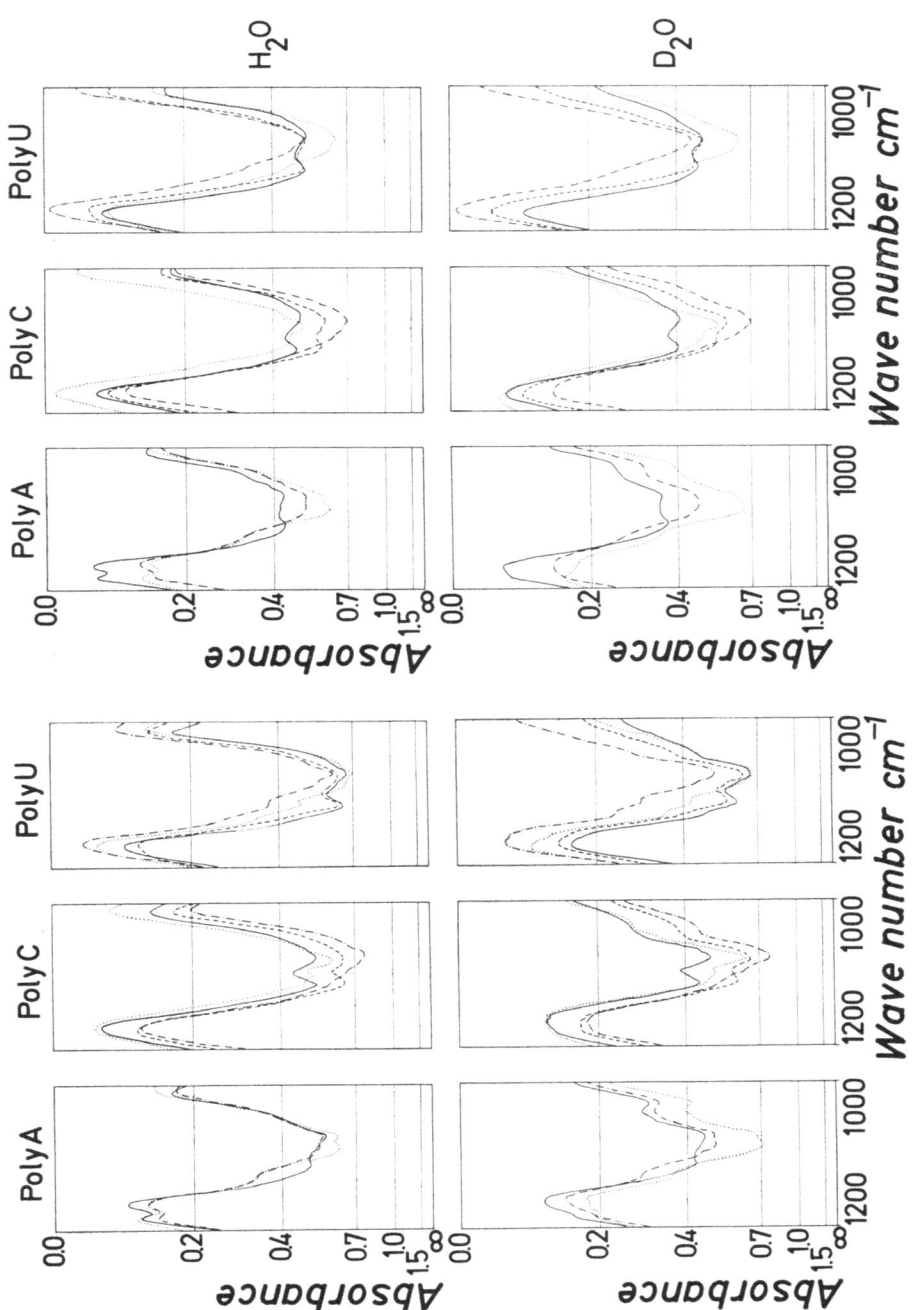

Fig. 4b.

compounds. It melts at higher temperatures and the melting curve is not steep. Thus the latter shows that in contrast to the structure formed by the non-methylated compound the cooperativity is not large. Hence the stabilizing forces of the structure of the 2'-O-methylated compound must be quite different from those which stabilize the structure of the non-methylated ones. Rabczenko and Shugar [14] suggested that hydrophobic interactions between the methyl groups and the base residues are of importance for the stabilization of the structure formed by the methylated compounds.

The observed backbone structure conforms well with the observation made by Melcher [15] on NMR investigation, namely that the 2'H of the ribose interacts with the π-electron system of the base, this as a result of hydrophobic interaction. Figure 3a shows that this 2'H group is turned toward the aromatic π-electron system.

The backbone of the RNA, in contrast to that of the DNA, can accordingly become more rigid through the formation of hydrogen bonds between the 2'OH groups and the O atom of the neighboring ribose residues. This result is of great significance for it shows that the difference between DNA and RNA, as far as this difference is due to the 2'OH group, is caused by the structure promotion of these hydrogen bonds.

2.1.1.1. *Dependence of the structure promotion on the type of base present.* Figure 4 shows these bands for poly (A), poly (C) and poly (U) in the presence of various cations. The spectra of the H_2O-hydrated samples are shown in the top left of the figure, respectively, those of the D_2O-hydrated sample below. The corresponding samples which have been extensively dried are represented on the right-hand side.

In the case of all D_2O-hydrated salts of the poly (A), the 2'OD bending vibration appears shifted toward smaller wave numbers as separate band at 1030 cm^{-1}. That is, the hydrogen bonds, which form the 2'OD groups with the ether O-atom of the neighboring ribose residues, are well formed with all salts of the poly (A), thus stiffening the backbone.

With the poly (C) the 2'OD bending vibration can only be recognized as a shoulder, that is, the band is no longer shifted so markedly toward smaller wave numbers due to the coupling, i.e., the tendency for the hydrogen bonds to form in the backbone has decreased from poly (A) to poly (C).

With poly (U) this shoulder is usually even weaker, but can be clearly recognized with the Mg^{2+} salt. The Li^+ salt, which exhibits a marked band at 1040 cm^{-1}, is an exception. Accordingly in this case the hydrogen bonds in the backbone are usually rather lacking in prominence, that is, with most salts of the poly (U) the stabilization of the backbone due to these hydrogen bonds plays only a minor role.

The tendency to form hydrogen bonds in the backbone accordingly increases in the order poly (U), poly (C), poly (A) [10]. This can be understood in the light of NMR investigations. It was shown by Wagner [16] that the tendency for the stacking of the base residues increases in the same sequence. This stacking favors formation of the secondary structure and hence hydrogen bonding in the backbone. The stabilization of the structure by hydrogen bonds and the stacking of the base residues favor each other, causing the observed dependence on the nature of the poly(nucleotide).

2.1.1.2. *Dependence of the structure formation on the degree of hydration.* Let us now compare the hydrated samples with the thoroughly dried ones in Figure 4. The band or shoulder, respectively, of the 2'OD bending vibration is far less marked and has sometimes disappeared completely in the thoroughly dried samples. The shift of this band toward smaller wave numbers due to the coupling is accordingly less, that is, the structure of the backbone breaks down on drying.

2.1.2. *Influence of Cations on the Conformation of the RNA Backbone in Neutral Medium*

We shall discuss the second question, which concerns how cations change the structure of the RNA backbone. Information on this problem is given by the band at about 1240 cm^{-1} [10]. With RNA this band sometimes shows a doublet structure* (Figure 5). Table I shows that the band observed on D$_2$O hydration lies between the two doublet maxima found on H$_2$O hydration. Thus no significant shift of the overall band complex is involved with the splitting. A similar band splitting is observed with natural RNAs, for instance with 23S r.RNA when Mg^{2+} ions are present [19].

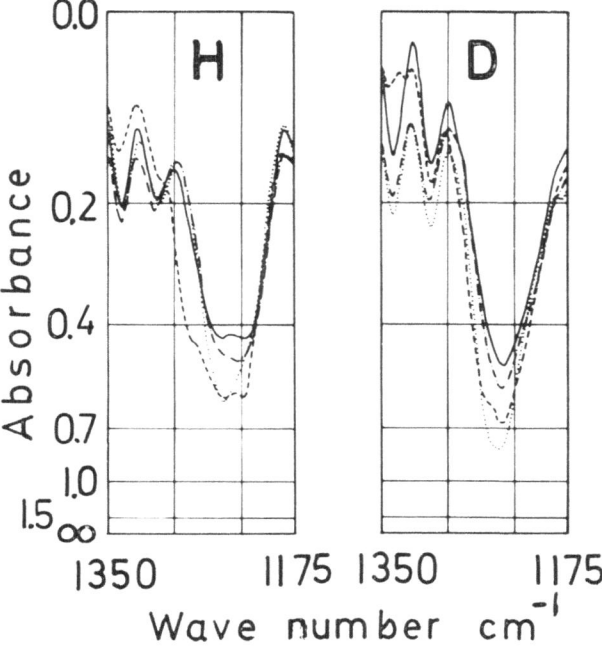

Fig. 5.   IR spectra of salts of poly (A) films hydrated at 75% relative air humidity. Left side: H$_2$O hydrated; Right side: D$_2$O hydrated. ———— Mg^{++}, – – – – Ca^{++}, ·–·–·– Ba^{++}, ····· Cs$^+$ salt.

* In the case of the poly (U) an additional band is observed on H$_2$O hydration at the slope of the former band toward larger wave numbers. This is probably the N glycosidic C-N stretching vibration, which on coupling with the NH bending vibration is shifted from 1300 cm^{-1}, where it is observed with the D$_2$O-hydrated sample, toward smaller wave numbers, that is, to 1266 cm^{-1} [17]. The bending vibration emerges on the transition from D$_2$O to H$_2$O hydration at 1422 cm^{-1} (see ref. [10]).

TABLE I

Splitting of the band at about 1240 cm^{-1}

Substance	Wave number cm^{-1}		
	H$_2$O		D$_2$O
Poly (A)			
Mg^{++}	1242	1224	1236
Ca^{++}	1242	1222	1235
Ba^{++}	1241	1227	1236
Li$^+$	1245	1220 sh	1237
Na$^+$	1241	1220 sh	1238
Poly (U)			
Mg^{++}	1242	1219 sh	1235

The band at about 1240 cm^{-1} is ascribed to the antisymmetric stretching vibration of the $>PO_2^-$ groups [2] [3]. The doublet structure, however, vanishes on H-D exchange. Hence a vibration which disappears on deuteration must participate in the coupling causing this splitting. In this range only OH and NH bending vibrations come into question. The only band common to all poly(ribonucleotides) and especially, too, to poly (A), is the in-plane bending vibration of the 2'OH group. It is known from investigations with secondary alcohols that in the region 1450–1000 cm^{-1} a band pair is observed in which the OH in-plane bending vibration participates [17]. Analogous bands in which the 2'OH in-plane bending vibration is involved are expected with poly(ribonucleotides). Thus the deuteration-sensitive splitting indicates that one of these two vibrations in which the 2'OH bending vibration is involved is contained in the band complex observed at about 1240 cm^{-1}.

Hence the doublet structure shows that the antisymmetric stretching vibration of the $>PO_2^-$ groups can couple with a vibration with which the 2'OH bending vibration is involved. This doublet structure, however, is only observed with some salts of the poly(ribonucleotides). Hence two different backbone conformations must exist, one whereby the transition moments of both vibrations are largely parallel and one whereby they lie perpendicular to each other. The transition moment of the antisymmetric $>PO_2^-$ stretching vibration is parallel to the line connecting the two non-ester-bonded oxygen atoms. The transition moment of the in-plane bending vibration of the 2'OH group is perpendicular to the OH bond and fixed in the furyl ring plane when – as previously discussed – the 2'OH groups cross-link the ribose residues.

Molecular models show how the groupings in the backbone must be oriented so that both transition moments are perpendicular or parallel, respectively, to each other. Figures 3c and d illustrate both cases. If the $>PO_2^-$ groups at the backbone are turned outward – as shown in Figure 3c – the transition moments of the antisymmetric $>PO_2^-$ stretching vibration and of the 2'OH bending vibration are perpendicular. No band splitting can be observed.

The $-CH_2-$group in 5'ribose position and the phosphate group, however, can be turned about the ester bonds in opposite directions. The $-CH_2-$group then turns outward whereas the phosphate group is turned toward the base, as illustrated in Figure 3d. The transition moments of the antisymmetric stretching vibration of the $>PO_2^-$ groups and those of the 2'OH bending vibration are oriented in parallel and hence couple. Conversely, the observed band splitting probably indicates this backbone conformation. This structure, where the bases are turned toward the phosphate groups, are relatively stiff mono-helices – as the molecular models show. These monohelices are right-handed screws. This structure of the backbone influences the structure formation of the double or triple helices formed by poly(ribonucleotides) [21] [22].

Similar band splitting at 1240 cm^{-1} was observed by Tsuboi *et al.* [1] [23] on investigating the double helical rice dwarf virus RNA. These authors were able to orient the macromolecules in the film and then to study the samples with polarized IR. They likewise found a band doublet on H_2O hydration. The one component appeared preferentially when the polarization plane was oriented parallel, the other when this plane was oriented perpendicularly to the helix axis. These authors assume that this splitting is the result of a coupling of the antisymmetric stretching vibration of all phosphate groups at the backbone. For Miyazawa's theory [24]–[26] states that due to such coupling, one vibration parallel to the helical axis and one vibration perpendicular to the helical axis is to be expected. As with this investigation, the band splitting, however, disappears on H → D exchange [1] [23]. This finding does not conform with the explanation given. A further problem arises, since Miyazawa's theory regarding the polynucleotides predicts not only a splitting but a considerable shift of the whole band complex, which is observed neither in the case of the rice dwarf virus nor with the poly(ribonucleotides) (Table I). On the other hand it can be assumed that dichroism would also be observed with the polymers we investigated if these could be oriented in the film.

The question now arises as to the possibility of interpreting these observations homogeneously and without contradictions.

A splitting is observed in the case of the DNA, too. The latter, however, is so slight that it appears merely as an extremely small band shift on investigation with an IR light oriented perpendicularly or parallel to the helix axis [7]. This splitting can be caused by the coupling of the antisymmetric stretching vibrations of the $>PO_2^-$ groups, since according to Miyazawa's theory in the case of polynucleotides, if the splitting is small the shift of the wole band complex is small, too.

The observations regarding the RNA could then be interpreted as follows: the relatively large splitting with RNA is caused by the coupling of the antisymmetric stretching vibration of the $>PO_2^-$ groups and of the vibration with which the in-plane bending vibration of the 2'OH groups is involved. In addition, these coupled vibrations of the individual nucleotides are, according to Miyazawa's theory, coupled mutually, too. This coupling causes additional splitting of the individual components, which, however, is not observed, since – as with the DNA – it is small. This additional coupling, which is increased by the previously discussed backbone stiffening in the monohelices, results, however, in the single doublet components obtaining a preferential direction. This interpretation would explain all experimental findings. Further studies, especially theoretical investigations, however, are necessary to provide a final interpretation of these coupling effects.

2.1.2.1. *The dependence of the orientation of the* $>PO_2^-$ *groups at the backbone on the cations present and on the degree of hydration.* Figure 6 shows that the splitting of the band at 1240 cm^{-1} is observed with the Mg^{2+} and Ba^{2+} salt of the poly (A). The second band appears as a marked shoulder at the slope of the band toward small wave numbers with the Li^+ salt of the poly(A), with the Mg^{2+} salt of the poly (U), and finally as a somewhat less marked shoulder with the Na^+ and K^+ salt of the poly(A). Similar band splitting is sometimes observed with natural RNAs when cations with strong fields are present [19]. According to the above, cations with strong fields probably turn the $>PO_2^-$ groups toward the base residues, thus

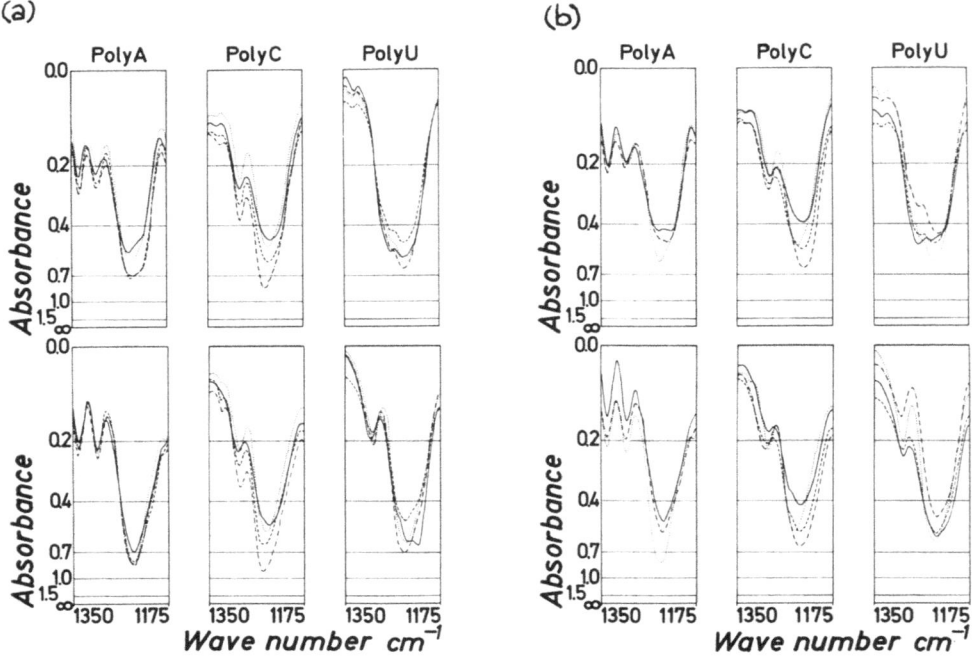

Fig. 6.   IR spectra of poly (A), poly (C) and poly (U), region 1350–1175 cm^{-1}. In each figure the upper spectra show the H$_2$O, the lower the D$_2$O hydrated samples. Films hydrated at 75% rel. humidity of the air. (a) ——— Li$^+$ – – – – – Na$^+$ ·—·—·— K$^+$ ..... Cs$^+$ salts; (b) ——— Mg^{++} ·—·—·— Ba$^+$ ..... Cs$^+$ salts.

inducing the stiff monohelical structure illustrated in Figure 3c. This is understand-able, for these cations can interact in this structure with the phosphate group. With this structure free enthalpy is gained by the interaction between cations and phosphate groups, and especially since the negative charges of the phosphate groups are screened to a large extent and do no longer repulse each other. In the case of cations with strong fields these free enthalpy gains evidently compensate for the lower free hydration enthal-py of the phosphate groups and the cations involved with this conformation. It is striking that this splitting is never observed with poly(C). According to this, the cations with poly(C) can never turn the $>$PO$_2^-$ groups toward the base residues.

The doublet structure disappears on drying. This is understandable, for, as we already know, the backbone structure is disturbed on removal of the hydration water.

Further information on the structure of the backbone and especially on structure changes caused by cations is given by other bands of the ribose residues, especially by a band at about 1130 cm^{-1} (Figure 6). This is an antisymmetric skeleton vibration of the $\begin{smallmatrix} C \\ C \end{smallmatrix}>C-$O group of the ribose residue [1] [3]. The intensity of this band is strongly dependent on the nature of the cations present. This dependence shows that the conformation of the ribose changes as a function of the cations present. The

conformation occurring in any specific case, however, can only be determined when
normal coordinate treatments are conducted for the various conformations.

2.3. SECONDARY STRUCTURES OF Mg^{2+} POLY(U) AND Ca^{2+} POLY(A)

Besides the discussed general effects with respect to the structure of the backbone,
very specific cation effects are also observed. I shall discuss two examples. Firstly, the
structure of poly(U), formed at low temperatures by Mg^{2+} ions. Secondly, a structure
of poly(A), induced by Ca^{2+} ions at neutral pH.

2.3.1. Mg^{2+} poly(U)

It is well known that at low temperatures Mg^{2+} ions form a secondary structure with
poly(U) [13] [27] [28]. This structure melts in the temperature range between
2 and 10°C. Figure 7 shows the two C=O stretching vibration bands of poly(U).
With the K^+ salt, on the left side, no change is observed due to temperature. With

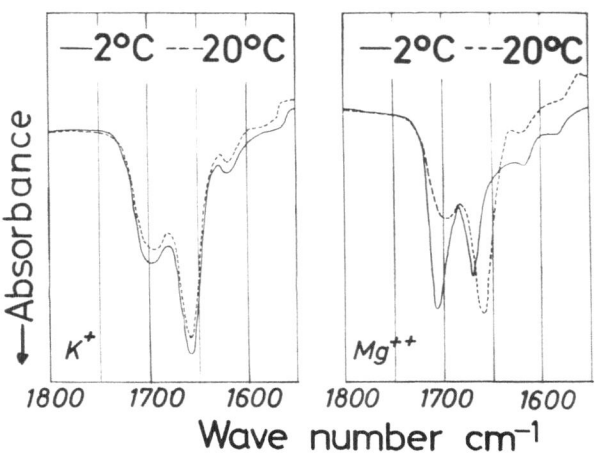

Fig. 7.   IR spectra of poly (U) ——— 2°C ----- 20°C; Left side: K⁺ salt; Right side: Mg⁺⁺ salt.

the Mg^{2+} salt, both bands shift a little toward larger wave numbers on cooling and
change their intensities. But, and this is very important, the distance between these
two bands is not increased and no additional splitting is observed [22].

Let us compare these results with results which were obtained by Howard et al. [4]
on investigating the base pairing of guanine and cytosine. These authors first investigated the addition of guanine monophosphate to poly(C). At low temperatures the
monomer is attached to the poly(C) in such a manner that a double helix is formed.
This structure melts when the temperature is increased. Figure 8 shows the IR spectra
of this system. At higher temperatures only one band of the two C=O stretching
vibrations of guanine and of cytosine is observed. With decreasing temperature this
band splits into two components when the structure is formed. Figure 9a shows that

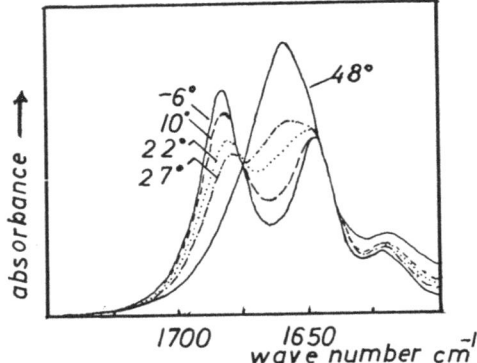

Fig. 8.   Temperature dependence of infrared spectrum of two-stranded helix formed between poly
(C) and 5'GMP. On cooling, the spectrum becomes essentially constant by 0°C (spectra taken from
ref. [4] with the kind permission of Prof. Miles).

(a)

(b)

(c)

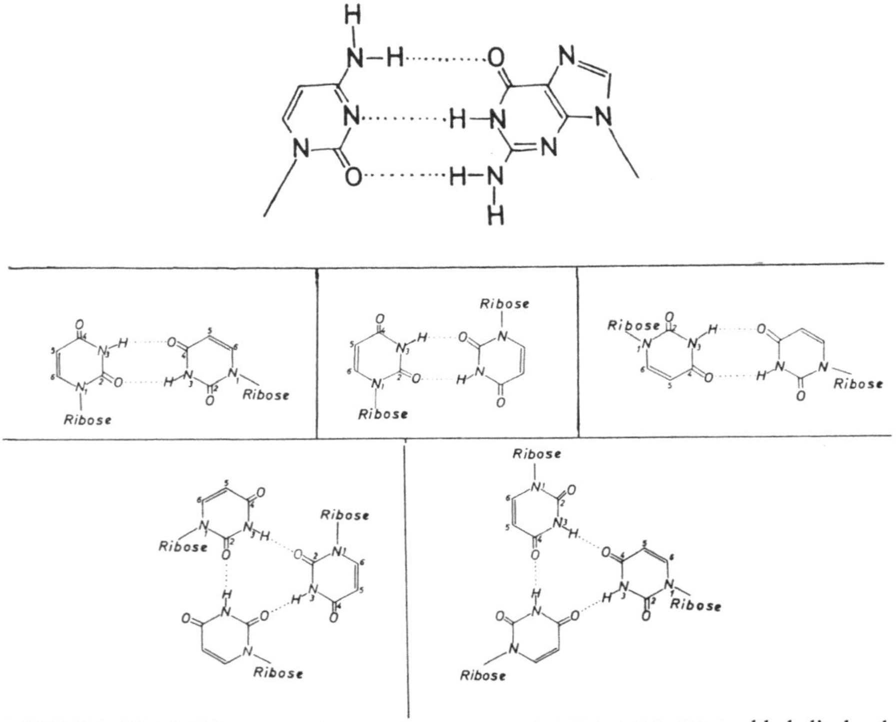

Fig. 9.   Schematic representation to explain the base pairing. (a) (G)+(C); (b) double helical poly
(U); (c) triple helical poly (U).

the transition moments of the stretching vibrations of the two C=O groups are
parallel in the base-paired structure. Similar splitting is observed in the case of the
(G+C) double helix [4].

The same should be observed with poly(U) if poly(U) were to form a double
helical structure, since with this structure formation transition moments of different

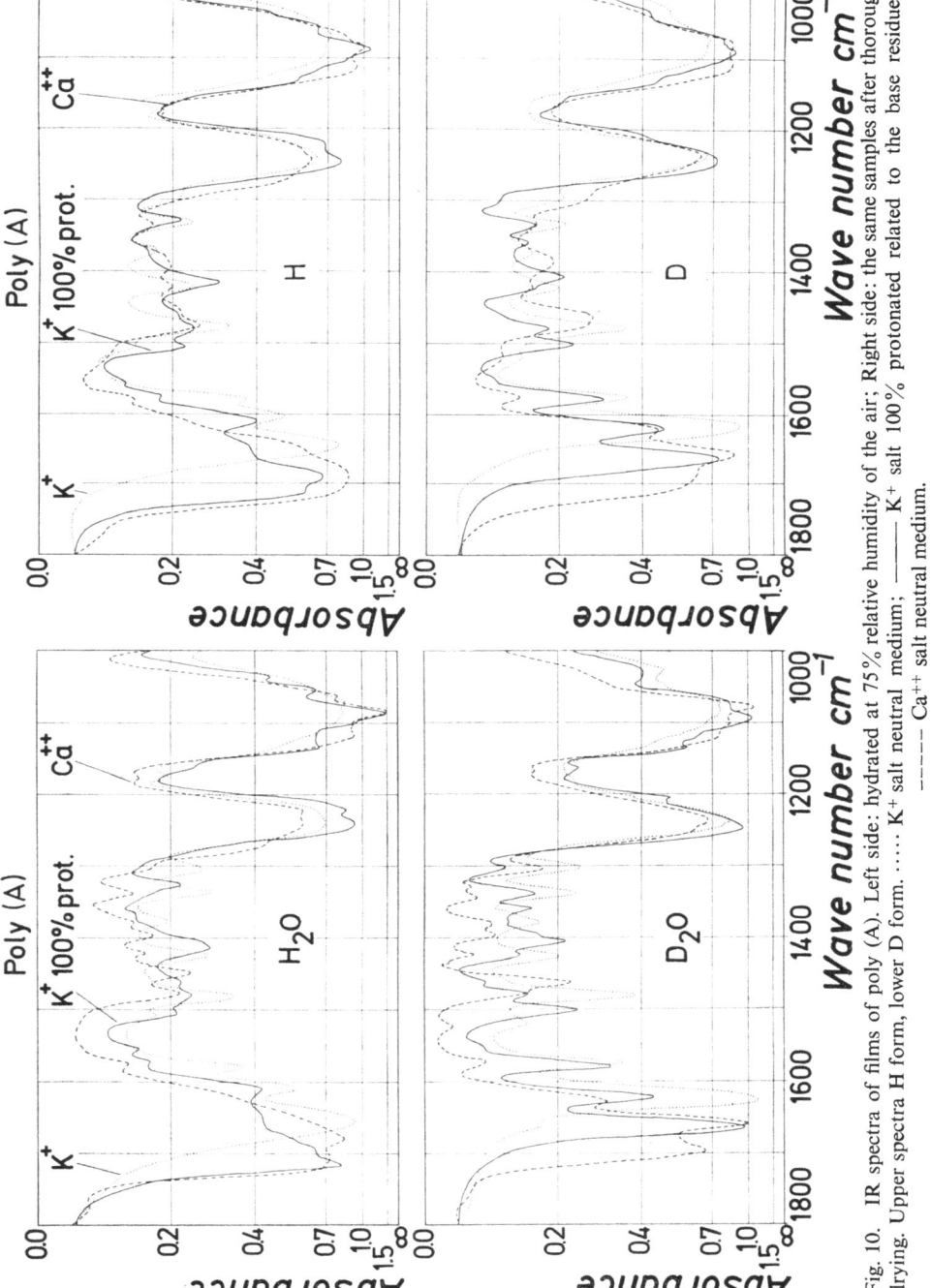

Fig. 10. IR spectra of films of poly (A). Left side: hydrated at 75% relative humidity of the air; Right side: the same samples after thorough drying. Upper spectra H form, lower D form. ..... K⁺ salt neutral medium; ——— K⁺ salt 100% protonated related to the base residues; ———— Ca⁺⁺ salt neutral medium.

C=O groups become parallel (Figure 9b). With Mg^{2+} poly(U), however, no additional splitting or shifting in opposite directions of the bands is observed.

It thus seems probable that the secondary structure formed by Mg^{2+} poly(U) is not a double but a triple helix [22], since in these cases the C=O groups would not be oriented either parallel or antiparallel to each other (Figure 9c). Hence, when a poly(U) triple helix is formed no band shifts or splittings caused by coupling should be observed. This corresponds very well with the observed results (for details see ref. [22]). A final decision, however, can only be made by determining the radius of gyration of the cross section by small angle X-ray scattering. Plotting Raman spectra would be useful, too. Plotting such spectra of polynucleotides can, however, only be effected with special equipment, Small and Peticolas [20].

2.3.2. $Ca^{2+}poly(A)$

In Figure 10 the dashed lines are spectra of the Ca^{2+} salt of poly(A). The dotted lines are spectra of the K^+ salt at neutral pH and the solid lines are spectra of the double helix formed in acidic medium, at the top with H_2O and below with D_2O hydration.

The spectra of the Ca^{2+} poly(A) differ fundamentally from those of the K^+ poly(A) in neutral medium. In some respects the spectra of Ca^{2+} poly(A) are similar to those of the double helix, which is formed by poly(A) in acidic medium. In other respects, however, these spectra are completely different [21].

Several findings in the spectrum of the Ca^{2+} salt show independently of one another that the Ca^{2+} ions transform the amino groups of the adenine base to imino groups. The Ca^{2+} ions shift the tautomeric amino-imino equilibrium toward the imino form.

The following facts prove this. Firstly, almost all ring vibrations change fundamentally. Secondly, the N glycosidic CN stretching vibration is found at the slope of the antisymmetric $>PO_2^-$ stretching vibration at 1284 cm^{-1}. In the case of the D_2O hydrated sample it is observed, as with the other polynucleotides, at 1303 cm^{-1}. This indicates that this band is shifted toward smaller wave numbers through coupling with the bending vibration of the NH group which is only present in the base residues of the imino form. The coupling effect is the same as with the protonated poly(A) and particularly with poly(U). In both cases an NH group is present in the base residue. Thirdly, with the Ca^{2+} salt, the band of the $-NH_2$ scissor vibration at 1580 cm^{-1} largely disappears. Fourthly, a ring vibration is formed at about 1700 cm^{-1} in the case of the Ca^{2+} salt as well as with the double helix. With the Ca^{2+} salt, the additional, extremely intense band at 1647 cm^{-1} is probably to be ascribed to the now present exocyclic C=N double bond. All these facts demonstrate that the Ca^{2+} ions induce the imino form of the base residues.

Let us now consider the bands of the sugar phosphate backbone. A doublet structure of the band at 1240 cm^{-1} is also observed. That is, here too, the Ca^{2+} ions turn the $>PO_2^-$ groups toward the base residues.

Hence the Ca^{2+} ions interact with the $>PO_2^-$ groups as well as with the base residues. The construction of molecular models illustrates that a mono-helical structure, whereby the Ca^{2+} ions interact with the phosphate and with the imino group at the same time, is not possible. All available findings point to the probability that a double helix of the type shown in Figure 3b forms. The Ca^{2+} ions induce the double helix, crosslinking the phosphate with the imino groups (for details see ref. [21]).

3. Na^+-K^+ Dependent Conformation Change of Proteins of Excitable Membranes

The IR spectroscopy is a very suitable method to decide with proteins between β-structures i.e. pleated sheet on the one side and α-helical structure and coil on the other. The proteins have many equal groupings in the backbone. The same groups, for instance, all $C=O$ groups in these peptide groupings have the same vibrations. The $C=O$ stretching vibration – the Amid I band – is observed near 1650 cm^{-1}. Miyazawa [24]–[26] has shown that all $C=O$ groups vibrate not independent from one another, if these groups are regularly arranged in secondary structures, whereby different components of the Amid I band are observed by polarized IR in the direction and perpendicular to the axis of the macromolecules (for details see [24]–[26] and [29]–[32]). Thus it is understandable that the exact position of these Amid I bands is determined by the secondary structure since the mutual orientation of the $C=O$ groups is completely different in β-structures and in α-helical structures. The Amid I band is observed with α-helical and coiled proteins at 1650 cm^{-1}. In contrast to this, this band is observed with proteins in β-conformation at 1630 cm^{-1} and when this β-structure is antiparallel a very weak but characteristical shoulder is observed at about 1690 cm^{-1} [29]–[32].

Based on these changes of bands dependent on the conformation we investigated the sodium-potassium dependent conformation change of proteins in excitable membranes [33].

The action potential is caused by a change of cation permeability of the axon membrane. In consequence of this permeability change sodium ions flow in and reverse the potential. The permeability then changes again, the resting potential is regenerated, and potassium ions flow out.

The question now arises: How does this permeability change come about? Possibly a conformation change of proteins in the excitable membrane is connected with the permeability change. It seems probable that these changes are induced by the cations themself.

To clarify this, we prepared membrane vesicles from the brains of young rats. From these, following the procedure given in [34], one can prepare a fraction of non-myelinated axon membranes. IR spectra were obtained from these strongly hydrated membrane vesicles in which K^+, Na^+ or Ca^{2+} ions were present. The results are shown in Figure 11. In the presence of the Na^+ and Ca^{2+} ions only one Amid I band is observed at 1652 cm^{-1}. In contrast to this result in the presence of K^+ ions, a strong Amid I band at 1630 cm^{-1} and a weak shoulder at 1695 cm^{-1} were observed.

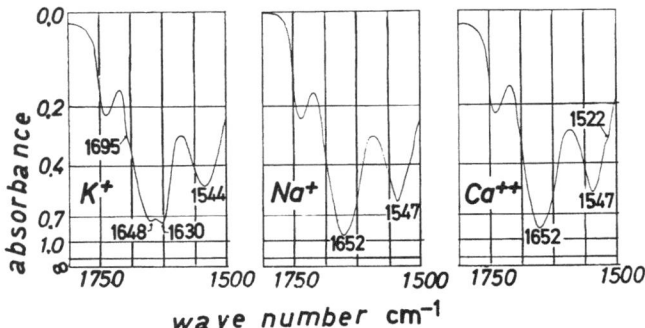

Fig. 11. Infrared spectra of K+, Na+ and Ca²+ form of axon membrane vesicles, samples H₂O hydrated at 90% relative air humidity.

This proves that the K⁺ ions cause a large portion of the membrane proteins to have β-structure. Further it is shown by this weak shoulder that this β-structure is at least partially antiparallel. In the presence of Na⁺ and Ca²⁺ ions no protein having β-structure was observed.

It cannot be determined from the IR spectra whether proteins in the presence of Na⁺ and Ca²⁺ ions are more α-helical or more coiled. This can be determined, however, by ORD and CD measurements. In the presence of Na⁺ ions in the membranes the ORD spectrum shows at 2330 Å a trough and the point of intersection with the zero line is observed at 2200 Å. Both of these observations are characteristic for proteins with α-helical structure.

Hence the IR investigations have shown that in the presence of K⁺ ions large parts of the proteins in the excitable membranes have antiparallel β-structure. In the presence of Na⁺ and Ca²⁺ ions, no β-structure is observed. The proteins are largely α-helical as shown by the ORD and CD measurements.

All these results suggest that relatively large parts of the membrane proteins change their conformation during the excitation process and that these changes can be induced by the cations themselves which flow through the membrane. Hence the permeability change during excitation of nerve membranes is probably strongly connected with such cation-induced conformation changes.

4. Cross Linking of Fixed Ions with Macromolecules by Polyvalent Counter-Ions

Another cation-induced change of bands in the IR spectra, which can provide information on the interaction of groups and especially on the structure, are band splittings caused by removal of degeneracy [35].

Groups with C_{3v} symmetry i.e. pyramid-shaped groups, have a degenerate antisymmetric stretching vibration. The degeneracy is, however, removed when cations are asymmetrically attached to these groups. In such cases a band splitting, caused by the removal of the degeneracy, is observed. This splitting is an indication of the interaction of cations, for instance, with negatively charged fixed ions.

We studied these interactions with artificial polyelectrolyte membranes, membranes from salts of polystyrene sulfonic acid [35] [36]. The fixed $-SO_3^-$ ions in these membranes have C_{3v} symmetry when they are isolated from their environment. In the membranes, however, the antisymmetric stretching vibration of these groups is split by removal of the degeneracy, caused by the interaction with the cations.

This is shown in Figure 12a. With increasing cation field the splitting of the antisymmetric stretching vibration of the $-SO_3^-$ groups increases by removal of the degeneracy [35] [36].

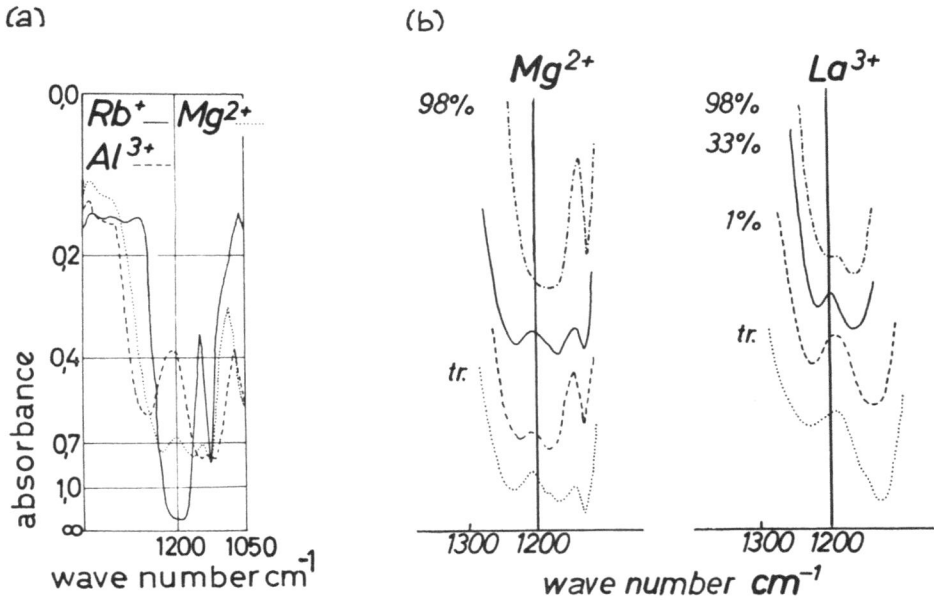

Fig. 12. Antisymmetric stretching vibration of the $-SO_3^-$ groups (doublet near 1200 cm^{-1}) with salts of polystyrene sulfonic acid to demonstrate the removal of degeneracy. (a) Rb$^+$, Mg^{2+} and Al^{3+} salt, dried samples; (b) Mg^{2+} and La^{3+} salt, dependent on the degree of hydration (films hydrated at 98%, 33%, 1% relative air humidity, tr thoroughly dried sample).

Figure 12b shows the dependence of this splitting on the degree of hydration [35] [37]. This splitting decreases with increasing degree of hydration. The cations become less firmly attached to the anions on the addition of water molecules.

Of special importance is the result that in the presence of polyvalent cations even at relatively large degrees of hydration no non-split band is observed other than the doublet (Figure 12b). This result shows that almost all divalent or trivalent cations cross-link two or three fixed anions, respectively (Figure 13).

4.1. CONCLUSIONS WITH REGARD TO BIOLOGICAL MEMBRANES

This cross-linkage of fixed anions by polyvalent cations, observed in the present system, may play an important role in some biological systems, especially in the

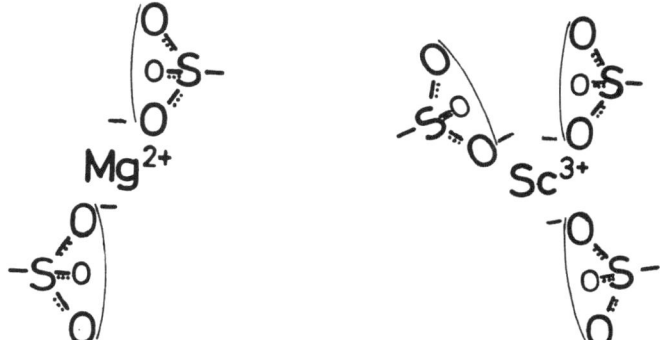

Fig. 13.   Schematic representation to demonstrate the cross-linking of fixed ions
by polyvalent counterions.

excitable membranes, too [35]. It is well-known that Ca^{2+} ions are of decisive importance for the formation of the resting and action potential with the excitable
membranes. As postulated by Tasaki [38] in his theory of the excitation processes,
the results with artificial membranes show that the Ca^{2+} ions cross-link fixed ions
in membranes. Thus they may regulate the permeability of these membranes in
cooperation with the Na^+-K^+ induced conformation changes of proteins in the
excitable membranes discussed in Section 3.

Acknowledgments

The spectra were plotted with the IR spectrophotometers model 221 and 325 supplied
by Bodenseewerk Perkin Elmer, Ueberlingen, W. Germany. Our thanks are due to
the Deutsche Forschungsgemeinschaft and to the Fonds der Chemischen Industrie
for providing the facilities for this work.

References

1. Tsuboi, M.: *Appl. Spect. Rev.* **3**, 45 (1969).
2. Tsuboi, M.: *J. Polymer Sci.* C7, 125 (1963).
3. Tsuboi, M., Matsuo, M., Shimanouchi, T., and Kyogoku, Y.: *Spectrochim. Acta* **19**, 1617 (1963).
4. Howard, F. B., Frazier, J., and Miles, H. T.: *Proc. Natl. Acad. Sci.* **64**, 451 (1969).
5. Zundel, G., Lubos, W. D., and Kölkenbeck, K.: *Biophys. J.* **12**, 1509 (1972).
6. Kyogoku, Y., Tsuboi, M., Shimanouchi, T., and Watanabe, I.: *Nature* **189**, 120 (1961).
7. Shimanouchi, T., Tsuboi, M., and Kyogoku, Y.: in *Advances in Chemical Physics* (ed. by J. Duchesne) **VII**, 435, Interscience Publ. N.Y., 1964.
8. Fritsche, H.: *Z. Chem.* **12**, 1 (1972).
9. Higuchi, S., Tsuboi, M., and Iitaka, Y.: *Biopolymers* **7**, 909 (1969).
10. Kölkenbeck, K. and Zundel, G.: *Biophysics of Structure and Mechanism* **1**, 203 (1975).
11. Tschirgadže, Ju. I., Sche, M., and Charitonenkov, I.: *Dokladi Akademii Nauk SSSR Biophysika* **203**, 959 (1972).
12. Rabczenko, A. and Shugar, D.: *Acta Biochim. Polonica* **19**, 89 (1972).
13. Zmudzka, B. and Shugar, D.: *Febs letters* **8**, 52 (1970).
14. Rabczenko, A. and Shugar, D.: *Acta Biochim. Polonica* **18**. 387 (1971).

15. Melcher, G.: *Biophysik* **7**, 29 (1970).
16. Wagner, K. G.: *Hoppe-Seyler's Z. Physiol. Chem.* **353**, 765 (1972).
17. Bellamy, L. J.: *The Infra-Red Spectra of Complex Molecules*, 2nd ed., Methuen, London, 1958.
18. Miles, H. T.: *Proc. Natl. Acad. Sci.* **51**, 1104 (1964).
19. Herbeck, R. and Zundel, G.: *Biochim. Biophys. Acta* **418**, 52 (1976).
20. Small, E. W. and Peticolas, W. C.: *Biopolymers* **10**, 69 (1971).
21. Kölkenbeck, K. and Zundel, G.: in preparation.
22. Herbeck, R. and Zundel, G.: in preparation.
23. Sato, T., Kyogoku, Y., Higuchi, S., Mitsui, Y., Iitaka, Y., Tsuboi, M., and Miura, K.: *J. Mol. Biol.* **16**, 180 (1966).
24. Miyazawa, T.: *J. Chem. Phys.* **32**, 1647 (1960).
25. Miyazawa, T.: *J. Chem. Phys.* **35**, 693 (1961).
26. Miyazawa, T., Ideguchi, Y., and Fukushima, K.: *J. Chem. Phys.* **38**, 2709 (1963).
27. Sazer, W.: *Biochem. Biophys. Res. Comm.* **20**, 182 (1965).
28. Rabczenko, A. and Shugar, D.: *Acta Biochim. Polonica* **19**, 89 (1972).
29. Miyazawa, T.: in G. D. Fasman, (ed.), *Poly-α-Amino Acids*, Dekker, New York, 1967.
30. Susi, H.: in S. N. Timasheff and G. D. Fasman (eds.), *Structure and Stability of Biological Macromolecules,* Dekker, New York, 1969.
31. Tschirgadže, Y. N.: *Infrared Spectra and Structure of Polypeptides and Proteins*, Akademia Nauk, SSSR, Moscow, 1965.
32. Parker, F. S.: *Infrared Spectroscopy*, Adam Hilger, London, 1971.
33. Papakostidis, G., Zundel, G., and Mehl, E.: *Biochim. Biophys. Acta* **288**, 277 (1972).
34. Lemkey-Johnston, N. and Dekirmenjian, H.: *Exp. Brain Res.* **2**, 392 (1970).
35. Zundel, G.: *Hydration and Intermolecular Interaction*, Academic Press, 1969, New York and Mir Moscow 1972.
36. Zundel, G. and Murr, A.: *Z. Naturforsch.* **21a**, 1640 (1966).
37. Zundel, G. and Murr, A.: *Electrochim. Acta* **12**, 1147 (1967).
38. Tasaki, I.: *Nerve Excitation*, Charles C. Thomas, Springfield Ill., 1968.

II

NON-ISOTHERMAL TRANSPORTS

NON-ISOTHERMAL TRANSPORT PHENOMENA
IN CHARGED GELS AND MEMBRANES

J. W. LORIMER

Dept. of Chemistry, University of Western Ontario, London, Ontario N6A 5B7, Canada

Abstract. The irreversible thermodynamic theory of discontinuous systems is applied to describe and classify the transport phenomena that can arise when a temperature difference exists between two reservoirs of electrolyte separated by a membrane and containing reversible electrodes through which current can enter and leave the system. The research that has been done on these phenomena is reviewed briefly. In the second part of the paper, the main features of some new results on thermo-osmosis in styrene-*p*-vinylbenzenesulfonate copolymer membranes are presented. The thermo-osmotic permeability depends strongly on the type of counterion and on temperature. A theoretical thermodynamic-hydrodynamic treatment of thermo-osmosis in charged membranes is outlined. The general theory is applied to a model consisting of charged polyelectrolyte rods in parallel arrays in domains; the domains are randomly oriented. Comparison of experiments and theory indicates that three main factors contribute to thermo-osmosis in charged membranes: polarization of solvent molecules in the electric field of the double layer, the heat of transport of the free ions in the membrane, and the mobility of these ions.

1. Introduction

Transport of ions and solvent molecules through charged membranes or gels due to a temperature difference has not been studied extensively, despite the possible import-ance of non-isothermal effects in both biological and industrial membrane processes. In addition, study of non-isothermal transport can provide fundamental information about the mechanisms of energy transfer in membranes.

In the first part of this paper, we survey the non-isothermal transport effects that are predicted to occur in charged membranes and review the existing experimental and theoretical work on these effects. In the second part, we summarize some recent experimental and theoretical work on thermo-osmosis.

2. Irreversible Thermodynamics of Non-Isothermal Transport Processes in Membranes

The basic equations of the irreversible thermodynamics of discontinuous non-isothermal systems have been discussed in several places [1–7]. It is useful, however, to write these equations in terms of experimentally-measurable quantities [5].

Consider a membrane separating two homogeneous solution phases I and II. The dissipation function for this system is [2]

$$\Phi = T\sigma = -j_q' dT/T - \sum_k j_k (d\mu_k)_T - \sum_k z_k j_k F d\psi \qquad (1)$$

where σ is the rate of entropy production due to irreversible processes taking place across the membrane and T is the temperature.

$$(d\mu_k)_T = (d\mu_k)_{T,P} + V_k dp \tag{2}$$

is the difference in chemical potential at constant temperature between solutions I and II, $(d\mu_k)_{T,P}$, dp and $d\psi$ are the corresponding differences in chemical potential (at constant T and pressure), in pressure and in electrical potential, V_k and z_k are the partial molar volume and charge number of species k, and F is the Faraday constant. A positive 'flux' $j_k = -dn_k^I/dt$ is the rate of decrease of the amount n_k of species k in phase I due to transport of k through the membrane. The 'reduced heat flux'

$$j_q' = j_q + \sum_k (H - H_k) j_k \tag{3}$$

is related by Equation (3) to the total heat flux j_q through the membrane and to the difference between the molar enthalpy H of the system and H_k, the partial molar enthalpy of species k. A positive flux j_q is the rate of loss of heat from phase I to phase II, while j_q' is the heat flux from the surroundings if phase I is considered to be closed. The significance of both these heat fluxes has been discussed thoroughly by Tyrrell [2].

Two features of Equation (1) that differ from customary notation should be noted. We write the dissipation function in terms of infinitesimal deviations of the state variables from their equilibrium values, rather than as finite but 'small' deviations [8]. In doing so, we take into account the possibility that the fluxes may be functions of the state variables of the two solution phases. The negative signs in Equation (1) coupled with the definitions of the fluxes guarantee that a negative flux of matter is from a higher to a lower chemical potential, and a negative heat flux is from a higher to a lower temperature. This choice agrees with the usual conventions in continuous systems, that fluxes of matter and heat are in a direction opposite to the respective gradients of chemical potential and temperature. It should also be noted that the 'fluxes' j_k and j_q are total flows rather than flows per unit area.

We now wish to express the dissipation function (1) in terms of overall flow quantities. These are the total electric current

$$I = F \sum_k z_k j_k \tag{4}$$

the total volume flow j_v, the salt flow j_s for a single salt $A_{v_1}B_{v_2}$ that ionizes completely in a solvent O to give v_1 cations of charge number z_1 and v_2 anions of charge number z_2, and a new heat flux j_q''. Clearly,

$$v_1 z_1 + v_2 z_2 = 0 \tag{5}$$

and, for the salt, the chemical potential is

$$\mu_s = v_1 \mu_1 + v_2 \mu_2 \tag{6}$$

and the partial molar volume is

$$V_s = v_1 V_1 + v_2 V_2 \tag{7}$$

The electric current enters and leaves the system via electrodes that are taken to be reversible to the anions, for convenience. If the difference in electrical potential between two wires at the differing temperatures of the electrodes is dE, then the difference in electrochemical potential of the anions in the two phases is

$$d\tilde{\mu}_2 = z_2 F dE = d\mu_2 + z_2 F d\psi$$
$$= (d\mu_2)_{T,P} - S_2 dT + V_2 dp + z_2 F d\psi \tag{8}$$

The salt flux through the membrane is

$$j_s = j_1/\nu_1 = j_2/\nu_2 - I/F z_2 \nu_2 \tag{9}$$

Substitution of Equations (2), (4), (8) and (9) into (1) and use of (6), (7) gives

$$\Phi = - \{ j_q' - TS_2 I/z_2 F \} dT/T - j_s (d\mu_s)_{T,P}$$
$$- j_0 (d\mu_0)_{T,P} - (j_s V_s + j_0 V_0) dp - IdE \tag{10}$$

Equation (11) contains the new heat flux

$$j_q'' = j_q' - TS_2 I/z_2 F \tag{11}$$

which clearly includes the heat flux that accompanies the electrode reactions, but *excludes* any heat flux associated with the electrode phases. For example, for silver-silver chloride electrodes, the reversible electrode reaction is

$$AgCl(s) + e^- = Ag(s) + Cl^-$$

and there is an accompanying entropy change

$$S_2 + S(Ag, s) - S(AgCl, s) - S(e^-) \tag{12}$$

Similarly,

$$j_v = j_s V_s + j_0 V_0 \tag{13}$$

is the volume flux, excluding any volume changes associated with the electrode phases. Weinstein and Caplan [9] (see also [10]) have discussed Equation (13) recently for isothermal transport. From the Gibbs-Duhem equation,

$$c_s (d\mu_s)_{T,P} + c_0 (d\mu_0)_{T,P} = 0 \tag{14}$$

where c_s, c_0 are the concentrations of salt and water. The osmotic pressure difference $d\Pi$ across the membrane is

$$d\Pi = - (d\mu_0)_{T,P}/V_0 = c_s (d\mu_s)_{T,P}/c_0 V_0 \tag{15}$$

Substitution of (12), (13) and (15) in (11) gives, finally,

$$\Phi = - j_s'' d\Pi/c_s - j_v d(P - \Pi) - IdE - j_q'' dT/T \tag{16}$$

where the relation $c_s V_s + c_0 V_0 = 1$ has been used. The significance of j_q'', j_v and j_s should be kept clearly in mind. In addition, it should be noted that dE is not the potential difference dE' between two isothermal wires connected to the electrodes, since dE' includes thermocouple effects. In fact, the system described here corresponds to a Class 1 thermocell for systems that do not contain a membrane. The theory of the e.m.f. of such cells has been described in detail by Agar [11]. (See also [2] Chap. 4.) The theory of the e.m.f. of non-isothermal membrane cells has been discussed by Hills et al. [12, 13]

3. The Phenomenological Equations

Linear phenomenological equations may now be constructed on the basis of the dissipation function, Equation (16). These are written in matrix form in order to display the matrix of phenomenological coefficients clearly:

$$
\begin{bmatrix} j_s \\ j_v \\ I \\ j_q'' \end{bmatrix} = - \begin{bmatrix} L_D & L_{DP} & L_{DE} & L_{DT} \\ L_{PD} & L_P & L_{PE} & L_{PT} \\ L_{ED} & L_{EP} & L_E & L_{ET} \\ L_{TD} & L_{TP} & L_{TE} & L_T \end{bmatrix} \begin{bmatrix} d\Pi/C_s \\ d(P-\Pi) \\ dE \\ dT/T \end{bmatrix}
\tag{17}
$$

The six Onsager relations $L_{DP}=L_{PD}$, etc. also hold, giving ten independent coefficients. Six of these are the customary isothermal coefficients, but L_{DT}, L_{PT}, L_{ET} and L_T are peculiar to non-isothermal systems. Clearly, L_T is connected with the thermal conductivity of the membrane, while the other terms describe coupling between heat flow and diffusive, bulk and charge flow.

4. Non-Isothermal Transport Processes

To discuss the variety of non-isothermal transport phenomena that can arise, it is convenient to solve Equation (17) for dE:

$$
dE = - \{I + L_{ED}d\Pi/C_s + L_{EP}d(P-\Pi) + L_{ET}dT/T\}/L_E
\tag{18}
$$

and substitute this equation into (17) to obtain

$$
j_s = L_{DE}I/L_E - M_D d\Pi/C_s - M_{DP}d(P - \Pi) - M_{DT}dT/T
\tag{19}
$$

$$
j_v = L_{PE}I/L_E - M_{DP}d\Pi/C_s - M_P d(P - \Pi) - M_{PT}dT/T
\tag{20}
$$

$$
j_q'' = L_{TE}I/L_E - M_{DT}d\Pi/C_s - M_{PT}d(P - \Pi) - M_T dT/T
\tag{21}
$$

where

$$
M_D = L_D - L_{ED}^2/L_E
\tag{22}
$$

$$
M_{DP} = L_{DP} - L_{DE}L_{EP}/L_E
\tag{23}
$$

$$
M_{DT} = L_{DT} - L_{DE}L_{ET}/L_E
\tag{24}
$$

$$
M_P = L_P - L_{PE}^2/L_E
\tag{25}
$$

$$
M_{PT} = L_{PT} - L_{PE}L_{ET}/L_E
\tag{26}
$$

$$
M_T = L_T - L_{ET}^2/L_E
\tag{27}
$$

All the M-coefficients have the same general structure: a leading term and a term resulting from the electrical potential difference. For example, M_{PT} contains a 'primary flow' L_P and a 'secondary flow' [14] that is an electro-osmotic flow induced by the thermal potential.

The flux equations may be written in yet another form by defining 'heats of transport' Q_s^*, Q_v^* and Q_I^* by the relations:

$$
\begin{bmatrix} L_{TD} \\ L_{TP} \\ L_{TE} \end{bmatrix} = \begin{bmatrix} L_D & L_{DP} & L_{DE} \\ L_{DP} & L_P & L_{PE} \\ L_{DE} & L_{PE} & L_E \end{bmatrix} \begin{bmatrix} Q_s^* \\ Q_v^* \\ Q_I^* \end{bmatrix} \tag{28}
$$

Substitution of Equations (28) in Equations (17) gives

$$
j_q'' = Q_s^* j_s + Q_v^* j_v + Q_I^* I - \{L_T - (Q_s^* L_{DT} + Q_v^* L_{PT} + Q_I^* L_{ET})\}\, dT/T \tag{29}
$$

These heats of transport are the heat fluxes j_q'' that accompany the fluxes of salt, volume and electric current at constant temperature.

There is a somewhat different, but equivalent way of obtaining the flux equations (19)–(21). Phenomenological equations relating j_q' and the j_k to their conjugate affinities may be written [8]. The M-coefficients then appear as combinations of phenomenological coefficients that refer to individual species and to heat. This approach, when used to derive equations for the e.m.f. of non-isothermal cells, leads to the introduction of heats of transport of ions and solvent [12, 13] as well as to the usual isothermal transport quantities [8]. However, all the transport quantities defined depend on the nature of the processes occurring inside the membrane.

Some of the various phenomena that can arise are given below. The list is not exhaustive, but covers the most important cases.

(a) Phenomena with $I = d\Pi = dP = 0$.

    Thermal diffusion: $\qquad\qquad M_{DT} = - Tj_s/dT \qquad\qquad\qquad$ (30)

    Thermo-osmosis: $\qquad\qquad M_{PT} = - Tj_v/dT \qquad\qquad\qquad$ (31)

    Thermal conductivity: $\qquad\ M_T \ = - Tj_q'/dT \qquad\qquad\qquad$ (32)

    Initial thermal potential: $\quad dE \ = - L_{ET} dT/TL_E \qquad\qquad$ (33)

(b) Phenomena with $I = dP = j_s = 0$.

    Soret effect: $\quad d\Pi/c_s dT = v d\mu_s/c_0 V_0 dT = $

$$
= - M_{DT}/T(M_D - c_s M_{DP}) \tag{34}
$$

    Steady-state thermal conductivity at zero salt flow:

$$
- Tj_q''/dT = M_T - M_{DT}^2/(M_D - c_s M_{DP}) \tag{35}
$$

    Steady-state thermal potential at zero salt flow:

$$
dE = \{M_{DT}(L_{DE} - c_s L_{PE})/(M_D - c_s M_{DP}) - L_{ET}\}/dT/TL_E \tag{36}
$$

(c) Phenomena with $I = d\Pi = j_v = 0$.

Thermo-osmotic pressure: $\quad dP = -M_{PT}dT/M_P T$ (37)

Steady-state thermal conductivity at zero volume flow:

$$-Tj_q''/dT = M_T - M_{TP}^2/M_P \tag{38}$$

Steady-state thermal potential at zero volume flow:

$$dE = \{M_{PT}L_{PE}/M_P - L_{ET}\}dT/TL_E \tag{39}$$

(d) Phenomena with $d\Pi = dP = dT = 0$.

Peltier effect: $\quad j_q''/I - L_{ET}/L_E \equiv Q_E^*$ (40)

(the Peltier heat)

Equation (33) for the initial thermal potential may be written, using (28) and (40), in terms of the Peltier heat:

$$\begin{aligned} -(TdE/dT)_{I=d\Pi=dp=0} &= Q_E^* \\ &= Q_I^* + L_{DE}Q_s^*/L_E + L_{PE}Q_v^*/L_E \end{aligned} \tag{41}$$

Other heats of transport can also be defined. One used frequently is

$$\begin{aligned} Q_p^* &= -(Tdp\ dT)_{I=d\Pi=j_v=0} \\ &= (j_q''/j_v)_{I=d\Pi=dT=0} = M_{PT}/M_P \end{aligned} \tag{42}$$

which is the thermo-osmotic pressure coefficient, Equation (37).

Of the effects listed above, only thermal conductivity, initial thermal potential, thermo-osmotic pressure, thermo-osmosis and one observation of the Peltier effect [15] appear to have been studied in charged membranes. The only data on thermal conductivity are some estimates by Haase [16] and some measurements on styrene-vinylbenzenesulfonate copolymer membranes, cellulose membranes and porous glass membranes by Thorsley [17]. Initial thermal potentials seem to have been observed first by Tyrrell et al. [18]. Further work has been reported by Lakshminarayanaiah [6], by Ikeda [19–22], by Tasaka et al. [23] and by Kobatake [24]. For a review of much of this work, see [6]. Thermo-osmotic pressure (Equation (37)) involves a combination of thermo-osmosis and mechanical (hydraulic) permeability, and has been measured by Haase and de Greiff [25] and by Rastogi and Singh [26]. Extensive but not very systematic work has been done on thermo-osmosis. The phenomenon was discovered by Lippmann in 1904, and investigated further by Aubert in 1912. The modern era of measurements begins with Carr and Sollner [27] and Haase et al. [16, 25, 28–31]. Haase [28] pointed out that many previous experimenters did not distinguish clearly between thermo-osmosis and thermal expansion effects. Carr and Sollner [27] gave reliable data at a fixed (but unknown) temperature difference for various collodion-based membranes. Rastogi et al. [26, 32, 33] found that thermo-osmotic flow through cellophane occurred from the warm to the cold side of the membrane, while Haase et al. [30] found that the direction of flow reversed at a higher temperature. Voellmy and Läuger [34] observed reversal of flow with 10^{-3} mol dm^{-3} aqueous KCl and a phenolsulfonate membrane, but not with 10^{-3} mol dm^{-3} aqueous HCl. Thermo-osmosis in copper (II) ferrocyanide membranes [29, 35] and in clay plugs [36–38] has also been studied. Several workers have interpreted their results in terms of higher-order phenomenological coefficients [37, 38], an approach which we consider to be misleading, since the thermo-osmotic permeability is expected to be a function of temperature even when first-order phenomenological coefficients are used. Finally, Ernst [39] has remarked on the possible importance of thermo-osmosis in biology.

The mechanism of thermo-osmosis in charged membranes was investigated by Kobatake and Fujita [40], who calculated the convective flow of solvent in the electrical double layer of a charged capillary when a thermal gradient was directed along the axis of the capillary. This theory accounts qualitatively for the dependence on salt concentration of thermo-osmosis in membranes of low charge, but does not predict the reversal of thermo-osmotic flow. Churaev et al. [14] carried out calculations similar to those of Kobatake and Fujita, but for a charged slit. They divided the thermo-osmotic flow into two parts, as mentioned above. The 'primary' flow is considered to arise from the change in enthalpy when solvent molecules are polarized in the electrical field of the double layer, while the 'secondary' flow depends on the thermal potential. The direction and magnitude of the flow depends on the sign of the fixed charge on the membrane and the balance between primary and secondary flow. Deryagin et al. [41] have presented observations of thermo-osmosis in thin layers of water in glass capillaries as evidence for their theory.

Thermo-osmosis is probably the easiest of the non-isothermal transport phenomena to observe, although experiments are tedious and require considerable care. In the next section, we summarize some recent experimental results and their interpretation [42]. The details of both experiments and theory will be published elsewhere.

5. Thermo-Osmosis in Ion-Exchange Membranes

Figure 1 shows thermo-osmotic volume flow j_v as a function of the temperature difference ΔT across a vinylbenzenesulfonate-styrene copolymer membrane [43] for various counterions and two different external salt concentrations. The 1.6 mm thick membrane contained 57 mole percent styrene, 40 mole percent p-vinylbenzenesulfonic acid and 3.1 mole percent divinylbenzene. It had a capacity of 2.7 mmol g^{-1} and a water/polymer mass ratio of 1.7, both based on the dry resin in hydrogen form. Thermo-osmotic flows were measured in a cell consisting of two identical water-jacketed all-glass cell compartments, each provided with a magnetically driven glass-enclosed stirrer. The membrane and two thermistors were mounted in a Plexiglass holder which was clamped between the cell halves. Flows could be measured in horizontal precision-bore capillaries attached to the cell with a reproducibility of one or two percent. Details of the cell and the measurements are given by Chan [42].

The sign convention for the volume flow has been discussed above: positive flow

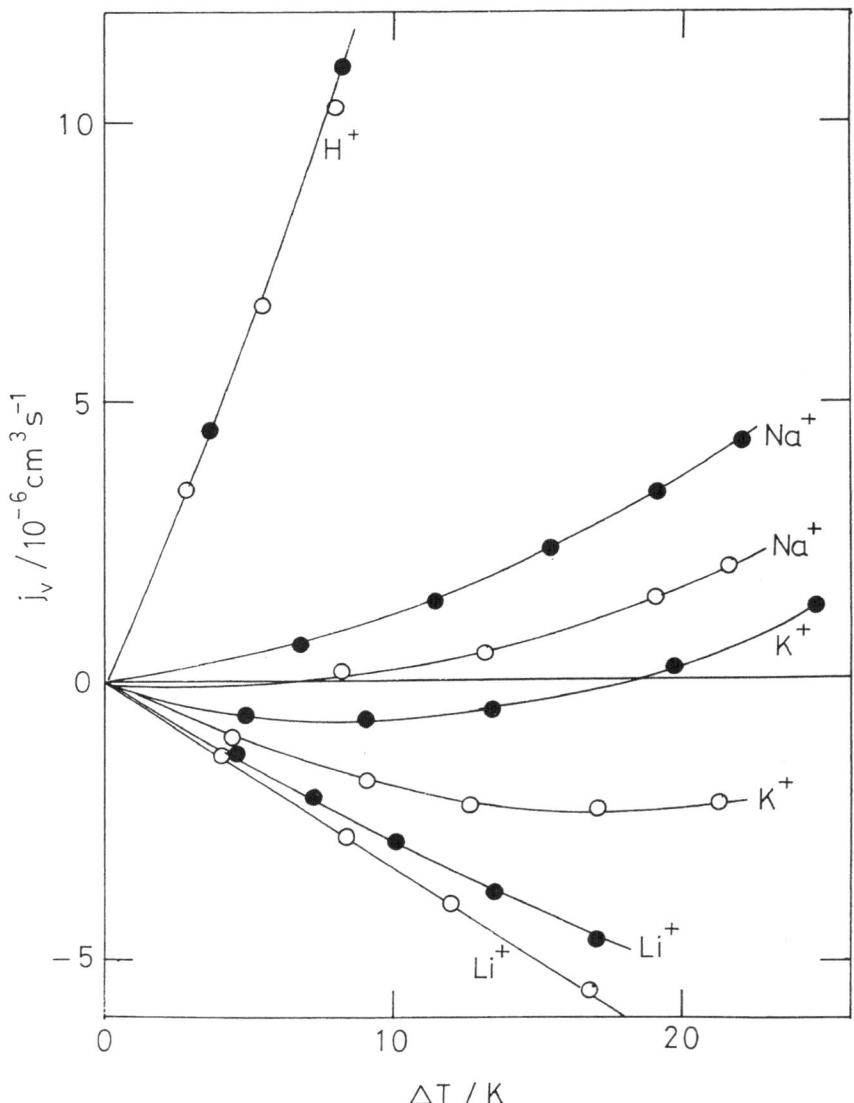

Fig. 1. Temperature dependence of thermo-osmotic flow for a styrene-*p*-vinylbenzenesulfonate copolymer membrane in H-, Li-, Na- and K- form at two concentrations: ○, water; ●, 0.15 mol dm^{-3} solution (0.17 mol dm^{-3} for HCl).

is from the cold to the warm side of the membrane. In general, the flow becomes more positive with increasing concentration. With H$^+$ as counterion, dependence of flow on concentration is small, and the flow is always positive. With Li$^+$ as counterion, the dependence on concentration is larger, but the flow is always negative. For Na$^+$ and K$^+$ as counterions, the dependence on concentration is still larger, and for K$^+$ reversal of flow is found at a temperature difference that depends markedly on concentration. While flows for pure water through a membrane in Na$^+$ form are difficult to observe below temperature differences of 10 K, analysis of the data suggests that reversal of flow also occurs in this case.

It is found (Figure 2) that the flow-temperature difference curves are all accurately parabolic; i.e., the mean thermo-osmotic permeability

$$\bar{M}_{PT} = - (Tj_v/\varDelta T)_{I = \mathrm{d}\varPi = \mathrm{d}p = 0} \qquad (43)$$

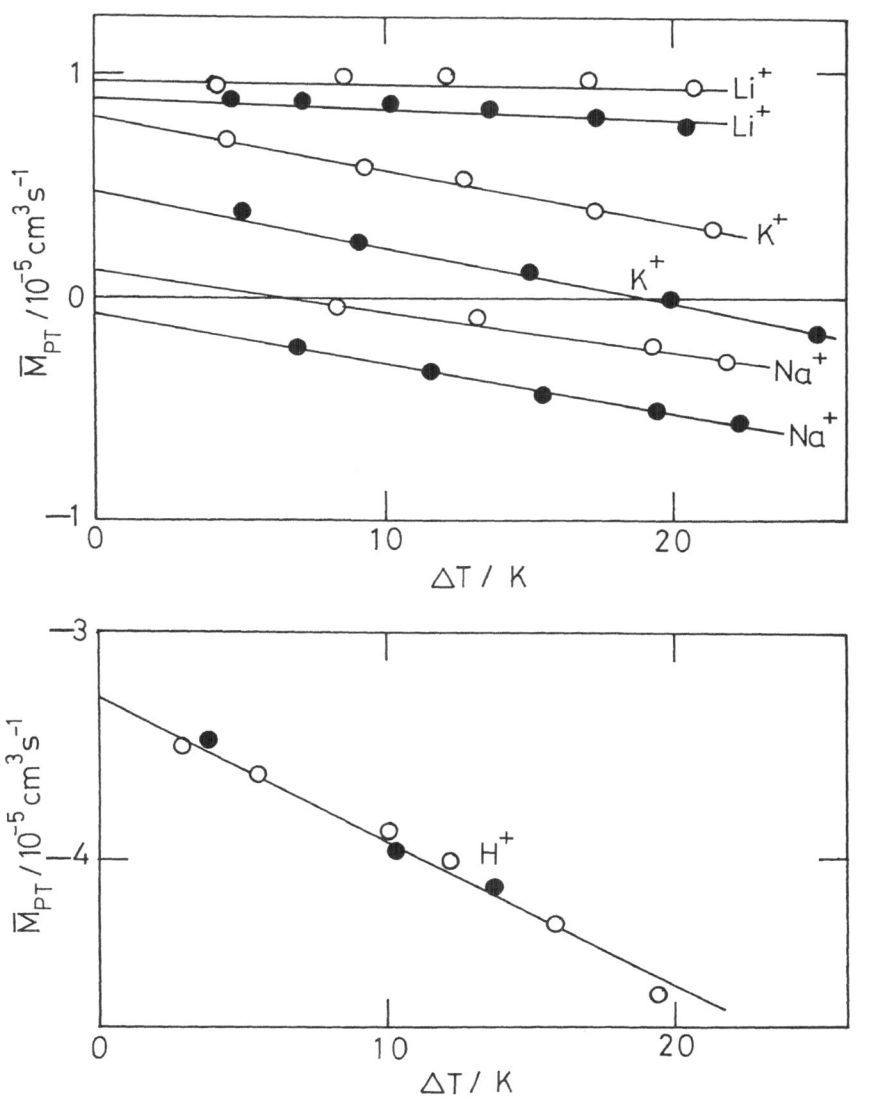

Fig. 2. Linear dependence of the average thermo-osmotic permeability $\bar{M}_{PT}$ on temperature difference: ○, water; ●, 0.15 mol dm^{-3} solution (0.17 mol dm^{-3} for HCl).

is a linear function of ΔT. This fact permits calculation of the thermo-osmotic permeability P_{PT} at a given temperature:

$$P_{PT} = M_{PT}l/A = (-Tj_vl/A\,\mathrm{d}T)_{I=\mathrm{d}\Pi=\mathrm{d}p=0} \tag{44}$$

where A is the area through which flow takes place and l is the thickness of the membrane. Note that a *positive* permeability corresponds to a *negative* flow, and *vice versa*. The thermo-osmotic permeabilities at 25°C increase in the order $Li^+ < K^+ < Na^+ < H^+$ which is the order for conventional heats of transport of ions in aqueous solution [11]. Mechanical permeabilities and electrical conductances were also measured as a function of concentration and temperature. The electrical conductance extrapolated to zero external salt concentration was a linear function of the counterion mobility at infinite dilution, λ_1°, in water, while the mechanical permeability was a linear function of the resistivity (reciprocal conductance) of the membrane.

To account for these results, we divide each local flux equation at a point in the membrane into two parts: a convective motion (of the centre of mass) and motions of the permeating species relative to the centre of mass. The velocity of the centre of mass is then calculated by application of the appropriate hydrodynamic equations to a rather general model of the membrane, and ionic motions relative to the centre of mass are taken to be the ionic motions in free solution. Theories of this type have been used before for both isothermal [44-48] and non-isothermal [14, 47] phenomena. However, for non-isothermal phenomena, calculations have not been done for highly-charged membranes, and have lacked rigour.

The membrane constitutes a convenient frame of reference for fluxes of ions and solvent inside the membrane, but whatever frame of reference is chosen, the dissipation function contains contributions from viscous flow. Mickulecky and Caplan [48] have shown how to avoid this difficulty in isothermal systems. In non-isothermal systems, it can be shown [42] by using the energy equation that their conclusions are still valid: for stationary flows in an isotropic membrane, the average dissipation function contains no viscous terms, and is given by

$$\langle \Phi \rangle = -\sum_{i-1}^{n-1} \mathbf{j}_i^m \cdot (\nabla \tilde{\mu}_i)_T - \langle \mathbf{j}_q' \rangle \cdot \nabla \ln T \tag{45}$$

where the averages (denoted by angular brackets) are taken over a thin slab parallel to the face of the membrane and perpendicular to the direction of average flow (the x-direction). The local flux $\mathbf{j}_i^m$ is measured relative to the stationary membrane, and is related to the flux $M_i\mathbf{j}_i$ relative to the centre of mass by

$$\mathbf{j}_i^m = c_i\mathbf{v}_i = \mathbf{j}_i - c_i\mathbf{v} \tag{46}$$

where c_i, $\mathbf{v}_i$ are the local molar concentration and velocity of species i, M_i is the molar mass of i, and $\mathbf{v}$ is the velocity of the centre of mass. The local reduced heat flux $\mathbf{j}'_q$ is independent of the choice of reference velocity, and is related to the local heat flux $\mathbf{j}_q$ by

$$\mathbf{j}'_q = \mathbf{j}_q - \sum_{i=1}^{n-1} H_i \mathbf{j}_i \tag{47}$$

All summations are over the $n-1$ permeating species; the membrane is species n. The chemical potential at constant temperature is defined by Equation (2).

We now write linear phenomenological equations for the fluxes $\mathbf{j}_i$ and $\mathbf{j}'_q$ and average these as in Equation (45). The results for a membrane in which there are no gradients of isothermal chemical potential are (cf. [48]):

$$\langle j_i^m \rangle = - \left\langle \sum_{j=1}^{n-1} V_j L_{ij} + c_i \alpha \right\rangle \partial P/\partial x - \left\langle \sum_{j=1}^{m-1} Z_j L_{ij} F + c_i \beta \right\rangle \partial \psi/\partial x$$
$$- \left\langle \sum_{j=1}^{n-1} q_j^* L_{ij} + c_i T \left(\alpha \partial P/\partial T + \beta \partial \psi/\partial T \right) \right\rangle \partial \ln T/\partial x \tag{48}$$

$$\langle j'_q \rangle = - \left\langle \sum_{i,\,j=1}^{m-1} q_i^* V_j L_{ij} + cq^* \alpha \right\rangle \partial P/\partial x$$
$$- \left\langle \sum_{i,\,j=1}^{n-1} q_i^* z_j L_{ij} F + cq^* \beta \right\rangle \partial \psi/\partial x$$
$$- \langle L_{uu} + cq^* T \left(\alpha \partial P/\partial T + \beta \partial \psi/\partial T \right) \rangle \partial \ln T/\partial x \tag{49}$$

In these equations, the x-component of the convective velocity $\mathbf{v}$ is given by (cf. [48])

$$v_x = - \alpha \partial P/\partial x - \beta \partial \psi/\partial x \tag{50}$$

where α, β are positive parameters that depend on the geometry and the distribution of electrical potential in the membrane. The gradients are functions of temperature [49]:

$$\partial/\partial x = (\partial/\partial x)_T + (\partial/\partial \ln T)(\partial \ln T/\partial x) \tag{51}$$

while the gradients of P and ψ in (48) are at constant temperature. Equation (50) holds for a large class of models for membranes, one of the most general of which is the 'domain model' illustrated in Figure 3. In each domain, uniformly-charged polyelectrolyte rods [50] all have the same orientation, but all orientations occur over all domains. Flow occurs parallel to the rods under the influence of the electrical double layers. It can be shown that the solution of the Navier-Stokes and Poisson equations for one domain applies to the whole model if the solution for one domain is multiplied by the geometrical factor 1/3 (cf. [50]). The appearance of convective terms in the equation for the reduced heat flux is to be expected, but the form of the equation is not obvious, and becomes apparent only after lengthy calculations. The average heat of transport q^* is given by

$$q^* = \sum_{i=1}^{n-1} c_i q_i^*/c \tag{52}$$

where c is the total concentration of permeating species.

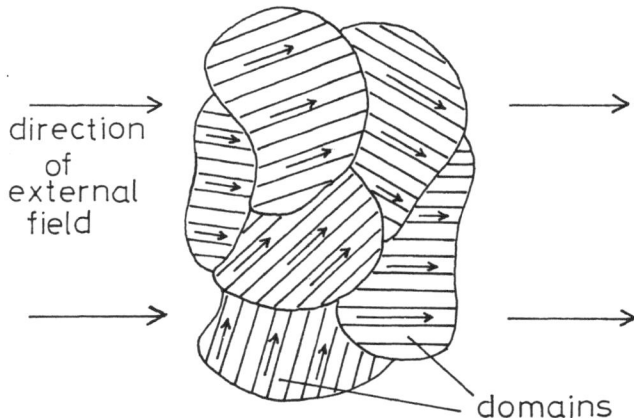

Fig. 3.   The domain model for a charged two-dimensional polyelectrolyte gel or membrane. The parallel lines in each domain are polyion rods along which ions and solvent molecules move as shown by the arrows.

The total volume flux (in the absence of electric current)

$$j_v = A \sum_{i=1}^{n-1} V_i \langle j_i^m \rangle \tag{53}$$

and the total electric current

$$I = AF \sum_{i=1}^{n-1} z_i \langle j_i^m \rangle \tag{54}$$

are calculated next, and the resulting stationary fluxes are integrated over the thickness l of the membrane to give

$$lj_v/A = - (L'_P + \langle \alpha \rangle)\, dp' - (L'_{PE} + \langle \beta \rangle)\, d\psi' - (L'_{PT} + \langle cq^*\alpha \rangle)\, d \ln T' \tag{55}$$

$$lI/A = - (L'_{PE} + \langle \beta \rangle)\, dp' - (L'_E + \langle \beta\alpha \rangle)\, d\psi' - (L'_{ET} + \langle cq^*\alpha \rangle)\, d \ln T' \tag{56}$$

$$lj'_q/A = - (L'_{PT} + \langle cq^*\alpha \rangle)\, dp' - (L'_{ET} + \langle cq^*\beta \rangle)\, d\psi' - L'_T\, d \ln T' \tag{57}$$

where

$$L'_P = \langle \sum_{i,j} V_i V_j L_{ij} \rangle, \quad i = 1, ..., n-1 \tag{58}$$

$$L'_{PE} = \langle \sum_{i,j} V_i z_j L_{ij} F \rangle \tag{59}$$

$$L'_{PT} = \langle \sum_{i,j} V_i q_j^* L_{ij} \rangle \tag{60}$$

$$L'_E = \langle \sum_{i,j} z_i z_j L_{ij} F^2 \rangle \tag{61}$$

$$L'_{ET} = \langle \sum_{i,j} z_i q_j^* L_{ij} F \rangle \tag{62}$$

L'_T is the coefficient of $\partial \ln T/\partial x$ in (49), and $\langle \beta \rangle = \langle \alpha F \sum_i z_i c_i \rangle = \langle \alpha \sigma \rangle$, $\sigma = F \sum_i z_i c_i$. The complicated quantities arising from terms in $\partial P/\partial T$ and $\partial \psi/\partial T$ in Equation (48) are expressed in simpler terms by assuming the Onsager relation $L^m_{PT} = L^m_{TP} = L'_{PT} + \langle cq^*\alpha \rangle$. The differences dp', $d\psi'$ and $d \ln T'$ in Equations (55)–(57) refer to differences at the membrane-solution boundaries, but inside the membrane. The general boundary conditions are given by equating the electrochemical potentials for each species in each phase at the phase boundaries, at which local equilibrium is assumed (cf. [30]):

$$- \Delta V_i \, dp + \Delta H_i \, d \ln T + z_i F d\psi = z_i F d\psi' \tag{63}$$

where dP, $d \ln T$, $d\psi$ are differences between the solution phases, and ΔV_i, ΔH_i are the changes in partial molar volume and enthalpy in transferring species i from the solution to the membrane.

A great simplification will now be made by applying the above equations to transport in a salt-free membrane [44]; that is, the membrane is in contact with pure water and contains no co-ions. This is the limiting case for highly-charged membranes, and for it the average convective terms in Equations (55)–(57) can be evaluated explicitly from the appropriate version of Equation (50) and the appropriate solution of the non-linear Poisson-Boltzmann equation. For a salt-free membrane, with counterions 1 and solvent 0, we have $L_{10} = L_{00} = 0$, since for the volume flux at zero electric current, the velocity of the centre of mass is the velocity of the solvent. The electric current is due only to the counter-ion flux. The overall transport parameters of interest are then the electrical conductivity

$$\begin{aligned} k &= - f(Il/Ad\psi)_{d\Pi = dp = dT = 0} \\ &= f \langle z_1^2 L_{11} F + z_1 c_1 \beta \rangle \end{aligned} \tag{64}$$

the mechanical permeability

$$\begin{aligned} P_p &= - f(j_v l/Adp)_{I = d\Pi = dT = 0} \\ &= f \langle \alpha \rangle - f^2 \langle \beta \rangle^2 / k \end{aligned} \tag{65}$$

and the thermo-osmotic permeability

$$\begin{aligned} P_{PT} &= - f(j_v l/Ad \ln T)_{I = d\Pi = dp = 0} \\ &= f \langle c_0 q_0^* \alpha \rangle - f \langle \beta \rangle \, q_1^*/z_1 F - f^2 \langle \beta \rangle \langle c_0 q_0^* \beta \rangle / k \end{aligned} \tag{66}$$

The factor f in these equations is the correction for the random orientation of domains. Ideally, $f = 1/3$. From Equations (42), (65) and (66), we note that $Q_P^* = P_{PT}/P_P$, and that Q_P^* cannot be interpreted as a simple heat of transport.

Since $\langle z_1 L_{11} F \rangle = \langle c_1 \lambda_1 \rangle$, where λ_1 is the conductance of the counter-ion relative to the solvent, Equation (64) predicts that the membrane conductance at infinite dilution should be a linear function of the counter-ion conductance in free solution. For membranes in which the ratio of charged to uncharged residues along the polymer chains is sufficiently low, the counter-ion conductance at infinite dilution will be appropriate. This condition appears to be satisfied for the membrane used in the experiments described above. Equation (65) predicts that the mechanical permeability should be linear in the reciprocal membrane conductance, in agreement with experiment. The quantity $f\langle\beta\rangle$ can be evaluated from the experimental data.

Individual ionic heats of transport in aqueous solution cannot be determined, but relative values $q_i^{*r}=q_{il}^*+q_r^*$ are available, using the choride ion as a reference ion [11], i.e., $q_r^*=0$ for chloride ion. Equation (66) can be written

$$P_{PT}+f\langle\beta\rangle q_1^{*r}/z_1F=f\langle c_0q_0^*\alpha\rangle$$
$$-f^2\langle\beta\rangle\langle c_0q_0^*\beta\rangle/k \tag{67}$$

The left-hand side of this equation can be computed from the experimental values of P_{PT} and P_P and tabulated values of q_1^{*r}, and should be linear in the reciprocal membrane conductance. Figure 4 shows that the left-hand side of (67) is indeed a linear function of the reciprocal conductance, within experimental error. The scatter of the values about the best straight line through the points can be attributed to the difficulty in measuring the small mechanical permeabilities accurately and to uncertainties in the values of q_1^{*r} [11]. The heat of transport of a liquid in an electrical double layer has been attributed to enthalpy changes due to polarization of the molecules [14]:

$$c_0q_0^* = (\varepsilon_0a/2)(\nabla\phi)^2 \tag{68}$$

where ε_0 is the permittivity of empty space, $\nabla\phi$ is the gradient of electrostatic potential in the double layer, and $a=D(1+\partial\ln D/\partial\ln T)$, where D is the dielectric constant of the liquid. This assumption is similar to that made in interpreting heats of transport

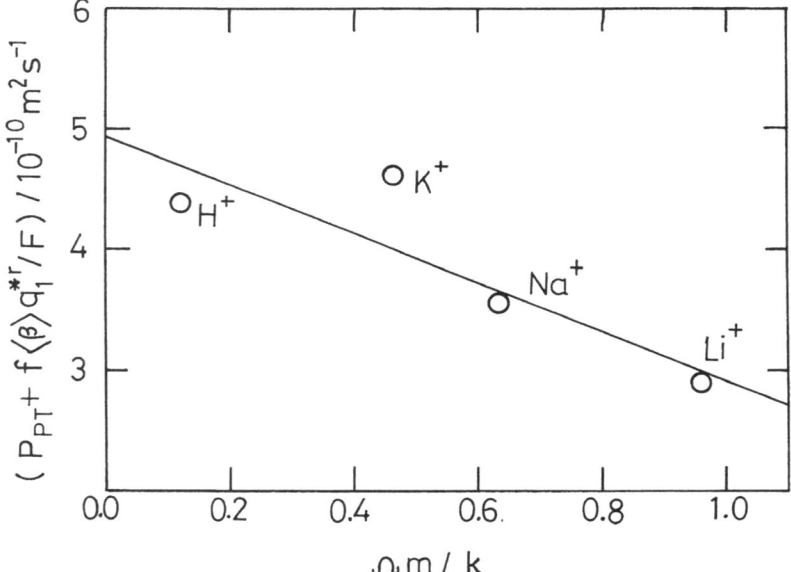

Fig. 4. $P_{PT}+f\langle\beta\rangle q_1^{*r}/z_1F$ as a function of the reciprocal limiting membrane conductance. The line is the least-squares line through the points.

in solution [11], but ignores any interactions of water molecules with the uncharged parts of the membrane matrix. For water, a is negative [14], and since α and β are positive, the averages $\langle c_0 q_0^* \alpha \rangle$ and $\langle c_0 q_0^* \beta \rangle$ should be negative. The slope of the plot in Figure 4 should therefore be positive, according to Equation (67). However, the sum of the two terms on the lefthand side of (67) is very sensitive to the value of q_1^{*r}. To investigate this point further requires an explicit calculation of the average convection terms. Ideally, these are found from the solution of the non-linear Poisson-Boltzmann equation for a charged rod [50] and equation (68), but these calculations are complicated mathematically. As an approximation, the potential distribution has been taken to be that in a charged slit of width $2h$:

$$\phi = \phi_0 + (2RT/z_1 F) \ln \cos \{ \kappa z \exp (-z_1 \phi_0 F/2RT) \sqrt{2} \} \qquad (69)$$

where ϕ is the potential at a distance z measured from the centre plane of the slit, ϕ_0 is the potential at $z=0$ and $\kappa^2 = z_1^2 c_1 F^2 / DRT\varepsilon_0$. The factor f and the average $\langle \sigma\beta \rangle$ can be evaluated from experimental conductance data using Equation (64). The value of h can then be found from the theoretical equation for $\langle \sigma\beta \rangle$ and the known value of κ. The equations for the averages and the details of the calculations are lengthy, but the results are reasonable: $2h = 1.8$ nm (the slit width), and $f = 0.20$ (ideal value 0.33). The other average quantities can now be calculated, and used along with the experimental values of the thermo-osmotic permeability and conductance to calculate q_1^* from Equation (66). These values are given in Table I, and

TABLE I
Ionic heats of transport from thermo-osmosis

Cation	q^*_1/kJ mol^{-1}		
	Eq. (65)	Free solution[a] (11)	Difference
H$^+$	8.2	12.7	4.5
Li$^+$	−2.9	−0.013	2.9
Na$^+$	0.77	2.9	2.1
K$^+$	−2.6	2.0	4.6

[a] relative to chloride ion.

are estimates of absolute single-ion heats of transport. They differ from the values in free solution by varying amounts, indicating that other factors in addition to the arbitrary choice of the reference ion for the values in free solution are important. Similar calculations for other temperatures have been carried out. A 'heat capacity for transport'

$$C_1^* = \{ q_1^*(T_2) - q_1^*(T_1) \}/(T_2 - T_1) \qquad (70)$$

can be defined [11]. Some results for $T_1 = 25°C$, $T_2 = 45°C$ are given in Table II. For comparison, values at $m = 0.05$ mol kg^{-1} in aqueous solution are given [11]. There is general agreement in the magnitude of the values for each ion. We conclude that

Equation (66) accounts for the dependence of thermo-osmosis in charged membranes on the nature of the counter-ion in a satisfactory way, and that solvent polarization in the double layer is the main factor that contributes to the heat of transport of water.

TABLE II

Heat capacity for transport from thermo-osmosis

Ion	$C*_1/JK^{-1} mol^{-1}$	
	Membrane	Free solution (11)
H^+	42	–
Li^+	71	97.9
Na^+	75	135
K^+	130	124

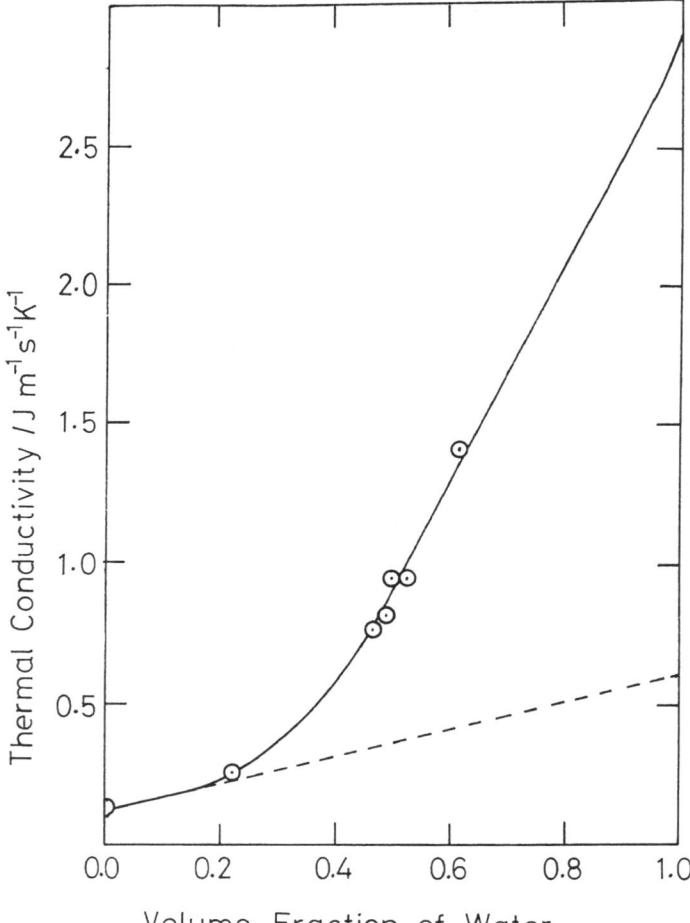

Fig. 5. Thermal conductivity of styrene-*p*-vinylbenzene copolymer membranes as a function of volume fraction of water in the membrane. The points are experimental; the dashed line joins values for pure water and pure polystyrene; the solid line is a best fit to theory (see text).

As a final piece of evidence concerning the nature of the solvent in charged membranes, Figure 5 shows the thermal conductivity of styrene-vinylbenzenesulfonate copolymer membranes of about the same charge density but different degrees of cross-linking as a function of water content [17]. The dashed line connects the thermal conductivities of pure water and pure polystyrene at 25°C, while the solid line is a calculated best fit to a theory of the thermal conductivity of composite media [51]. The large value of the thermal conductivity of water in the membrane that is required to account for these results is apparent; it corresponds to the thermal conductivity of ice at about −50°C [52]. The solvent must indeed be highly polarized in the membrane.

It appears that the mechanism of thermo-osmosis in membranes such as cellophane or copper (II) ferrocyanide with small or negligible charge density differs from that discussed above for highly-charged membranes. Haase *et al.* [16, 30] have given a thermodynamic theory of thermo-osmosis in solubility membranes. From experiments, they conclude that cellophane membranes conform neither to a pure solubility model nor a pore model, but to a model that has features of both.

Acknowledgment

The work described in this paper was supported by the Defence Research Board and the National Research Council of Canada.

References

1. de Groot, S. R.: *Thermodynamics of Irreversible Processes*, North-Holland Publishing Co., Amsterdam, 1952. Chap. 5, Sect. 67.
2. Tyrrell, H. J. V.: *Diffusion and Heat Flow in Liquids*, Butterworths, London, 1961, pp. 13–18, 24–28, Chap. 4.
3. de Groot, S. R. and Mazur, P.: *Non-Equilibrium Thermodynamics*, North-Holland Publishing Co., Amsterdam, 1962, Chap. XV.
4. Haase, R.: *Thermodynamik der Irreversiblen Prozesse*, Dr. Dietrich Steinkopff Verlag, Darmstadt, 1963, Sections 3.9–3.14.
5. Katchalsky, A. and Curran, P. F.: *Non-Equilibrium Thermodynamics in Biophysics*, Harvard University Press, Cambridge, 1965, Chap. 13.
6. Lakshminarayanaiah, N.: *Transport Phenomena in Membranes*, Academic Press, New York and London, 1969, Section 3.20, Chap. 7.
7. Hanley, H. J. M.: in Hanley, H. J. M. (ed.) *Transport Phenomena in Fluids*, Marcel Dekker, New York, London, 1969, Chap. 4; Mickulecky, D. C.: *ibid.*, Chap. 12.
8. Lorimer, J. W., Boterenbrood, E. I., and Hermans, J. J.: *Faraday Soc. Discussions* **21**, 141 (1956).
9. Weinstein, J. N. and Caplan, S. R.: *J. Phys. Chem.* **77**, 2710 (1973).
10. Kedem, O.: *J. Phys. Chem.* **77**, 2711 (1973).
11. Agar, J. N.: in Delahay, P. (ed.), *Advances in Electrochemistry and Electrochemical Engineering*, Interscience Publishers, New York, London, 1963, p. 31.
12. Hills, G. J., Jacobs, P. W. M., and Lakshminarayanaiah, N.: *Nature* **179**, 96 (1957).
13. Hills, G. J., Jacobs, P. W. M., and Lakshminarayanaiah, N.: *Proc. Roy. Soc.* **A262**, 246 (1961).
14. Churaev, N. V., Deryagin, B. V., and Zolotarev, P. P.: *Dokl. Akad. Nauk SSSR* **183**, 1139 (1968). (*Doklady Phys. Chem.* **183**, 935 (1968).)
15. Shaffer, L. H.: *Nature* **191**, 591 (1961).
16. Haase, R.: *Z. Physik. Chem. N.F.* **51**, 315 (1966).

17. Thorsley, J. D.: Ph. D. Thesis University of Western Ontario, London, Canada, 1972. See also Lorimer, J. W. and Thorsley, J. D.: International Symposium on Macromolecules, Helsinki, 1972, abstract II-40.

18. Tyrrell, H. J. V., Taylor, D. A., and Williams, C. M.: *Nature* 177, 668 (1956).

19. Ikeda, T.: *J. Chem. Phys.* 28, 166 (1958); 31, 267 (1959).

20. Ikeda, T. and Tsuchiya, M.: *Kagaku* (Tokyo) 30, 364, 481 (1960).

21. Ikeda, T.: *Repts. Liberal Arts. Fac. Shizuoka Univ., Nat. Sci.* 2, 237 (1961).

22. Ikeda, T., Tsuchiya, M., and Nakano, M.: *Bull. Chem. Soc. Japan* 37, 1482 (1964).

23. Tasaka, M., Morita, S., and Nagasawa, M.: *J. Phys. Chem.* 69, 4191 (1965).

24. Kobatake, Y.: *Bull. Tokyo Inst. Technol. Ser. B*, 97 (1959).

25. Haase, R. and de Greiff, H. J.: *Z. Physik. Chem. N. F.* 44, 301 (1965).

26. Rastogi, R. P. and Singh, K.: *Trans. Faraday Soc.* 62, 1754 (1966).

27. Carr, C. W. and Sollner, K.: *J. Electrochem. Soc.* 109, 616 (1962).

28. Haase, R.: *Zeit. Physik. Chem. N. F.* 21, 244 (1959).

29. Haase, R. and Steinert, C.: *Zeit. Physik. Chem. N. F.* 21, 270 (1959).

30. Haase, R., de Greiff, H. J., and Buchner, H.-J.: *Z. Naturforsch.* 25a, 1080 (1970).

31. Haase, R. and de Greiff, H. J.: *Z. Naturforsch.* 26a, 1773 (1972).

32. Rastogi, R. P., Blokhra, R. L., and Agarwal, R. K.: *Trans. Faraday Soc.* 60, 1386 (1964); *Indian J. Chem.* 2, 166 (1964).

33. Rastogi, R. P., Singh, K., and Misra, B. M.: *Desalination* 3, 32 (1967).

34. Voellmy, H. and Läuger, P.: *Ber. Bunsenges. Physik. Chem.* 70, 165 (1966).

35. Vetö, P.: *Acta Biochim. Biophys. Acad. Sci. Hung.* 2, 441 (1967).

36. Dirksen, C.: *Proc. Soil Sci. Amer.* 33, 821 (1969).

37. Srivastava, R. C. and Abrol, I. P.: *Indian J. Chem.* 7, 1121 (1969).

38. Singh, K. and Singh, H. P.: *Indian J. Chem.* 7, 690 (1969).

39. Ernst, E.: *Acta Biochim. Biophys. Acad. Sci. Hung.* 1, 211 (1966).

40. Kobatake, Y. and Fujita, H.: *J. Chem. Phys.* 41, 2963 (1964).

41. Deryagin, B. V., Ershov, A. P., and Churaev, N. V.: *Poverkh. Sily Tonkikh Plenakh Dispersnykh Sist., Sb. Dokl. Konf. Poverkh. Silam*, 4th 1969; *Chem. Abs.* 78, 8225h (1973).

42. Chan, S. H.: Ph. D. Thesis, University of Western Ontario, London, Canada, 1973. See also Lorimer, J. W. and Chan, S. H.: *International Symposium on Macromolecules*, Helsinki, 1972, abstract II-39.

43. Brydges, T. G., Dawson, D. G., and Lorimer, J. W.: *J. Polymer Sci. A-1* 6, 1009 (1968).

44. Dresner, L.: *J. Phys. Chem.* 67, 1635 (1963).

45. Kobatake, Y. and Fujita, H.: *J. Chem. Phys.* 40, 2212, 2219 (1964); *Kolloid-Z.* 196, 58 (1964).

46. Läuger, P. and Kuhn, W.: *Ber. Bunsenges. Physik. Chem.* 68, 4 (1964).

47. Churaev, N. V. and Deryagin, B. V.: *Dokl. Akad. Nauk SSSR* 169, 396 (1966).

48. Mickulecky, D. C. and Caplan, S. R.: *J. Phys. Chem.* 70, 3049 (1966).

49. Bearman, R. J. and Kirkwood, J. G.: *J. Chem. Phys.* 28, 136 (1958).

50. Katchalsky, A., Alexandrowicz, A. and Kedem, O.: in B. E. Conway and R. G. Barradas (eds.): *Chemical Physics of Ionic Solutions*, Wiley, New York, 1966, p. 295.

51. Brailsford, A. D. and Major, K. G.: *Brit. J. Appl. Phys.* 15, 313 (1964).

52. Ratcliffe, E. H.: *Phil. Mag.* 1, 1197 (1962).

THE EFFECT OF TEMPERATURE AND OUABAIN ON THE MEMBRANE PROPERTIES OF A GIANT ALGAL CELL

ANITRA THORHAUG

Dept. of Microbiology, School of Medicine, University of Medicine,
University of Miami, Miami, Fla., U.S.A.

Abstract. This investigation treats the effect of temperature on a series of membrane properties of an ideal model cell, *Valonia*, which is a primitive, marine algal cell found throughout the world's tropics. The effect of temperature on the electrical properties of the *Valonia* membrane system including bioelectric potential difference, A.C. conductance, chord conductance, and short-circuit current are reported and discussed. Ion relations of potassium and sodium with respect to temperature were also investigated. The *Valonia* membrane system appears to have a fairly stable region between 15 and 30°C, where $d\varphi/dT$ is close to zero and other electrical properties are linear. Above 30°C and below 15°C there are abrupt discontinuities in electrical properties of the membrane system. Ionic relations for potassium and sodium show these same temperature regions for long-term (3 days) incubation periods, but not for short-term (45 min) periods. Standard deviations of ionic concentrations between cells are high. Ouabain showed a marked effect on bioelectric potential and short-circuit current at concentrations of $10^{-3}M$ and $10^{-5}M$ with no effect at $10^{-3}M$ or $10^{-6}M$ or below. Implications of these data on transport processes in plant cells are discussed.

1. Introduction

1.1. TEMPERATURE EFFECTS ON MEMBRANES

Although a great deal of effort has been directed toward the effect of physical or chemical factors on living membranes, most workers have ignored temperature or have treated its effect superficially. Most of the organisms in the world are poikilothermic, meaning that they do not control their temperature and thus are at the mercy of their environmental temperature. This is changing daily, in fact hourly, and therefore temperature must have a profound effect on the membrane transport processes in the vast majority of organisms on this planet.

On the other hand, temperature has been of interest to physical chemists who have done a series of studies on oil or lipid membranes. Rosano *et al.* (1961), using a butanol membrane system, measured the transport rate of sodium chloride and potassium chloride. Ting *et al.* (1966) also looked at the effects of temperature on a butanol membrane for the transport of rubidium and saw an anomalous transport rate at 15°C. A series of investigators examined temperature effects on bimolecular phospholipid membranes: Thompson (1964) and D. Chapman (1967), as well as Bean and Chan (1969). Johnson and Bangham (1969) have studied the effect of anesthesic agents on phospholipid membranes as a function of temperature. Among their results are transport rates in terms of apparent energy of activation. Steinert and Haase (1969) observed in treated cellulose membranes apparent heats of transport anomalies near 32°C. These authors noted that previously there had not been such an anomaly reported for cellulose itself at this temperature.

Martin-Lof and Soremark (1969a, b, c) have looked at the effects of temperature on cellulose membranes using various techniques and they too saw an anomaly near 35°C; using an NMR linewidth in hemicellulose they found an anomaly near 30°C. Earlier studies by Ramiah and Goring (1965) studied the water cellulose interaction and found anomalous changes in the expansion of cellulose-hemicellulose and lignum caused by the hydrophylic surfaces of the woody macromolecules. This has pertinence to *plant membrane* studies since plants are surrounded by a cellulose cell wall.

There have been a series of *animal membrane* studies concerning temperature, one of the most extensive of which is that by Wright and Diamond (1969); also Diamond and Wright (1969). They studied the membrane permeability of various non-electrolytes in a comparative study of permeability coefficients. They were concerned as to the role of the unstirred layer and point out that the unstirred layer may play a dominant, in fact a destructive, influence on the interpretation of their investigative data aimed at calculating activation energies of permeation. However, in their particular study, perhaps the effect of the unstirred layer plays a less crucial role because the unstirred layer simply shifts the absolute reflection coefficient due to the fact that their method is a comparative one.

Several reports exist on the effect of temperature on erythrocytes. Good (1960), and Coleman and Good (1968, 1969) studied the effect of temperature on the hemolysis and transport in erythrocytes and also the effects of various pharmacons on their hemolysis. Solomon and coworkers have investigated the hydraulic permeability of red blood cells of humans and dogs as a function of temperature. They concluded that the hydraulic conductivity was constant over the temperature range explored from 5 to 39°C. Therefore, temperature induced no restraints on equivalent pores. Their data could be interpreted as above 27°C their hydraulic conductivity is more or less independent of temperature.

Caffier and Kuchler (1972) studied the effects of temperature on frog skeletal muscles, finding between 20 and 45°C a different Q_{10} between 20 and 30°C than that between 30 and 41°C. In the latter, higher range the temperature did not appear to affect the membrane potential, whereas the Q_{10} below 30 was about 1.2 with a reversible effect. Above 42°C the temperature caused irreversible depolarization of the membrane.

Several workers have investigated the temperature dependence of ATPase from animal cells by showing changes in activity. Other studies include Dalton and Hendrix (1962) on the lobster axon; Lippold *et al.* (1960); Hensel *et al.* (1960); Hensen (1963); Zucker (1973); and Zotterman (1959).

Plant cells have enjoyed far less attention than animal cells. Thus, it is not surprising that there are fewer studies on the effect of temperature on plant cells. The membrane studies are complicated in plants by the fact that there are two membranes. The *plasmalemma*, or outer membrane, and the *tonoplast*, or the membrane separating the internal vacuole from the cytoplasm. The problems of dealing with plant cells have been discussed in detail by Dainty (1963) and Dainty (1971).

Unlike animal cells, plant cells have never developed the ability to control their temperature. Therefore, all plant cells are poikilothermic; in addition to which some of the plant cells are our oldest living fossils known on the earth, especially the marine plant cells. Thus, it is interesting to look at a very simple temperature sensitive organism to see if we can understand the underlying mechanism of temperature on a membrane.

Blinks (1942) was one of the first to look at the effects of temperature on a plant cell membrane. The potential, in fact, was not smooth linear function of temperature. Rather, below about 15°C the potential difference increased, and above

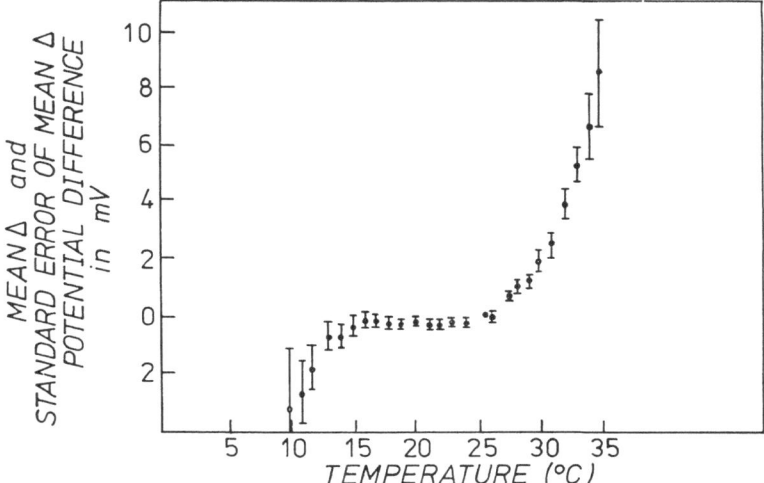

Fig. 1. Change of mean potential (open dots) from resting potential at 25° and standard error of mean potential (vertical lines) across the protoplasm of 25 *Valonia macrophysa* cells versus temperature ($E = \Psi_i - \Psi_0$).

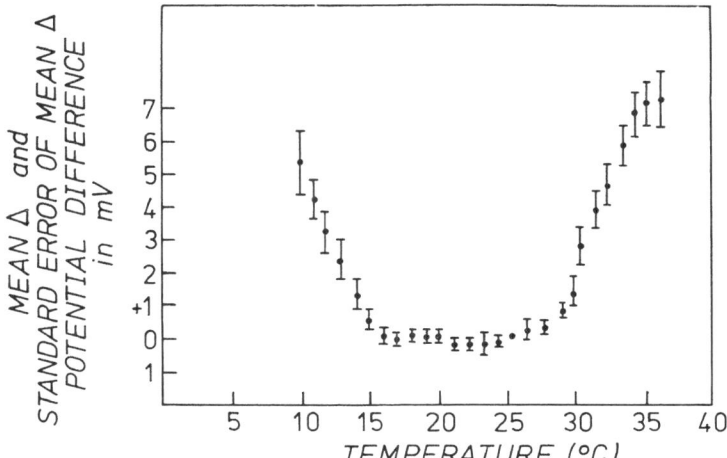

Fig. 2. Change of mean potential (closed dots) from resting potential at 25° and standard error of mean potential (vertical lines) across the protoplasm of 30 *Valonia ventricosa* cells versus temperature ($E = \Psi_i - \Psi_0$).

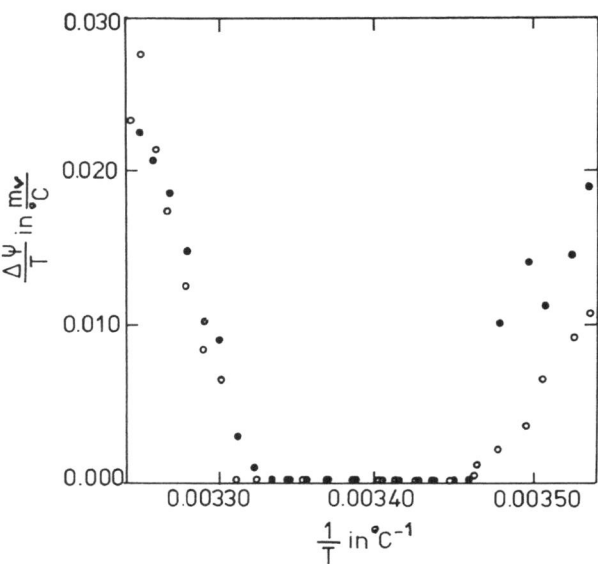

Fig. 3. Plot of data from Figure 2 (black dots) and 1 (open dots) for the potential difference across *Valonia* membrane system as a function of temperature. Change of potential per degree celsius is plotted versus reciprocal temperature. (From Thorhaug, 1971b.)

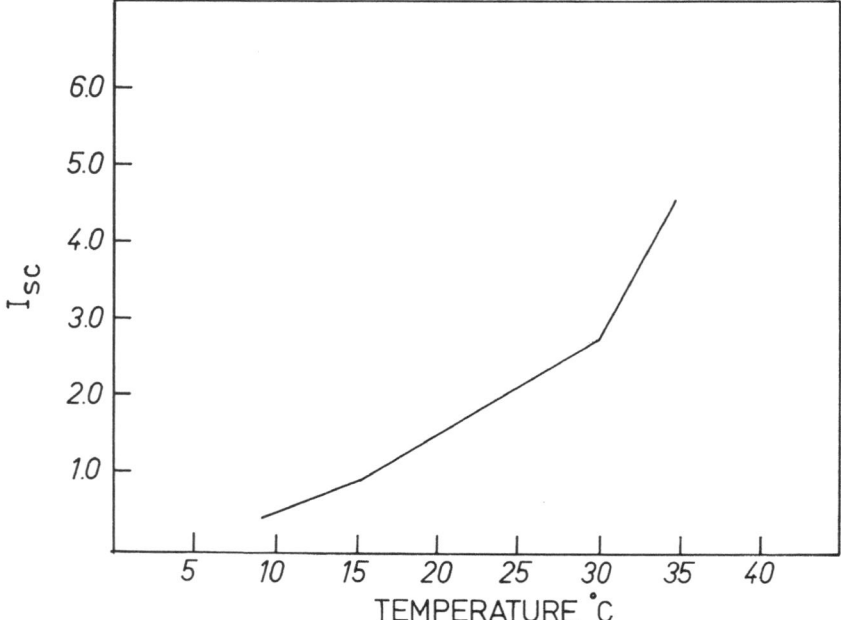

Fig. 4. Short-circuit current in amp/cm² versus temperature for 15 cells of *Valonia ventricosa* under high potassium, low sodium conditions.

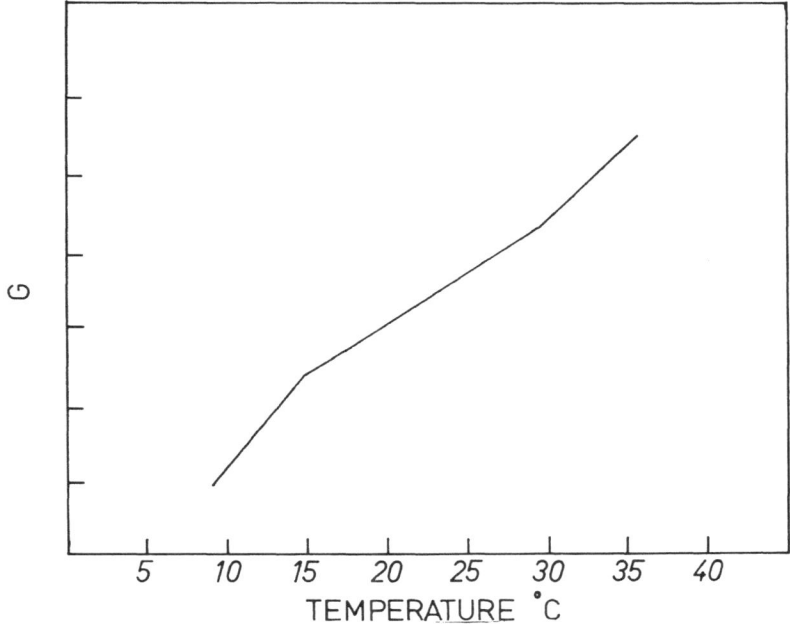

Fig. 5. Preliminary data on A. C. conductance versus temperature under normal conditions (seawater, sap).

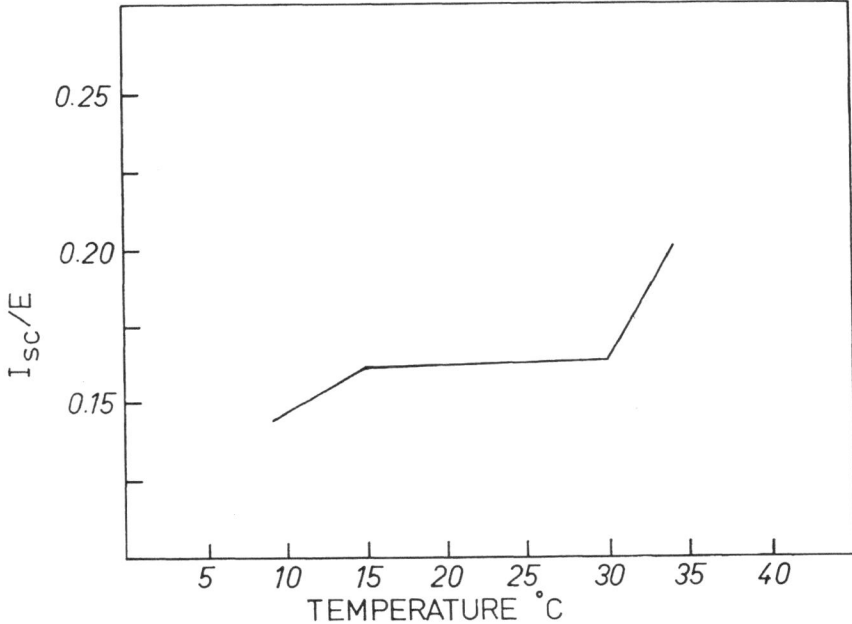

Fig. 6. Chord conductance taken from short-circuit current divided by potential versus temperature for 15 cells of *Valonia ventricosa* under high potassium and low sodium conditions.

approximately 31 °C the potential difference also increased with a fairly stable area between 15 and 30 °C although this work was done on one cell and on rather wide temperature intervals. Hogg *et al.* (1968) looked at the effect of temperature on *Nitella*, showing a linear dependency in potential and conductance in the range of about 5 to 17 °C and then erratic behavior above this temperature. Raven (1967) showed a rather strong temperature dependency on the active uptake of *Nitella*, while the passive or ouabain poisoned uptake of potassium was only very slightly temperature dependent. Thorhaug and Drost-Hansen (1967), using *Valonia utricularis*, and Thorhaug (1971a, b), and Thorhaug and Fernandez (1973) investigated two other species of *Valonia*.

In these studies of two species of algae, *Valonia ventricosa* and *Valonia macrophysa*, a non-linear relationship between temperature and a bioelectric potential was seen (Figures 1 and 2). An Arrhenius plot of the same information (Thorhaug, 1971b) is shown in Figure 3 and subsequently interpreted in the framework of irreversible thermodynamics. A relationship to the flow of active transport with temperature was postulated. It is important to see that in the area between 15 °C and 30 °C there is a surprisingly stable potential. And $d\psi/dT$ is close to zero. However, it changes markedly below and above these temperatures. This was done with 25 cells of *Valonia macrophysa* and 30 cells of *Valonia ventricosa*. Possible reasons for this effect have been discussed in the literature cited. The temperature dependence of the short-circuit current of *Valonia* (Figure 4), the A.C. conductance under normal conditions (seawater external and vacuolar sap internal) (Figure 5), as well as the D.C. chord conductance (Figure 6), have been studied (Thorhaug, Fernandez and De Diego, in press).

1.2. POTASSIUM AND SODIUM CONCENTRATION VERSUS TEMPERATURE

Since Meyer (1891) showed that potassium was concentrated in cellular fluids of *Valonia*, many types of cells have been used for experimental purposes. The regulatory nature of the cell membrane is readily demonstrated by the marked difference between cellular and extracellular concentrations of potassium and sodium. These are the two most abundant cations in cellular fluids. Due to their obviously discriminatory requirements, erythrocytes and muscle fibers have been among those selected; other examples include giant nerve cells of squid as well as frog skin. Due to its large size, attaining a diameter of 10 cm or more, *Valonia* has been one of the cells most frequently used in membrane studies. Its large size dictated and abundance of internal fluids, making possible direct analysis of the internal ion concentration. Experimental results using *Valonia* have been perplexing because, unlike some cells, they showed a wide variation in potassium and sodium concentration (Table I).

Steward (1937) measured the potassium and sodium content of the vacuolar fluids of *Valonia* and concluded that the ionic concentration varied with certain environmental conditions. He estimated that under 'normal' conditions the potassium content ranged from 0.498 to 0.520 gm equiv l^{-1} in *V. macrophysa* and from 0.587 to 0.595 gm equiv l^{-1} in *V. ventricosa*. The cells he examined came from the Dry Tortugas. Although

TABLE I

Potassium and sodium content of *Valonia* cell sap

Author	Location	K+ in moles/l	Na+ in moles/l	T°	Species
Osterhout (1922)	Bermuda	0.492	0.090	–	*V. macrophysa*
Brooks (1933)	Naples	0.465	0.164	12°	
Brooks (1933)	Tortugas	0.494	0.097	–	
Brooks (1933)	Bermuda	0.517	0.090	25°	
Steward (1937)	Tortugas	0.509 ±0.011	0.113 ±0.010	–	
Cooper and Blinks (1928)	Tortugas fresh	0.587	0.035		
Cooper and Blinks (1928)	Tortugas 3 weeks	0.537	0.057	25°	*V. ventricosa*
Brooks (1933)	Tortugas	0.562	0.046	–	
Brooks (1933)	Tonga	0.545	0.055	–	
Steward (1937)	Tortugas	0.591 ±0.004	0.043 ±0.003	–	
Aikman and Dainty (1966)		0.550 ±0.040	–	–	
Gutknecht (1966)		0.625 ±0.005	0.012	–	
Camlong and Genevois (1930)	Naples	0.368	0.020		*V. utricularis*
Pantanelli (1918)	–	0.291	0.152	–	
Brooks (1933)	–	0.372	0.266	18°	

Steward considered many factors which could affect the vacuolar fluid composition of both species, he did not make direct temperature observations. He merely quoted Osterhout (1933a, 1933b), noting that the potassium ion content of the fluids decreased when the cells were kept for a period of days. More recent work shows similar variations; Aikman and Dainty (1966) reported large variations for *V. ventricosa* cells, while Gutknecht (1966) gave somewhat smaller variations.

The *Valonia* cell may be an excellent experimental entity with which to investigate the effects of various parameters and potassium and sodium transfer. This is due to a number of factors, including: (a) a large amount of cellular fluid, (b) a greater concentration of potassium in the cell fluid than in most other cells, and (c) a wide variation in the ionic content of the cell fluid (some lower animal cells seldom exceed a concentration variation of ± 0.0005 M for potassium and sodium).

One of the environmental parameters is temperature. Although no previous work has been done to show the magnitude of the temperature effect on internal ion content, a review of past work causes one to suspect that the concentration variation of ions in cellular fluids is at least in part temperature dependent. This view is strengthened by the data obtained in the potential measurement experiments.

The Nernst equation, which supposedly expresses the potential across the *Valonia* membrane system, shows that the potential is directly proportional to the log of the ratio of the concentration inside to the concentration outside. Aikman and Dainty (1966) stated that when considering *Valonia* potential, potassium is the dominating ion. In flux measurements, Gutknecht (1966) found evidence which would indicate a system with an active potassium and sodium flux and a passive chloride flux.

It has been shown that the PD in *Valonia* has abrupt discontinuities at 15 and 30 °C (Figures 1–3). If the relations described in the Nernst equation hold, it follows that discontinuities might arise at 15 and 30 °C in the internal potassium concentration of the cell. Obviously, this requires a rather large change in potassium content to effect an appreciable change in the potential. Within the limits of the variability of potassium concentration given by various authors, the potential is almost unchanged (0.3 mV is the greatest difference). Natural fluctuations (biological noise) in bioelectric potential under a given set of conditions have been estimated to be about ± 1 mV. This natural fluctuation is greater than theoretical calculations, due to the variability of potassium ion content.

Since the effect of temperature on the sodium and potassium ion content of the *Valonia* vacuole was unknown and since these concentrations directly entered into the theoretical calculations of the potential difference, it seemed relevant to examine the effect of temperature on ionic content over the range of the PD measurements.

1.3. The effect of ouabain

The enzymatic basis for active transport across living cell membrane or the so-called 'pump' has been intensively explored during the past ten years, chiefly utilizing such animal material as nerve, muscle and red blood cells. A number of important discoveries have been made about the enzyme systems which transport ions and other materials against an electrochemical gradient. In brief, it has been shown that energy for the active transport of sodium and potassium is linked (or coupled) and that ATP is required for the transport action. It has also been shown that this enzyme system is located in the plasma membrane. Excellent review of the recent work can be found in Albers (1967), Skou (1964 and 1966), and Hokin and Hokin (1963). One of the most characteristic tests for determining the presence of the ATPase transport system in animal cell is by inhibition of the system using cardiac glycosides such as ouabain (Wetherall, 1967). When cardiac glycosides are administered to cells they characteristically gain sodium and lose potassium. This indicates a loss of regulatory ability and is reflected in the decay of the membrane potential.

Despite the voluminous literature concerning the effect of cardiac glycosides on animal cells and biochemically interesting compounds, little interest or enthusiasm has been shown in exploring their effects on plant and, in particular on algal cells. Although a great many of the early classical studies of membrane biophysics employed algal cells, little has been done using algal cells to test the 'pump' or active transport system in comparison to animal cells. Very recently Wins and Kleinzeller (unpublished) have isolated a sodium and potassium ATPase system from the algal cell *Hydrodictyon* which is sensitive to ouabain on concentrations of about 10^{-3} M, while being insensitive at concentrations of 10^{-4} M and lower. This enzyme system has some of the characteristics of animal cell ATPase systems. It is important to note that the observed concentrations of ouabain which inhibit the ATPase system in the algal cells is of much higher concentrations than necessary to inhibit certain animal cell transport, i.e. 10^{-3} M vs 10^{-6} M. However, preparations of mouse and rat tissue have sensitivities in the area of 10^{-4} M.

Besides a biochemical study by Wins and Kleinzeller (1967), there are only a few other published studies concerning the effect of cardiac glycosides on algal cells: MacRobbie (1963) used the fresh-water species *Nitella* and found that potassium influx was decreased by ouabain concentrations of 5×10^{-5} M. After long treatments at this concentration there appeared to be a tendency in some cells for the potassium influx to increase. The potential showed little or no change, however, the time period of the potential measurements was not mentioned. Spanswick and Williams (1964) reported no effects of ouabain at concentrations of 10^{-5} M on the membrane potential across the plasmalemma and the tonoplast on exposures up to 4 h. Janacek and Rybova (1966) examined higher concentrations of ouabain (10^{-3} and 10^{-4}) on *Hydrodictyon reticulatum*. They found a gain in potassium chloride and a swelling of the cells after three days. However, in examining the potential across the membrane system after one, two and three days, they saw no change in bioelectric potential. More recently, Gutknecht (1967) used 10^{-4} M ouabain on *Valonia ventricosa* with no change in potential or short-circuit current.

Raven's study (1967) was previously mentioned. His ouabain concentrations were in the order of 10^{-4} M to 10^{-3} M.

In summary, there appears to be evidence that the potassium content of algal cells appears to be affected at high concentrations of ouabain. This cardiac glycoside is produced from the bark and the root of the ouabain tree and from the seeds of *Stropinthus gratus*. It has been used as an arrow and dart poison by various African tribes. Since it is produced from a plant itself, one might reason that plants have less sensitivity to it. It should be noted, however, that except for one *Valonia* study, all investigations utilized fresh water algae. The recent work of Slayman (in press) emphasizes the danger of assuming that at these high concentrations, ouabain acts specifically on the cationic pump.

In the following study, the effect of temperature and the effect of ouabain have been investigated on the concentration of ions potential and the short-circuit current.

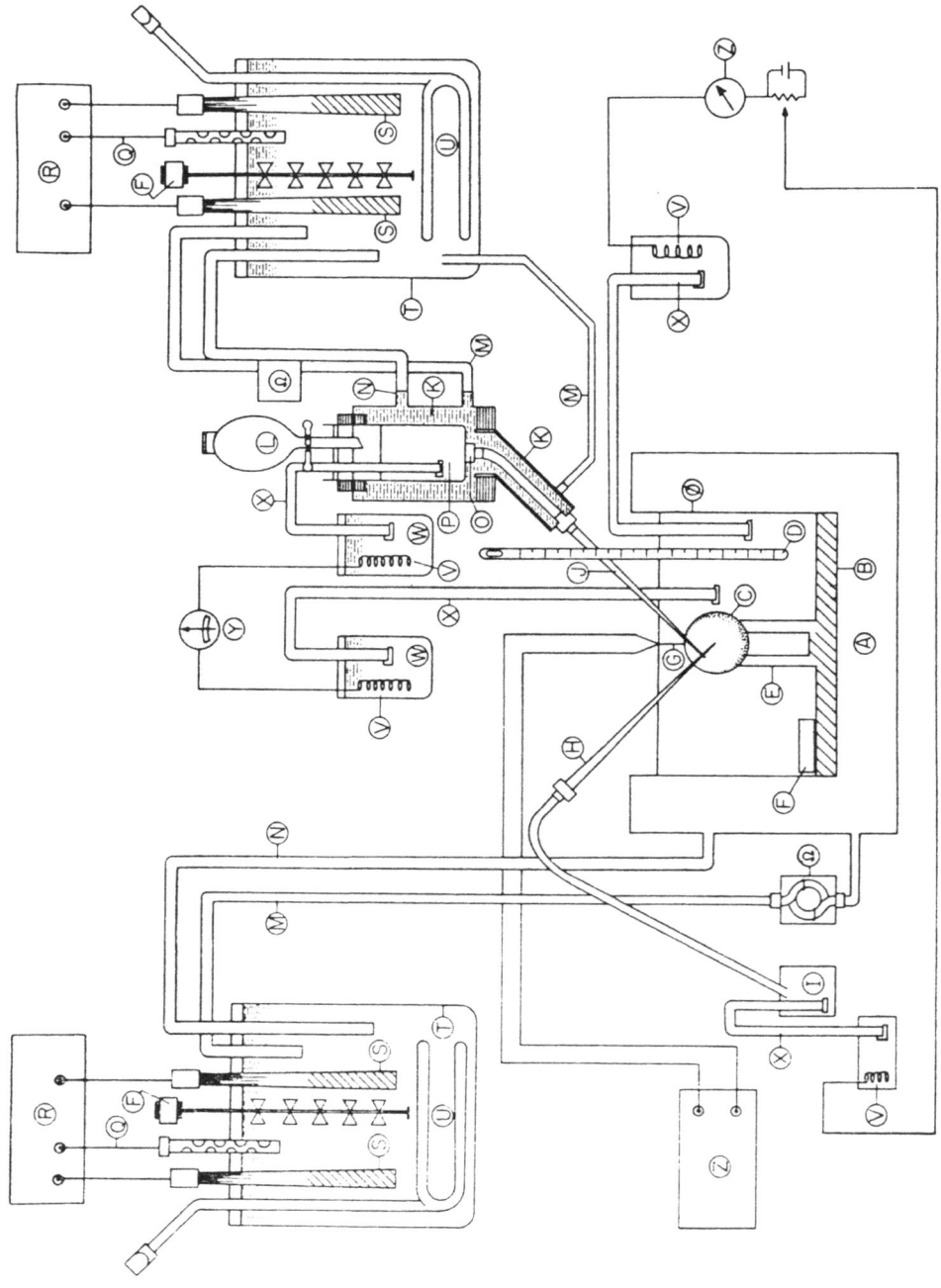

2. Methods

2.1. CULTURE AND HANDLING OF *Valonia*

Specimens have been collected from the Florida Keys and maintained for several years in out of doors running seawater tanks covered with plastic screening permitting of penetration of 5% normal daylight. This allows optimum culture conditions (Thorhaug, 1969 and 1971a) and insures healthy cells, and is a vital point since *Valonia* are relatively delicate in comparison to plants such as *Nitella*, *Chara* or *Acetabularia*.

Blinks (personal communication) has shown that it is essential for the cells to be in perfect health in order to obtain reproducible results; thus, several tests to detect and eliminate cells which were dying or injured were employed. Light-dark responses were measured by placing the cell in the dark for one minute. In healthy cells, this caused a drop in potential; upon sudden exposure to light there is a rapid increase in potential, then a return to the resting potential. Other tests used were turgor pressure, lack of spots of aplanospores and appearance of shiny green color.

2.2. PERFUSION TECHNIQUE

This method is basically similar to that of Blinks (1933), modified by Gutknecht (1967a). However, we have modified it to include temperature control, rapid and thorough mixing, complete view of cell and several other important features.

A single *Valonia* cell, approximately 1 cm in diameter is seated on the glass support, fixed to the bottom of an absorption dish (Figure 7 from Thorhaug and Katchalsky, 1972). Two micropipets are inserted through the cell wall and underlying protoplasm into the cell's vacuole. The insertion is accomplished in such a manner that the pipets are crossed. Dye experiments have shown that this configuration accomplished optimal mixing of the inflow fluid within the vacuole, which is essential for a homogeneous vacuole temperature.

After a recovery period of approximately five hours, during which time the protoplasm forms a tight seal about the portion of the pipet in contact with it, perfusion with an artificial cell sap is begun at the rate of approximately 1 ml min. When the cell potential returns to normal, the appearance of the cell as viewed through the microscope and bottom and side mirrors appears normal, and the flow rate of the perfusion is optimum, the experiment is begun. The sap continually drips into the syringe via a separatory funnel with a needle valve for fine adjustment of dropping rate.

←

Fig. 7. Assembly for *Valonia* perfusion with temperature control. A. Aluminum block. B. Mirror fitted to bottom of dish. C. *Valonia* cell. D. Thermometer. E. Glass stand for supporting cell. F. Stirring bar. G. Thermistor needle impaled into cell. H. Efflux pipet. I. KCl solution. J. Influx pipette. K. Temperature jacket for influx fluid. L. Reservoir for inflow sap. M. Connecting inflow tubing from temperature control bath. N. Connecting outflow tubing to temperature controlled bath. O. Teflon adapter for syringe containing inflow fluid. P. Artifical sap. Q. Thermistor for temperature control apparatus. R. Temperature control apparatus. S. Knife blade heater. T. Temperature bath. U. Cooling coils. V. Electrode. W. KCl solution. X. Salt bridge with vycor tips. Y. Voltmeter. Z. Conductance bridge. Ω-Pump. ϕ-Absorption dish.

2.3. ELECTRICAL MEASUREMENTS

The potential difference between the vacuole and seawater was measured with a high impedance electrometer, which makes an electric contact with the inflow reservoir and external seawater surrounding the cell via potassium chloride salt bridges (Y). The bridges are fitted with Vycor sintered glass plugs containing 0.5 M KCl and connected via 1 M KCl to calomel electrodes (X, Y, Z).

Resistance is measured by a conductance bridge via a KCl salt bridge connected to the outflow tube which communicates through the membrane, and another salt bridge with calomel electrode connected to the bathing seawater (Z, V, X, I). Short-circuit current was accomplished as in Gutknecht (1967b).

2.4. ION CONCENTRATION

The cells of *Valonia macrophysa* employed in this study were grown from aplanospores, from a few single cells and raised under identical environmental conditions. Those used were 0.7 cm diameter. The cells were placed in a polythermostat which is basically a precision-bored aluminum block heated at one end and cooled at the other so that a temperature gradient is maintained down the length of the bar. The bar has been fitted with two replicate cuvettes at each temperature cell. After the polythermostat had been brought to the desired temperatures, the seawater was placed in tubes and three cells were placed in each tube. Salinity and pH were measured before and after experiments. In three days, running they did not vary more than ± 0.5 ppTh and ± 0.2 pH. Light in the tubes was less than one foot candle. The experiment consisted of allowing the cell to remain at the given temperature for a net period which was 45 min, corresponding to the average time interval used in conducting the bio-electric potential studies and a second period of three days. The cell fluid was analyzed. Cells removed from the polythermostat were immediately blotted thoroughly on absorbent tissue and punctured with a clean needle to allow the fluid to drain into plastic specterphotometer cups using a calibrated eppendorf micropipet with disposable plastic tips. 15 μl of fluid was drawn from the cup and placed into two ml distilled water for sodium analysis. 5 μl fluid was placed into 5 ml of distilled water for fluid in analysis. The samples were stored in plastic containers until analyzed with the absorption spectrophotometer.

2.5. OUABAIN

The first group of experiments consisted of measuring the potential difference across the membrane system of *Valonia macrophysa* and *V. ventricosa* using the bioelectric methods described in the previous section. First, a light response was measured to determine the general health of the cell. Only cells showing over 1 mV change were used for the ouabain experiments; a time course for $\frac{1}{2}$ h was plotted to see the cells' variability. The ouabain was then introduced into 200 ml of filtered seawater used as the bathing solution.

A continuous plot of potential versus time was carried out on a Moseley Autograph

7100B strip chart recorder. Ouabain in concentrations of 10^{-3}, 10^{-4}, or 10^{-5} M and lower were then administered to the cells. The temperature was held at 25.0°C (± 0.01 °C).

Dilution effects were tested by adding an amount of water equal to the total ouabain solution to various cells. Changes in either the composition or effects of ouabain on the cells when held for a period of 3 days at 25°C in seawater were tested by placing cells into the ouabain seawater solution after various time periods, and comparing the effects with these cells having freshly administered ouabain.

3. Results

3.1. RESULTS OF TEMPERATURE ON ION CONTENT

The first group of experiments for determining the temperature effect on the potassium and sodium ion content of *Valonia* vacuolar fluid was conducted by holding the cells in the polythermostat for 45 min. Four sets of three cells were held at 19 different temperatures ranging from 8.0 to 42.1 °C with intervals varying from 1.3 to 2.9 °C. The results as determined from analysis of 228 cells of *V. macrophysa* are given in Figures 8 and 9. The concentration of potassium in individual cells varies from 70

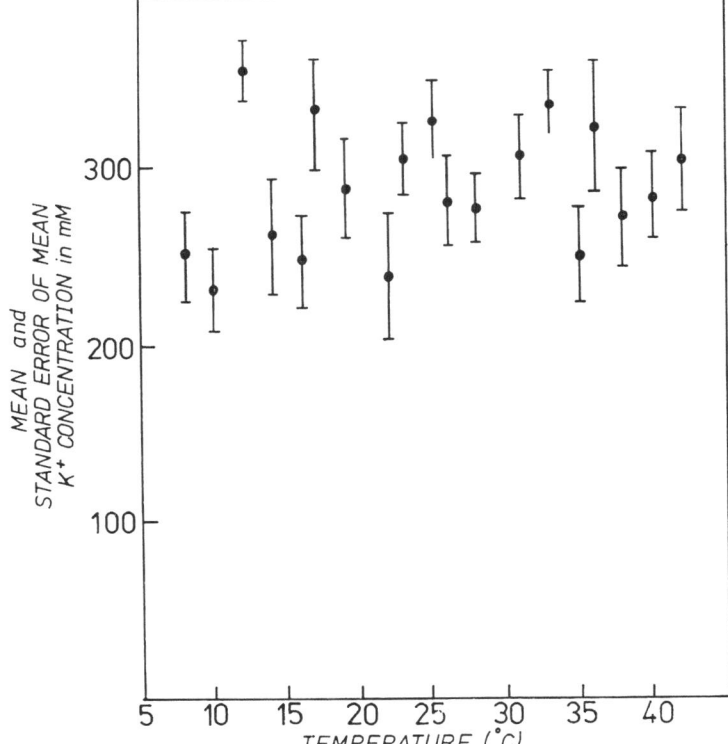

Fig. 8. Mean and standard error of mean potassium concentration in vacuolar sap of *Valonia macrophysa* kept at the given temperature for 45 min. Each point represents the mean of 12 cells.

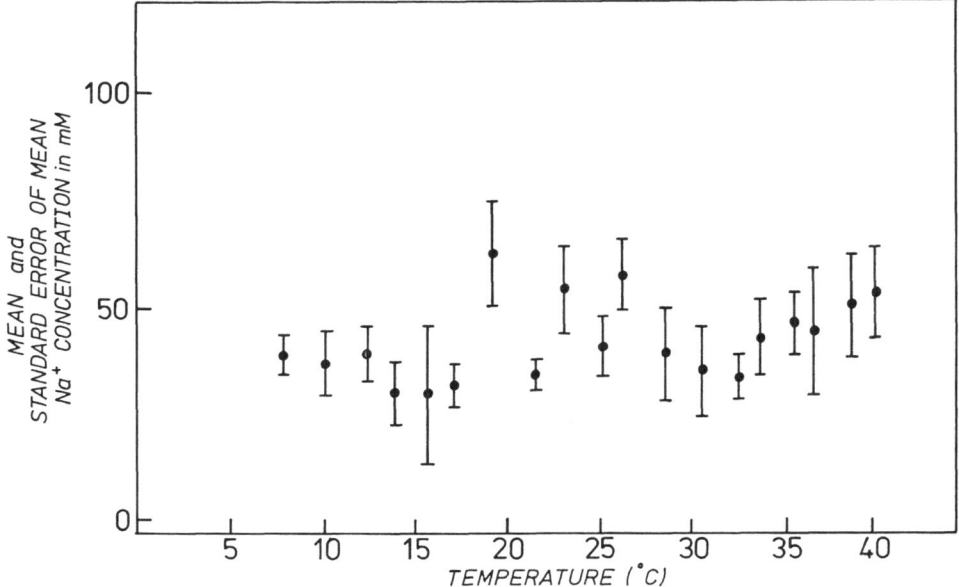

Fig. 9.   Mean and standard error of mean sodium concentration in vacuolar sap of *Valonia macrophysa* kept at the given temperature for 45 min. Each point represents the mean of 12 cells.

to 540 mM. For sodium the range is 19 to 96 mM. The range of the mean as shown is from 231 to 357 mM for a maximum variation of 32 mM for sodium. It is immediately apparent that the scatter far exceeds any noticeable temperature effect.

The standard error, ranges from ±17.0 to ±38.0 mM, for an average of about ±25 mM. This corresponds with the ±40 mM reported by Aikman and Dainty (1966). The standard error values are about 10 per cent of the mean potassium value. Gutknecht (1966) claimed a standard error of ±5 mM; however, his experiments utilized a maximum of 13 cells. Steward's (1937) standard errors of ±11 and ±4 mM are quite puzzling, since he used as many as 50 cells per determination. However, he speaks of taking 'composite' samples which obviously would make standard error very low (Figures 8 and 9).

The second group of experiments was conducted by holding the cells in the poly-thermostat for three days. The same numbers of cells were held at temperatures ranging from 7.0 to 39.6 °C with intervals varying from 0.7 to 2.1 °C. The potassium and sodium concentrations are given in Figures 10 and 11.

In contrast with the first experiment, the results of this experiment are striking and readily demonstrate a marked discontinuity of potassium concentration at about 13 and 33 °C. Actually, the potassium content between 15 and 32 °C was in the range of 345 to 421 mM, while that below 13 °C and above 33 °C was an order of magnitude different. The potassium values determined in this experiment for the range of 15 to 32 °C are somewhat lower than those given by other investigators. However, it must be emphasized that prior studies were made with very few cells, in some cases as few

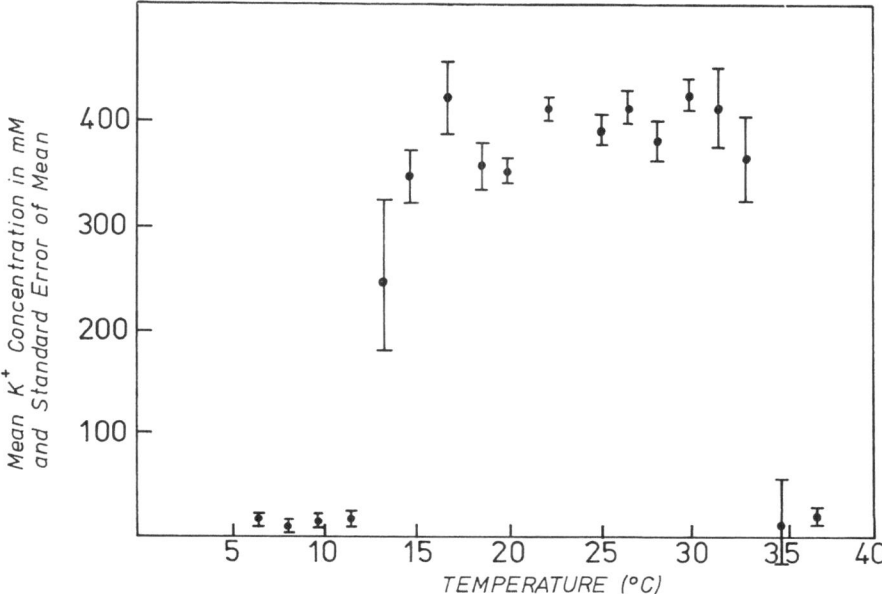

Fig. 10.   Mean and standard error of mean potassium concentration in vacuolar sap of *Valonia macrophysa* kept at the given temperature for three days. Each point represents the mean of 12 cells.

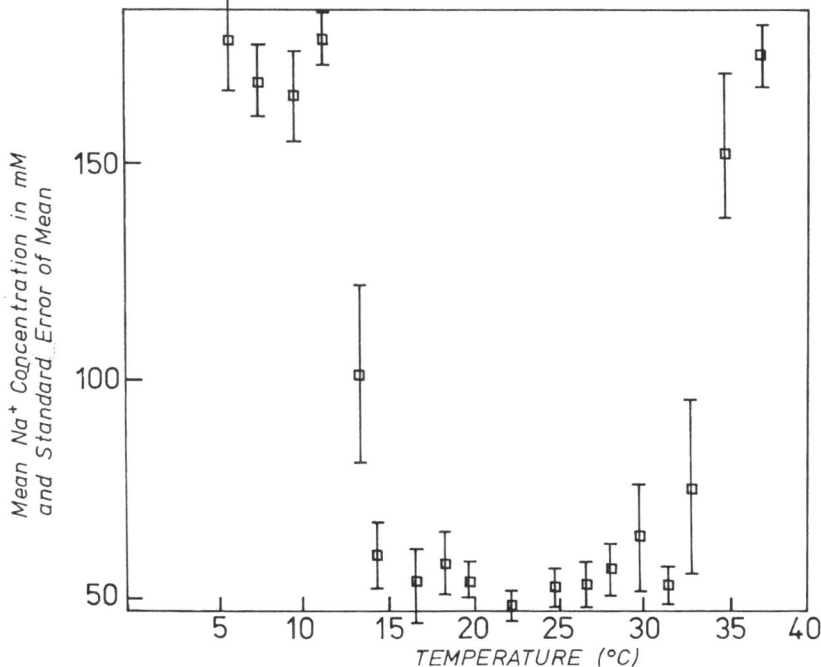

Fig. 11.   Mean and standard error of mean sodium concentration in vacuolar sap of *Valonia macrophysa* kept at the given temperature for three days. Each point represents the mean of 12 cells.

as four. Then, too, with the numerous cells used in this investigation, the maximum variation between any two cells often exceeded 100 mM. Brooks (1933) claims that in his study the temperature was not far from 12°C; his value given for *Valonia* in the Gulf of Naples could easily have been 14 or 15°C, thus comparable to the data of this investigation. Gutknecht (1966) conducted his experiments in artificial seawater; thus his results are not comparable.

Between 15 and 33°C, the maximum standard error recorded was ±38 mM (see Table II). This corresponds very well with the ±40 mM of Aikman and Dainty (1966). This is similar to the standard error in the first experiment just described. Gutknecht

TABLE II

Standard error of mean potassium and sodium content for *Valonia macrophysa* after 45 min exposure to a given temperature

Temperature	K+ SE of mean in mM	Na+ SE of mean in mM
8.0	±26.1	±4.5
10.2	24.4	7.7
12.1	17.8	7.3
13.9	32.7	7.0
15.8	26.0	7.9
17.1	30.2	5.0
19.4	28.7	10.2
21.5	35.8	4.3
23.1	19.5	10.0
25.0	23.0	6.1
26.4	25.8	6.2
27.8	20.0	8.3
30.7	24.0	4.6
32.5	18.4	5.5
34.6	26.8	10.8
36.3	38.0	6.8
37.9	27.8	9.5
40.0	24.9	10.1
42.1	29.7	9.0

(1966) reported a standard error of ±5 mM; however, he used a maximum of 13 cells for any one measurement. Steward's (1939) standard error of ±11 and ±4 mM is quite puzzling, since he used as many as 50 cells per determination.

By far the most interesting point to observe about the standard error is that at 13.3 and 33.5°C it is markedly different than the standard error, respectively, below and above these temperatures. This abrupt change is identical with the striking discontinuity in potassium and sodium concentrations at 13.3 and 33.5°C. It should be noted that, as the temperature decreases, the standard error increases to a maximum of ±89.50 mM at 13.3°C. Likewise, as the temperature is increased to 33.5°C, the standard error reaches a maximum of ±37 mM. Between 20 and 25°C, the average standard error is a low ±14.6 mM.

It is also essential to point out that the percentage of the mean represented by the standard error is quite different than in the 45 min experiment. Except for the transition zones of 13 to 15°C and 32 to 34°C, the standard error values are from 3% to 10% of the mean potassium value. In the transition zones the values are 20% and 32%.

The sodium ion content of the cells used in this three day experiment is similar to potassium in that it appears to be highly temperature dependent. However, as one would expect from our knowledge of cation relations in cells, the sodium and potassium are inversely related. Between 15 and 33°C the sodium content of the *Valonia* vacuolar fluid ranges from 50.5 to 80.0 mM for an average of about 60 mM. Just as with the potassium concentrations, there are two abrupt changes. Although the points of temperature change are not identical, they are very similar to that occurring between 31.8 and 35.2°C and that between 13.3 and 15.0°C. Above 33.5°C the average

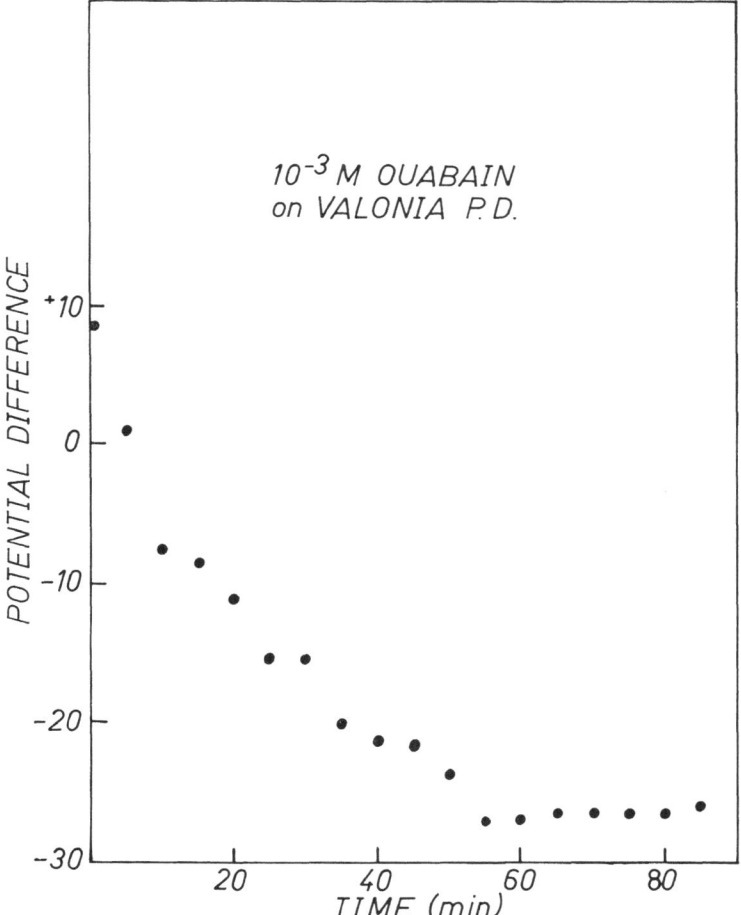

Fig. 12. Results of the effect of 10^{-3} M ouabain on *Valonia ventricosa* P.D. versus time.

sodium concentration is shown to be 171 mM, while below 15 °C the concentration is about 163 mM.

The standard error of the mean sodium concentration ranges from ± 1.15 to ± 21.5 mM. However, as the transition zones described earlier are approached, the standard error or variation in vacuolar sodium concentration becomes erratic. From 31.8 to 15.0 °C, the average variation in the sodium content is ± 6.8 mM. From 11.2 to 7.0 °C, this value is a comparable ± 6.5 mM. At 36.9 °C it is about ± 5.0 mM. The transition zones appear to fall between 33 to 35 °C and around 13 °C, where the standard errors are approximately ± 17 and ± 21 mM, respectively.

Considering the previously reported variability of sodium in *Valonia* vacuolar fluid, the values in this experiment are remarkably consistent except for the transition zones.

3.2. THE EFFECT OF OUABAIN ON BIOELECTRIC POTENTIAL

The results of twelve cells tested with 10^{-3} ouabain are seen in Figure 12, which shows potential difference versus time. The potential decayed within the first 40 min to high negative values. The mean value was near -27 mV. However, there were several occasions on which potential for individual cells dropped to -50 or -70 mV. The standard deviations were quite large in some cases. The time sequence of decay of potential varied markedly from cell to cell. In some cells the potential dropped within the first five minutes, while other cells had no change for up to 20 min after which their potential dropped markedly. In addition, there were some cells which did not react to ouabain. These numbered approximately 20% of those cells tested. Control cells, which were administered the same solution without ouabain, showed no change in potential. Other control cells from the same batch were run for equal time periods showing ± 1 mV variation in 90 min. One interesting observation was that after $3\frac{1}{2}$ to 4 h, the cells' potential returned to the values near the original resting potential and remained there even though the cell was allowed to remain in the ouabain bathing solution for up to 3 days. If a second cell was placed in the same solution the new cell would react in a typical manner of having a rapid negative potential.

Most of the ouabain literature does not examine long term time effects of ouabain on membrane potentials or other membrane properties. This time variability would be a fruitful path for further study.

The result of 7 cells run at 10^{-4} M ouabain showed very little change in potential. The change was within the time course variability, i.e. several millivolts.

Cells run at 10^{-5} M ouabain showed that within 60 min the average cell decreased 53 mV. The time course for this decrease over long periods of time – such as 3 h or longer – has not been determined at present. Lower concentrations of ouabain, down to 10^{-8} M were applied and did not appear to affect the membrane potential of *Valonia*.

The effect of ouabain at 10^{-3} M on the vacuolar side of the plasma membrane was negligible. Eight cells were subjected to a perfusate containing 10^{-3} M ouabain and the potential remained within the normal variability over the time periods of up to three hours.

3.3. OUABAIN AND SHORT-CIRCUIT CURRENT

The effect of 10^{-3} M ouabain was tested on cells perfused with the short-circuit current. The short-circuit current decayed to zero within 35 min of application of ouabain. Seven cells were tested. The mean shortcircuit current was 1.76 μ amp/cm². The mean potential was 10.6 mV at the start of the experiment. The potential went to a very low value under the short-circuit conditions within a mean time of 20 min. As the short-circuit current decayed it went to zero after more than 30 min. Thus, the decay rate of the potential and the short-circuit current were not necessarily parallel. Ouabain placed on the vacuolar side of the plasma membrane at concentrations of 10^{-3} M did not effect the short circuit current or the potential.

4. Discussion

Many properties of the *Valonia* membrane system show anomalous behavior near 15 and 30°C which were presented in the Introduction and Results section. This may be interpreted as phase transitions in one, or perhaps, several membrane properties. Both rate reactions, such as active transport (manifested in the short-circuit current), and electrical properties manifesting structural characteristics, such as conductance, show these transitions.

For short time periods the transitions are reversible with little apparent hysteresis as compared to the studies of Tasaki (this volume). However, for longer time periods, the transitions result in cellular death. Water transport increases markedly above 33°C and decreases markedly below 10°C (Thorhaug et al., 1974) which may indicate opening and closing of pores. Gutknecht (1967) stated *Valonia* did not show evidence of pores at 25°C, but this would not preclude pores at other temperatures. The water permeability itself is sensitive to temperature (Thorhaug et al., 1974).

It appears probable that both structural and enzymatic changes occur simultaneously at temperatures which exceed the viability limits. However, an exact proof of the mechanism of thermal death requires that each system which could undergo a thermal transition in the cell membrane be examined. Various hypotheses have been put forth, such as protein denaturation (Brandts, 1967), lipid melting (Bean and Chan, 1969), higher order phase transitions in water structure (Drost-Hansen, 1965).

One interesting result is the remarkable lack of temperature dependence between 15 and 30°C of several properties such as bioelectric potential and chord conductance. The double membrane system may provide a balance so that the activities at one membrane have a feedback loop to the other membrane and thus maintain a steady state within the viability limits.

Despite the extensive literature on the accumulation of salts in *Valonia* cells and despite the many attempts to elucidate the factors which affect salt accumulation, these data on ion content are the first to demonstrate the effect of temperature on the ion content of *Valonia* vacuolar fluid. Over a short time period (45 min) the cell to cell variation in potassium and sodium was greater than any discernible temperature effect. Over experimental periods of three days, however, there is a striking relationship between the potassium and sodium content and the temperature.

Let us first discuss the experiment which exposed the cells to a set of temperatures for 45 min. This showed that the variability of the mean range of potassium was from 231 to 357 mM. Individual potassium values ranged from 70 to 540 mM for similar cells. Previous workers have reported values ranging from 291 to 540 mM for three species (*V. ventricosa*, *V. macrophysa* and *V. utricularis*). (Gutknecht reported 625 mM with artifical media, which probably caused increased concentration). Brooks (1933) suggested that the sap composition may be of some taxonomic value. It is now clear, however, that the variation between similar cells of a species by far exceeds the intraspecific differences. Consequently, sap analysis for taxonomic purposes could be extremely misleading. Sodium was likewise variable, although consistently lower in concentration. The mean content of 12 cells ranged from 31 to 63 mM, while individual cells ranged from 19 to 96 mM. The standard error was about ± 7 mM.

The investigation was carried out primarily to test the assumption that the Nernst equation adequately described the potential across the membrane system of *Valonia* in terms of the ion content. The dominating ion supposedly is potassium. However, since the potential undergoes abrupt changes at 15 and 30°C, and the only cellular variable in the equation is the concentration of potassium ion, one may then expect a sharp change in potassium concentration at 15 and 30°C, if the relationship between PD and concentration is as described by the Nernst equation. But when the concentration values found during this experiment are substituted into the Nernst equation, the widely varying concentrations of potassium ions between 15 and 30°C would indicate that the potential should fluctuate widely between 15 and 30°C; yet, it has been shown to be relatively constant in this range. In addition, when the potassium concentration values above 30°C are substituted into the Nernst equation, no marked increase in PD can be seen as previously shown in the bioelectric studies cf *V. macrophysa* nor can a marked decrease in PD below 15°C be ascertained. If sodium ion concentrations are used as C_0/C_i in the Nernst equation, the bioelectric potential varies in a manner which does not follow the measured bioelectric potential data. The bioelectric potential behavior of *V. macrophysa* appears not to be attributed to changes in potassium ion content, nor to changes in sodium ion content. Thus, the concentration of the major cations in the vacuole does not describe the potential in the *Valonia* cell with respect to temperature. It can be argued that the vacuolar concentration of potassium does not change markedly, but the protoplasmic content, the plasmalemma content and/or the cellulose wall content of potassium may change markedly.

An investigation of concentration of potassium ions in the protoplasm as a function of temperature would thus be highly interesting for a thorough examination of the Goldman equation. There may be a sudden increase of potassium ions at the plasma membrane which for this short duration is not reflected in the sap. Analysis of the protoplasm itself might be a more meaningful measure of the concentration value of potassium described in the Nernst equation. Yet, it may be argued that the vacuolar sap should rapidly reflect an increase of potassium in the protoplasm for the following reasons: (a) rapid increase in potential of *Valonia* cells when placed directly into a high temperature seawater in which the potential may rise 10 mV within 10 seconds (Blinks, 1942; Marsh, 1939); (b) rapid rate of ionic diffusion, about 10^{-5} cm^2 s; and (c) thinness of the protoplasm layer, less than 10 μ. The increase of potassium ions in the vacuole would have to be quite large to produce the 10 mV change seen above 30°C in *V. macrophysa*. The validity of the Nernst or Goldman equation when applied to the relatively complex *Valonia* membrane system should be questioned. It may be possible that the Nernst equation applies to such systems as frog sartorius muscles, but does not adequately describe other systems. The increased complexity caused by a tonoplast membrane may be a partial explanation.

Examining the three day experiment, one finds a definite concentration difference as a function of temperature. Below 15°C and above 32°C, the potassium concentration is an order of magnitude below the values found between 15 and 32°C. The sodium ion concentration also showed a great change at precisely the same temperatures; sodium ion concentration is higher at the extremes, however, and lower between 15 and 30°C. The obvious conclusion from these data is that the cell is able to regulate normally between 15 and 30°C, regulate weakly at the limits of these values (13.3 and 33°C), but is unable to regulate (after three days) at the temperature extremes. Decrease in concentration of potassium and sodium ions is not necessarily the cause of the irreversible plasmolysis which has occurred. This process is no doubt the effect of a complex process occurring within the cell three day period.

It is important to compare the standard error of the three day experiment with that of the 45 min experiment. The standard error of the potassium concentration is within the same range (± 15 to ± 38 mM) from 15 to 31°C in the three day experiment, as in the entire range of the 45 min experiment. This would substantiate the previous statement that there is a large biological variability, since 456 cells were used for the total measurements. Steward, who had a much lower standard error, did not use such a large number of cells. Moreover, he sampled the cells directly in the field. The cells used in the described experiments were grown at the laboratory and sampled under laboratory conditions. This is important because most results concerning *Valonia* potential experiments have used cells grown in the laboratory and handled under laboratory, rather than field, conditions. If the measurements of variability of cell content from Steward are correct, the cells in their natural environment have far less variation than cells under laboratory conditions. The fact that Gutknecht obtained appreciably higher values than any previous investigator for potassium ion concentration in *V. ventricosa*, after raising the cells on an artificial medium, may indicate that the cells are sensitive on a long term basis to the concentration in which they are raised.

Sodium ions have previously been found to be markedly more variable (Table I) than potassium ions. This present study also finds that the amount of variability (SE) of sodium demonstrated by the 45 minute cells and the SE of cells between 15 and 32°C in the three day experiment was markedly greater in percentage of total ion value than was the potassium. It may indicate that the sodium pump does not regulate as effectively as the potassium pump in *Valonia* cells. This would be in contrast to the pumps of certain other cells. It may be possible, on the other hand, that the sodium pump is highly regulated and that the large variability is due to the effect of some other ionic constituent. For instance, the sodium pump may compensate for changes inside the vacuole of the *Valonia* cell. However, such a mechanism would make the sodium ion pump more highly regulatory than the potassium ion pump. To elucidate this problem, further studies should include investigations of the Na^+-K^+-ATP-ase system in *Valonia* cells. Furthermore, the greater variability of the sodium ion concentration than that of potassium may indicate a lack of 'one-to-one' coupling of sodium and potassium as postulated by Scott and Haywood (1955) chiefly from a comparison with data obtained using *Ulva*.

The effect of ouabain on the bioelectric potential and short-circuit current was definite and reproducible. There has been controversy (Slayman, in press) as to the mechanism of action of ouabain on plant cells due to the high concentrations evidently necessary to produce the desired effect. It is not clear that ouabain acts as a specific inhibitor of the Na^+-K^+-ATPase system at these high levels; rather the mechanism may be one of a surfactant. Our results at 10^{-3} M do not elucidate any information bearing on this argument without further simultaneous information. However, the results at 10^{-5} M along with the short-circuit current measurements would preliminarily indicate a specific Na^+-K^+-ATPase inhibition in *Valonia* due to ouabain. These results are in contradiction to those of Gutknecht (1969) who found no effect of ouabain at concentrations of 10^{-5} and 10^{-4} on the bioelectric potential and short-circuit current. We have found the preparation of ouabain to be temperature light and time dependent so that proof of its action must be used with very fresh ouabain kept cold and protected from light.

The effect of ouabain on the plasmalemma would indicate an active pump at this membrane. However, Gutknecht (1969) has indicated that the major pump for ouabain is located at the tonoplast. Until the mechanism of 10^{-5} M ouabain has been established, it is difficult to further interpret these results. Nevertheless, it is interesting that ouabain at 10^{-3} M does not affect the short-circuit current or bioelectric potential of the plasmalemma.

The ouabain results yielded two distinct time patterns for action. Many of the cells, both with short-circuit current conditions and normal conditions showed a drop in potential only after 15 to 20 min. This would argue against an electrogenic pump. However, other cells responded within two to 5 min to ouabain. Further investigation into the origin of the membrane pump of *Valonia* should be undertaken.

Acknowledgements

I wish to acknowledge the support of the U.S. National Science Foundation Grant

P2B0689. Technical assistance of Dr. Marcela Fernandez and Steven Bach was appreciated. Stimulating discussions with the late Professor Aharon Katchalsky and with Drs. Dainty, Gutknecht and Kleinzeller were valuable in assessment of the data.

Literature Cited

Aikman, D. P. and Dainty, J.: 1966, in H. Barner (ed.) *Some Contemporary Studies in Marine Science*, Allen and Unwin Ltd., London, pp. 37–43.

Albers, R. W.: 1967, *Ann. Rev. Biochem.* **36**, Part II, 727–756.

Bean, R. and Chan, H.: 1969, in C. D. Tosteton (ed.), *The Molecular Basis of Membrane Function*, Prentice-Hall, p. 133.

Blinks, L. R.: 1933, *J. Gen. Physiol.* **17**, 109–125.

Blinks, L. R.: 1942, *J. Gen. Physiol.* **25**, 905–916.

Brandts, J. F.: 1967, in A. Rose (ed.) *Thermobiology*, Academic Press, pp. 25–72.

Brooks, S. C.: 1933, *Science* **77**, 221–222.

Caffier, G. and Kuchler, G.: 1972, *Actabiol. Med. Germ.* **29**, 109–118.

Chapman, D.: 1967, in *Ordered Fluids and Liquid Crystals*, pp. 157–166, Advances in Chemistry Series, No. 63, Washington, D.C.

Coldman, M. F. and Good, W.: 1969, *Biochem. Biophys. Acta* **183**, 346–349.

Dainty, J.: 1963, *Adv. Bot. Res.* **1**, 279–326.

Dainty, J.: 1969, in M. C. Wilkins (ed.), *The Physiology of Plant Growth and Development*, McGraw-Hill, London, pp. 453–506.

Dalton, J. C. and Hendrix, D. E.: 1962, *Am. J. Physiol.* **202**, 491.

Diamond, J. M. and Wright, E. M.: 1969, *Proc. Roy. Soc. B.* **172** (1028), 273–316.

Drost-Hansen, W.: 1965a, *Ann. New York Acad. Sci.* **125**, 471–501.

Drost-Hansen, W. and Thorhaug, A.: 1967, *Nature* **215** (5100), 506–508.

Good, W.: 1960, *Biochim. Biophys. Acta* **44**, 130–143.

Gutknecht, J.: 1966, *Biol. Bull.* **130**, 331–334.

Gutknecht, J.: 1967a, *Sci.* **158** (3802), 787–788.

Gutknecht, J.: 1967b, *J. Gen. Physiol.* **50**, 1821–1834.

Hensel, H., Iggo, A., and Witt, I.: 1960, *J. Physiol.* **153**, 113–126.

Hensel, H.: 1963, in Ch. M. Herzfeld (ed.), *Temperature*, Vol. 3, part 3 (Biology and Medicine), Reinhold Publishing Corp., New York.

Hogg, J., Williams, E. J., and Johnston, R. J.: 1968, *Biochim. Biophys. Acta* **150**, 640–648.

Hokin, L. E. and Hokin, M.: 1963, *Ann. Rev. Biochem.* **32**, 553–578.

Johnson, S. M. and Bangham, A. D.: 1969, *Biochim. Biophys. Acta* **193**, 92–104.

Lippold, O. C., Nicholls, J. G., and Redfearn, J. W. T.: 1960, *J. Physiol.* **153**, 218–231.

MacRobbie, E. A. C.: 1963, *J. Gen. Physiol.* **45**, 861–876.

MacRobbie, E. A. C.: 1970, *Quart. Rev. Biophys.* **3**, 352–281.

Marsh, G.: 1939, Carnegie Inst., Wash., *Papers from Tortugas Lab.* **33**, 67–84.

Martin-Lof, S. and Soremark, C.: 1969a, *Svenska Traforshningsinstitutet Series B* (8), 15 pp.

Martin-Lof, S. and Soremark, C.: 1969b, *Svenska Traforshningsinstitutet Series B* (1), 1–15.

Martin-Lof, S. and Soremark, C.: 1969c, *Svenska Papperstidning*, **72**, 193–194.

Meyer, A.: 1891, *Ber. Deutsch. Bot. Ges.* **9**, 77–79.

Oppenheimer, C. H. and Drost-Hansen, W.: 1960, *J. Bact.* **80**, 21–24.

Osterhout, W. J. V.: 1933a, *Ergebnisse Physiol.* **35**, 967–1021.

Osterhout, W. J. V.: 1933b, *Cold Spring Harbor Symp. on Quant. Biol.* **1**, 1–4.

Ramiah, M. V. and Goring, D. A. I.: 1965, *J. Polymer Sci. Part C* (11), 27–48.

Raven, J. A.: 1967, *J. Gen. Physiol.* **50**, 1607–1625.

Rosano, H. L., Duby, P., and Scholman, J. H.: 1961, *Phys. Chem.* **65**, 1704–1708.

Scott, G. T. and Hayward, H. R.: 1955, in A. M. Shanes (ed.), *Electrolytes in Biological Systems*, Am. Physiol. Soc., Washington, pp. 35–64.

Skou, J. C.: 1964, *Prog. Biophys. Chem.* **14**, 131–166.

Slayman, C. L.: 'Proton Pumping and Generalized Energetics of Transport: A Review', in press.

Steward, F. C. and Martin, J. C.: 1937, *Carnegie Inst. Wash. Publ.* **475** 87–170.

Thompson, T. E.: 1964, in M. Lucke (ed.), *Cellular Membranes in Development*, Academic Press, New York, pp. 83–96.

Thorhaug, A.: 1969, 'Temperature Effects on Membrane Phenomena', Ph. D. Dissertation. University of Miami, Coral Gables, Florida, 165 pp.

Thorhaug, A.: 1971a, *Biochim. Biophys. Acta* **225**, 151–158.

Thorhaug, A.: 1971b, *Proc. First European Biophysics Congress*, EVIII/49, 419–428.

Thorhaug, A. and Fernandez, M.: 1973, *Bull. Mar. Poll.* **4**, 70–73.

Thorhaug, A. and Katchalsky, A.: 1972, *Proc. 7th International Seaweed Symp.*, pp. 279–285.

Thorhaug, A. Fernandez, M., and de Diego, A.: 1974, 'The Effect of Temperature and Temperature Gradients on Electrical Measurements of *Valonia* Membrane Properties', in J. Dainty and U. Zimmerman (eds.), in *Giant Plant Cell, Water and Ion Relations*, in press.

Ting, H. P., Bertrand, G. L., and Sears, D. F.: 1966, *Biophys. J.* **6**, 813–823.

Wright, E. M. and Diamond, J. M.: 1969, *Proc. Roy. Soc. B.* **172**, 227–271.

Zucker, R. S.: 1973, *J. Physiol.* **229**, 787–810.

III

CHARGED MOSAIC AND ARRAYS

CHARGED MEMBRANE ARRAYS: MOSAICS AND STACKS

S. ROY CAPLAN

Weizmann Institute of Science, Rehovot, Israel

Abstract. Membrane arrays built of unlike elements, both in parallel and in series, may possess important properties entirely due to the interactions between constituent elements. The simplest and perhaps most interesting examples consist of arrays of cation-selective and anion-selective elements. Such systems are particularly amenable to nonequilibrium thermodynamic analysis, from which considerable insight may be obtained into their behavior. Parallel arrays or charge-mosaics are characterised by electrical circulation, and exhibit various regimes of operation according to the electrical resistance in the bathing solutions. They have a number of potential applications, most strikingly in desalination by 'piezodialysis'. An illustrative calculation is given of piezodialysis with cylindrical tubes or hollow fibers made from a hypothetical charge-mosaic, similar in properties to the Zeo-karb 315 cation-exchange membrane described by Meares and co-workers. Other applications include the determination of salt activities in solutions containing nonelectrolyte. Series arrays, or multimembrane stacks, have unexpected symmetry properties with respect to the passage of electric current. Certain classes of intermembrane compartments may be identified whose concentrations change similarly under different steady state conditions. Among these is a set whose concentrations are independent of the steady state current. For the case of a closed stack of paired membranes containing alternately positive and negative fixed charges, with indefinitely large end reservoirs, a set of desalinated intermembrane compartments may exist in the steady state, or a set of salinated compartments, but never both simultaneously. Such a stack may act as an efficient current rectifier.

1. Introduction

As with so many other membrane phenomena, the essential properties of mosaic membranes were predicted and demonstrated by Sollner many years ago. This fact reminds me that in common with Dr. Gregor I also spent time in Dr. Sollner's laboratory at an early stage of my career, so it is perhaps not inappropriate to recount at least one of my experiences there. I came to Bethesda in the late 50's after working in an England where austerity was still very much the rule, where envelopes were re-used until they fell apart, and where I was used to regenerating the mixed-bed ion-exchange columns used for preparing the de-ionized water every week. One day while I was going through this fairly lengthy procedure Dr. Sollner came up to me and said "Dr. Caplan, why *exactly* are you doing this?" When I explained, he rather vehemently compared the cost of my time (to the American tax-payer) with the cost of a new column available commercially, and almost succeeded in persuading me that throwing a batch of perfectly good ion-exchange resin down the sink was a desirable thing to do. Economically, of course, the argument made excellent sense; but as we all know today, ecologically it's a disaster. Now, it might be thought that this story is hardly apropos of my topic, but in fact it has a rather direct bearing on it. Mixed-bed ion-exchange chromatography is essentially a batch process. Imagine an attempt to convert this technique into a continuous process, and at the same time save resin, by utilizing a 'mixed-bed membrane' equivalent to a thin slice of the column. Would such a procedure succeed? On simple intuitive grounds one might suppose that the possibility exists of extruding, under pressure, a concentrated salt solution through this type of membrane, leaving a more dilute solution on the upstream side. This might occur because of the very favorable partition coefficient regulating the distribution of salt between the solution and the membrane: the effective concentration of salt in the membrane would be several molar. However, it is apparent that anyone trying to use this process for desalination would be defeated by that selfsame partition coefficient. The membrane would be extremely reluctant to release salt on the downstream side, and consequently no perceptible desalination of the upstream water would take place. The problem here is the randomness of the mixed bed, i.e. the total lack of organization, which means that no cooperativity with respect to flow can exist between the two kinds of ion-exchange particle.

Figure 1 shows a cross-sectional representation of several different possible types of array of anion and cation exchange elements in a membrane. Only two have

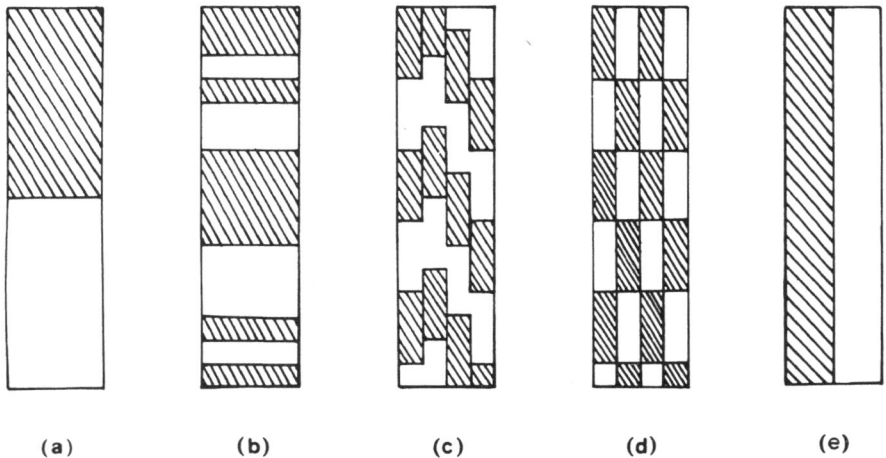

(a)   (b)   (c)   (d)   (e)

Fig. 1.   Cross-sectional representations of composite membranes of varying degrees of organization (see text).

properties of practical interest: the balanced mosaic array at (a) and the series or stack array at (e); these correspond to the two extremes of organization. I cannot refrain from recalling at this point Dr. Boyd's opening comments concerning the Yin and the Yang. It was Dr. John Weinstein, who was responsible for most of the studies of mosaic membranes in my laboratory, who pointed out to me the tremendous relevance in this field of the dualistic cosmic theory of the Yang and the Yin. The positive cosmic principle Yang, and the negative cosmic principle Yin, were considered to control the so-called 'blood current', and in the 3rd century B.C. it had already been claimed that under this influence "the blood current flows continuously in a circle and never stops". It will be seen that this description bears a striking resemblance to that of a functioning charge-mosaic membrane.

2. Mosaics

A charge-mosaic membrane consists of an array of anion and cation exchange elements arranged in parallel, each element providing a continuous pathway from one bathing solution to the other. When a gradient of electrolyte concentration is established across the membrane, anions and cations flow in parallel through their respective pathways without violating macroscopic electroneutrality. In electrical terms these flows appear as a circulation of current between the individual ion-exchange elements. This phenomenon was predicted by Sollner [1] and later demonstrated experimentally in systems constructed of two separate membranes of opposite fixed charge by Sollner and co-workers [2–4]. A diagrammatic representation of a

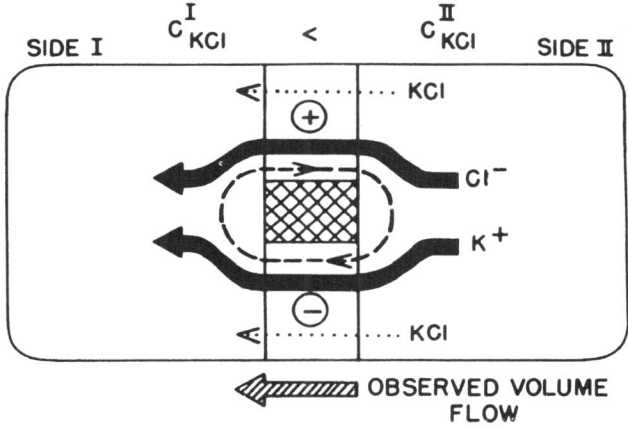

Fig. 2.   Diagrammatic representation of a functioning charge-mosaic membrane showing the circulating electric current.

functioning charge-mosaic membrane is shown in Figure 2, in which the circulation of electric current is indicated. Membranes consisting of only one kind of element, in which no circulation current can flow, will permit only a relatively minor leakage flow of salt to pass.

The phenomenon of current circulation enables a charge-mosaic to show a salt permeability orders of magnitude greater than the intrinsic salt permeability of its constituent parts, and indeed also greater than its permeability to nonelectrolytes of comparable size. This high salt permeability can be attributed to two factors: (i) the high effective concentration of electrolytes in the membrane as a whole, and (ii) the short-circuiting of the membrane potentials of the individual elements by the circulating current. In addition, a mosaic may exhibit a number of other unusual transport phenomena, the most notable being negative osmosis.

The simplest theoretical treatment of a charge-mosaic is an electrical circuit model in which one calculates the factors determining the 'loop current' circulating between the elements [5, 6]. In this simple electrical view the membrane elements are analogous to batteries with internal resistances, which reinforce each other in driving current around a circuit containing load resistances. The load resistances, of course, represent the part of the current pathway which lies in the bathing solutions on either side of the membrane. If the internal resistances of the elements predominate, intrinsic membrane properties such as thickness, fixed charge density, and porosity control the current flow. On the other hand, if the load resistances predominate, then solution phase properties such as electrolyte concentration and geometry of the conductance paths control the current. This model, therefore, delineates two of the regimes in which a mosaic can operate: 'membrane control' and 'solution control'. Although this conceptual picture is instructive, it takes into account the circulating current only. If we wish to have a more complete description which includes co-ion and volume flows as well as their coupling to the circulating current, we must turn to non-equillibrium thermodynamics.

The first nonequilibrium thermodynamic analysis of a charge-mosaic expressing the transport properties of the whole membrane in terms of those of its constituent parts was given by Kedem and Katchalsky [7]. The treatment of Kedem and Katchalsky, however, took no account of the specific geometry of the mosaic pattern, and tacitly assumed that the elements of the mosaic were short-circuited together, i.e. that the membrane was necessarily operating in a regime of membrane control. This treatment was extended beyond the regime of membrane control by Weinstein et al. [6], who took explicit account of membrane geometry and the nature of the solution compartments. It is necessary to give a broad outline of some essential components of the treatment in order to indicate what practical conclusions can be drawn.

As is usual in nonequilibrium thermodynamic treatments, one starts with the dissipation function Φ in terms of volume flow J_v, salt flow J_s, and electric current I, as derived by Kedem and Katchalsky [8] for membrane phenomena:

$$\Phi = J_v(\Delta p - \Delta \pi) + J_s \Delta \mu_s^c + IE \tag{1}$$

Here Δp and $\Delta \pi$ are the hydrostatic and osmotic pressure differences, $\Delta \mu_s^c$ is the concentration dependent part of the chemical potential difference of the solute, and E is the emf (i.e. the potential difference measured with anion reversible electrodes). Flows are considered positive if they proceed from side I to side II, and $\Delta X = X_1 - X_{11}$ for all thermodynamic potentials. X. E is positive if it tends to drive current through the membrane from side I to side II. The corresponding linear phenomenological equations, in terms of resistance coefficients, are

$$\Delta p - \Delta \pi = R_{11}J_v + R_{12}J_s + R_{13}I \tag{2}$$

$$\Delta \mu_s^c = R_{21}J_v + R_{22}J_s + R_{23}I \tag{3}$$

$$E = R_{31}J_v + R_{32}J_s + R_{33}I \tag{4}$$

where the subscripts 1, 2, and 3 denote volume, salt, and charge flow processes respectively. Onsager's reciprocal relations indicate that in all cases $R_{ij} = R_{ji}$. Those who are familiar with Kedem and Katchalsky's work know that this set of equations is transformable into more practical sets of equations, the set most useful in treatment of mosaics being their 'second set':

$$J_v = L_p(\Delta p - \Delta \pi) + c_s L p (1 - \sigma) \Delta \mu_s^c + \beta I \tag{5}$$

$$J_s = c_s L_p (1 - \sigma)(\Delta p - \Delta \pi) + c_s \omega' \Delta \mu_s^c + (\tau'_+/v_+ z_+ F)I \tag{6}$$

$$E = - \beta(\Delta p - \Delta \pi) - (\tau'_+/v_+ z_+ F) \Delta \mu_s^c + (1/\kappa')I \tag{7}$$

The six independent 'practical' coefficients appearing are the filtration coefficient L_p, the reflection coefficient σ, the electroosmotic permeability β, the solute permeability ω', the cation transport number τ'_+, and the electrical conductance κ'. The logarithmic average concentration c_s is defined by the expression

$$c_s = (c_I - c_{II})/\ln(c_1/c_{11}) \tag{8}$$

The stoichiometric and charge numbers of the cation of the permeable salt are given by v_+ and z_+ respectively, F being the Faraday constant.

In the format given above the coefficients of the independent variables are all 'complete' coefficients in the terminology of Weinstein *et al.* [6], i.e. they are all simple algebraic combinations of the original R coefficients. For example,

$$c_s(1 - \sigma) = (J_s/J_v)_{\Delta \mu^c_s = 0, \, I = 0} = - R_{12}/R_{22} \tag{9}$$

$$c_s \omega' = (J_s/\Delta \mu_s^c)_{\Delta p - \Delta \pi = 0, \, I = 0} = (R_{22} - R_{12}^2/R_{11})^{-1} \tag{10}$$

Since the osmotic pressure difference $\Delta \pi_s$ due to the permeable solute is related simply to the chemical potential difference $\Delta \mu_s^c$ by

$$\Delta \pi_s = c_s \Delta \mu_s^c$$

the equations are readily written in terms of $\Delta \pi_s$.

Conductances in parallel may of course be summed, hence the technique used by Kedem and Katchalsky to obtain the overall coefficients of a mosaic array of membrane elements was to express the coefficients of the individual elements in terms of conductances and then to subject them to a summation and averaging process. For example, for the electrical conductance of the mosaic one obtains

$$\kappa'_m = \gamma_a \kappa'_a + \gamma_c \kappa'_c \tag{11}$$

where m denotes a property of the mosaic as a whole and the quantities γ_a and γ_c denote the fraction of active membrane area occupied by the anion or cation exchange regions respectively. However, not all the expressions are simply additive as in the case of electrical conductivity. Indeed, in the case of the coefficients of most interest with regard to the mosaic, namely ω', σ, and L_p, each overall expression contains two 'additive terms' representing the contributions of the separate membrane elements, and in addition a 'non-additive term' associated with the circulation of current. These results are collected together in Table I, where it will be seen that each one of

TABLE I

Transport properties of a charge-mosaic in terms of the properties of its constituent resin elements the Geometry of the mosaic pattern, and the properties of the bathing solutions

$\kappa'_m \qquad = \gamma_a \kappa'_{ta} + \gamma_c \kappa'_{tc}$

$$\frac{1}{\kappa'_{tr}} = \frac{1}{\kappa'_{1r}} + \frac{1}{\kappa'_r} + \frac{1}{\kappa'_{11r}}$$

$$\alpha_{tr} = \gamma_r \kappa'_{tr} / \kappa'_m$$

$\tau'_{+m} \qquad = \alpha_{ta} \tau'_{+a} + \alpha_{tc} \tau'_{+c}$

$$\frac{1}{\kappa'_{loop}} = \frac{1}{\gamma_a \kappa'_{ta}} + \frac{1}{\gamma_c \kappa'_{tc}}$$

$\beta_m \qquad = \alpha_{ta} \beta_a + \alpha_{tc} \beta_c$

$c_s \omega'_m \qquad = \gamma_a c_s \omega'_a + \gamma_c c_s \omega'_c + \kappa'_{loop} \left(\dfrac{\tau'_{+a} - \tau'_{+c}}{\nu_+ z_+ F} \right)^2$

$L^m_p \qquad = \gamma_a L^a_p + \gamma_c L^c_p + \kappa'_{loop} (\beta_a - \beta_c)^2$

$c_s L^m_p (1 - \sigma_m) = \gamma_a c_s L^a_p (1 - \sigma_a) + \gamma_c c_s L^c_p (1 - \sigma_c) + \kappa'_{loop} \left(\dfrac{\tau'_{+a} - \tau'_{+c}}{\nu_+ z_+ F} \right) (\beta_a - \beta_c)$

the non-additive terms contains the quantity κ'_{loop}. This quantity is the series con-ductance of the *entire circuit* including the path of the current through the solution on either side of the membrane. The loop conductivity was introduced by Weinstein *et al.* who showed how it could be computed for a mosaic pattern of given geometry, in particular a square array of alternating circular cation and anion exchange elements embedded in an impermeable supporting matrix.

The transport equations given above are deceptively simple, as the coefficients are functions of state and may have a pronounced dependence on the concentration profile through the membrane. One may assume that linearity and Onsager reciprocity hold *locally* for the salt, volume, and current transport at each point in the membrane. If all the local resistance coefficients were independent both of position and of all state properties which might vary across the membrane, then the integrations required to arrive at a set of 'global' coefficients in the steady state would be trivial. Unfortunately, this is not the case. However, Weinstein *et al.* were able to demonstrate that certain of the coefficients are essentially independent of concentration, and that it is just these coefficients which enter into the description of processes in which the mosaic membranes may be expected to have some practical utility. Since none of the applications envisaged involve the passage of *net* current across the membrane, one need only concern oneself with the two-process phenomenological equations for salt and volume flow. For a good quality charge-mosaic one may write

$$J_v = L_p^m (\Delta p - \Delta \pi) + c_s L_p^m (1 - \sigma_m) \Delta \mu_s^c \tag{12}$$

$$J_s = c_s L_p^m (1 - \sigma_m) (\Delta p - \Delta \pi) + c_s \omega_m' \Delta \mu_s^c \tag{13}$$

and use the approximation

$$c_s \omega' \approx \kappa_{loop}' \{(\tau_{+a}' - \tau_{+c}')/\nu_+ z_+ F\}^2 \approx \kappa_{loop}'/F^2 \tag{14}$$

since

$$(\tau_{+a}' - \tau_{+c}') \approx - 1 \tag{15}$$

If the anion and cation exchange materials are highly permselective, then

$$\sigma_a \approx \sigma_c \approx 1 \tag{16}$$

The transport equations therefore take the form

$$J_v = L_p^m (\Delta p - \Delta \pi) - (\kappa_{loop}'/F) (\beta_a - \beta_c) \Delta \mu_s^c \tag{17}$$

$$J_s = - (\kappa_{loop}'/F) (\beta_a - \beta_c) (\Delta p - \Delta \pi) + (\kappa_{loop}'/F^2) \Delta \mu_s^c \tag{18}$$

It is important to note that none of the parameters in these equations contain the average concentration c_s and none of the coefficients are dependent on local concentration, at least to a good level of approximation. In effect, these equations neglect the

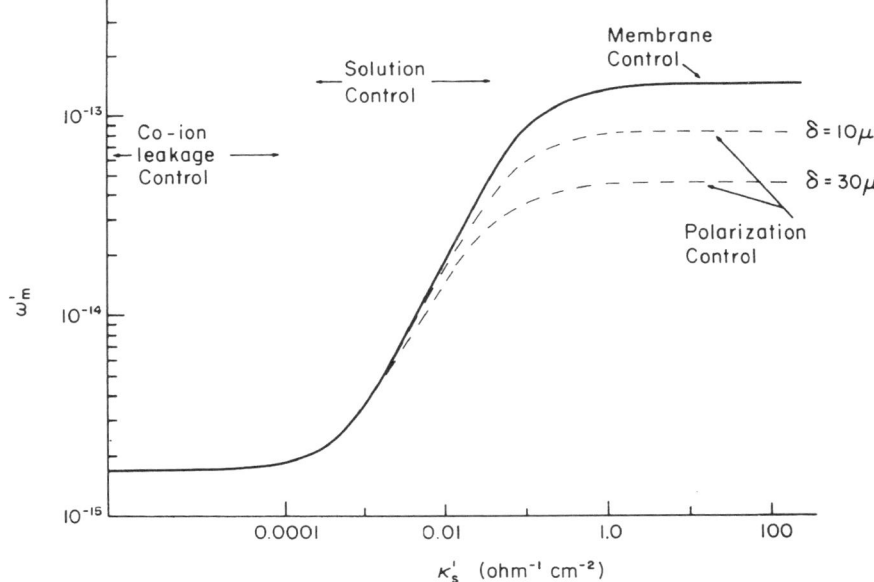

Fig. 3.   Salt permeability in the four regimes of charge-mosaic operation. The solid curve was calculated from a set of typical parameter values used by Kedem and Katchalsky [7]. The dashed lines represent the apparent salt permeability which would be found in unstirred layers of 10 μ and 30 μ, assuming the mosaic elements to be very large compared with the thickness of the unstirred layers. The concentrations are held constant, κ_s' represents the solution path conductance. After Weinstein *et al.* [6].

terms arising from leakage of co-ions across the homogenous membrane elements. Provided a high quality mosaic membrane is used this will generally be a good approximation for the treatment of brackish water.

As an example of this treatment some experimental results obtained by Weinstein et al. [9] for the reflection coefficient of a charge-mosaic are given in Figure 3, where a comparison is made with the theoretical predictions. As can be seen, negative values of the reflection coefficient are readily obtained, even though the experiments were performed with the electrochemically symmetrical salt KCl. This means that the non-additive term in the expression given in Table I far outweighs the sum of the two additive terms. This is true not only of the expression for the reflection coefficient but also of the expression for the solute permeability of the mosaic, although in general it is not true of the filtration coefficient (i.e., the non-additive term is usually quite small in this case). Since the ratio J_s/J_v is essentially the concentration of the salt solution emerging on the downstream side of a membrane under pressure, it can be seen immediately from Equation (9) that a negative value of the reflection coefficient is associated with an enrichment of the salt in the emerging solution.

The four regimes of operation of a charge-mosaic are shown in Figure 4. In this representation the solution conductivity κ_s' is assumed to be varied by altering the spacing between the ion-exchange elements. It can be seen that in the regime of co-ion leakage control the loop resistance has become so large that the distinctive circulating currents of a charge-mosaic are reduced to insignificance and the various

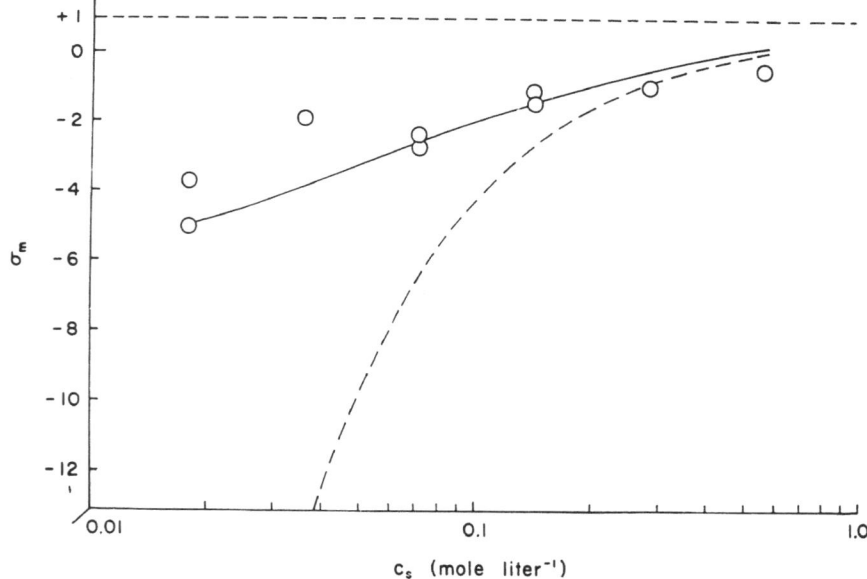

Fig. 4.   The reflection coefficient σ_m measured at a series of 2:1 concentration ratios. The solid curve was calculated by taking into account the solution conductances (see text). The dashed curve was calculated from the unmodified Kedem-Katchalsky treatment and corresponds to the membrane control limit. After Weinstein et al. [9].

charge elements function independently. If a mosaic is made progressively thinner it passes successively from membrane control into solution or polarization control. Any further decrease in thickness will have no additional effect on the circulating current and tend only to degrade the membrane's properties in other ways. Thus it is not advantageous to produce a mosaic membrane of active thickness in the micron range with charged regions in the millimeter range. The ratio between the conductivity of the bathing solutions and that of the charged elements determines how fine the mosaic pattern must be before a decrease in membrane thickness will influence circulating current. Generally speaking, if the ratio is of the order of unity then the ratio of membrane thickness to nearest neighbour distance should also be of the order of unity.

The ability of a charge-mosaic to distinguish between electrolytes and nonelectrolytes of low molecular weight suggests the following applications which have been

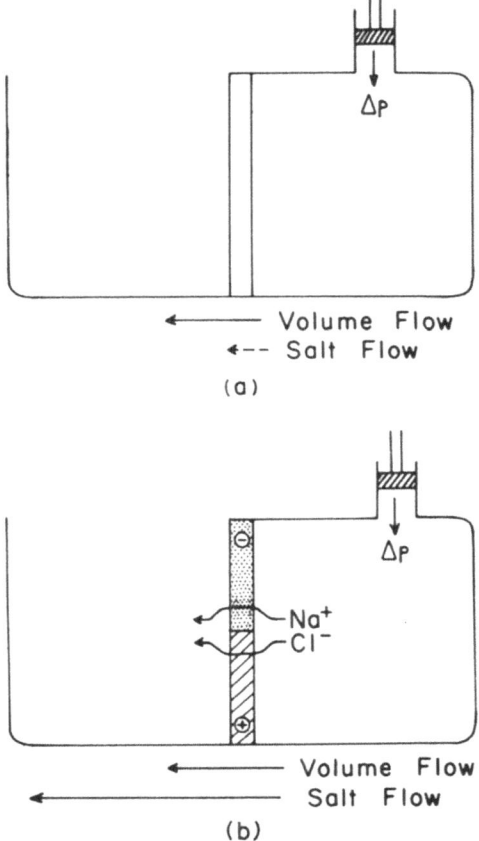

Fig. 5. Schematic comparison of (a) reverse osmosis, and (b) piezodialysis desalination. In actual practice, the membranes for reverse osmosis are usually cast as tubes or hollow fibers through or past which feed solution is forced under pressure. A similar form would probably prove most practical for piezo-dialysis as well. After Gardner et al. [13].

already demonstrated in the laboratory: (i) the desalting and concentration of solutions of small nonelectrolytes, i.e. "charge-mosaic dialysis" [10] and (ii) the determination of binding constants and activity coefficients of electrolytes in the presence of small nonelectrolytes by means of equilibrium dialysis [11]. It has also been suggested [6] that they could be used in the adjustment of ionic constituents in body fluids by dialysis, or by filtration as in a suggested modification of a portable artificial kidney proposed by Brown and Kramer [12]. The most interesting potential use, however, is in 'piezodialysis', which is the desalination technique whereby pressure is applied to one side of a charge-mosaic, the resulting streaming potentials tending to be short-circulated by the circulating current. Consequently, the overall reflection coefficient is negative and the solution emerging on the low pressure side can indeed be enriched in salt, as envisaged in the Introduction. A schematic comparison of two membrane desalination processes, piezodialysis and reverse osmosis, is shown in Figure 5. It should be noted that although the thermodynamic minimum energy requirement for the production of a given volume of potable water would in principle be the same for either process, frictional losses should be lower in piezodialysis since only a small volume of highly concentrated salt solution need be driven through the membrane. In a large-scale operation piezodialysis would probably be performed in much the same way as reverse osmosis is, using tubular membranes or hollow fibers. However, in piezodialysis the product water remains on the feed side.

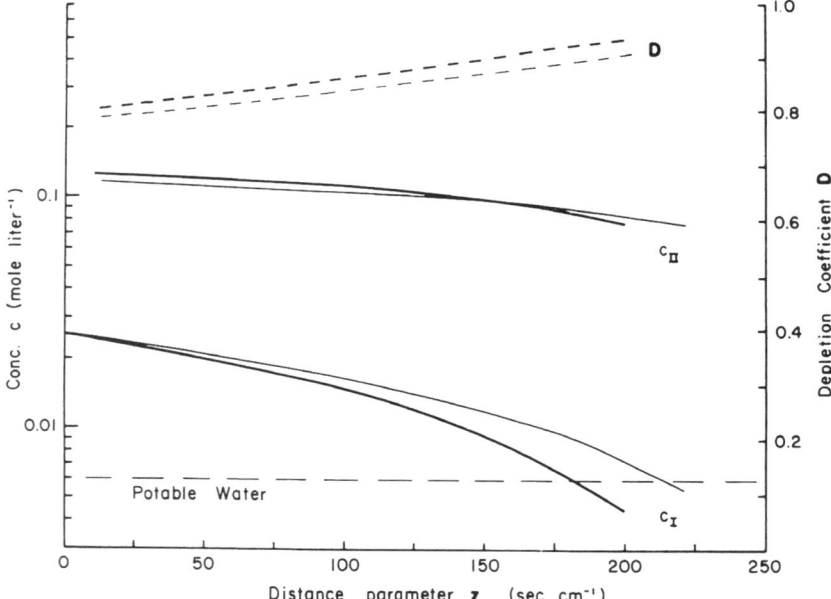

Fig. 6.   Course of a hypothetical piezodialysis process using the 'Zeo-karb mosaic'. The solid curves represent values of the concentrations c_I and c_{II} on the high and low pressure sides respectively. The dashed curves represent the depletion coefficient D. Heavy lines are used for results calculated neglecting co-ion leakage, while lighter lines include the effects of co-ion leakage and the concentration dependences of the coefficients. All are plotted as functions of the parameter z, which is proportional to the distance travelled down the tube. After Gardner *et al.* [13].

An analysis of a continuous-flow piezodialysis desalination system, based on the transport equations given above, has been given by Gardner *et al.* [13]. These workers treat a simple first approximation case of ideal radial mixing in a cylindrical tube, with no axial diffusion or axial mixing. The calculation is based on data obtained by Meares *et al.* [14–17] with the cation-exchanger Zeo-karb 315, and it assumes the existence of an anion-exchanger with identical properties except for the sign of the fixed charge. This 'Zeo-karb mosaic' was taken to have a thickness of 1 mm. The results of the calculation are shown in Figure 6. The distance parameter z in this figure is given by the relation

$$z = y/du_0 \tag{19}$$

where y is the distance measured down the tube in the direction of flow from its entrance, d is the internal diameter of the tube, and u_0 is the inlet velocity. The 'depletion coefficient' for piezodialysis D was defined as

$$D = 1 - c_1/c_{11} \tag{20}$$

It is related to the conventional rejection coefficient R by the expression

$$D = R/(R-1) \tag{21}$$

and consequently positive values of D correspond to negative rejection. This calculation showed that under an applied pressure difference of 100 atmospheres an 81% recovery may be expected of potable water of 350 ppm from a brackish water feed of 1500 ppm, the rate of production being 23 gpd/ft^2. A thinner mosaic of correspondingly reduced pattern size would give a proportionately higher production rate with no loss in recovery, provided that significant polarization could be prevented. This calculation sets upper limits on the performance of a given membrane by assuming that its properties, rather than the hydrodynamic conditions, are rate controlling.

3. Stacks

We turn now to what might be termed the 'dual' of the mosaic array. Dr. Avraham Naparstek, working at first with the late Prof. Aharon Katchalsky and subsequently in my laboratory, showed that series arrays, or multimembrane stacks, have rather unexpected symmetry properties with respect to the passage of electric current [18]. The approach followed is based once again on the Kedem and Katchalsky practical equations, but in this case on the 'first set'; in other words, it is assumed that volume flow rather than the difference between hydrostatic and osmotic pressure differences is an independent variable. In this treatment, volume flow is assumed to be zero or negligible.

A series array is considered of n rigidly-supported ion-exchange membranes of varying properties (and in any given sequence) separated by closed compartments containing a solution of 1:1 electrolyte, as shown in Figure 7. Electric current is passed through a pair of reversible electrodes situated in 'infinite' terminal reservoirs

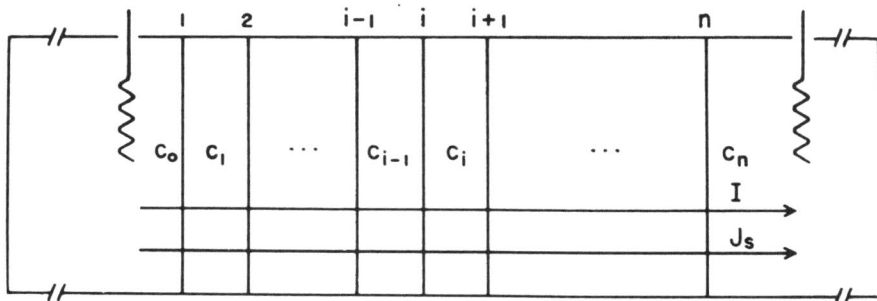

Fig. 7. Schematic diagram of an array of n ion-exchange membranes separated by closed compartments containing electrolyte solution of varying concentrations c_i. The directions of solute flow J_s and of electric current flow I are shown. After Naparstek *et al.* [18].

at the same pressure. Under stationary-state conditions the salt flux is the same in each of the membranes and we can write

$$J_s = \omega_i \Delta \pi_i + (\tau_i/F)I \tag{22}$$

where the subscript i represents the ith membrane and the transport number refers to the cation. In order to express the dependence of J_s on the total osmotic pressure difference, Equation (22) must be summed over all the membranes:

$$J_s = \left\{ \sum_{i=1}^{m} (1/\omega_i) \right\}^{-1} \left[\Delta \pi_{tot} + (I/F) \sum_{i=1}^{m} (\tau_i/\omega_i) \right] \tag{23}$$

where

$$\Delta \pi_{tot} = \sum_{i=1}^{m} \Delta \pi_i = 2RT(c_0 - c_n) \tag{24}$$

From these equations one obtains the following relation between $\Delta \pi_i$ and $\Delta \pi_{tot}$:

$$\Delta \pi_i = \left\{ \omega_i \sum_{i=1}^{m} (1/\omega_i) \right\}^{-1} \left[\Delta \pi_{tot} + (I/F) \sum_{i=1}^{m} (\tau_i/\omega_i) \right] - (I/F)(\tau_i/\omega_i) \tag{25}$$

This relation leads in turn to a condition for the identical current-dependence of two intermembrane concentrations c_k and $c_{k'}$. If k is greater than k',

$$c_k - c_k' = \frac{c_0 - c_n}{\sum\limits_{i=1}^{n} (1/\omega_i)} \sum_{j=k'+1}^{k} (1/\omega_j) - \frac{I}{2RTF \sum\limits_{i=1}^{n} (1/\omega_i)} \sum_{i=1}^{n} \sum_{j=k'+1}^{k} (\tau_i - \tau_j)/\omega_i \omega_j \tag{26}$$

From Equation (26) it follows that

$$c_k - c_{k'} = \frac{c_0 - c_n}{\sum\limits_{i=1}^{n} (1/\omega_i)} \sum_{j=k'+1}^{k} (1/\omega_j) = \text{constant (for any given } k) \tag{27}$$

if the following condition is obeyed:

$$\sum_{i=1}^{n} \frac{\tau_i}{\omega_i} \bigg/ \sum_{i=1}^{n} \frac{1}{\omega_i} = \sum_{i=k'+1}^{k} \frac{\tau_i}{\omega_i} \bigg/ \sum_{i=k'+1}^{k} \frac{1}{\omega_i} \tag{28}$$

If this condition holds for any given sequence of membranes in the 'inner' group $(k'+1$ to $k)$ and the 'outer' group $(1$ to k', $k+1$ to $n)$, it will hold for any other sequence of membranes within the two groups, including sequences in which the inner group is shifted in position relative to the outer group. When this condition is obeyed the inner group may be considered 'representative' of the whole system.

To demonstrate the significance of these results it is useful to examine a model system consisting of $n/2$ pairs of alternating anionic and cationic membranes as used in electrodialysis. The stack is closed except at the end compartments which are maintained at identical concentrations. Typical steady-state concentration profiles for such a system are shown in Figure 8. It can be seen that $n/2$ intermembrane compartments are either salinated or desalinated according to the direction of the current, while the remaining compartments are all at the same concentration level as the external reservoirs.

The *approach* to the stationary state has been computed for a four-membrane version of the model: the result of the computation is shown in Figure 9. Practical electrodialysis cells are designed to perform continuous desalination and hence the

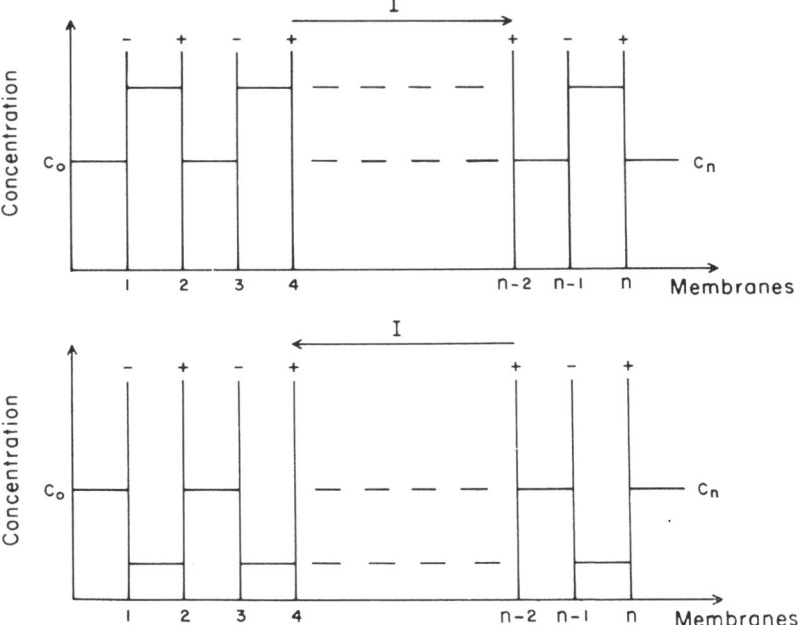

Fig. 8.   Schematic diagram of a model electrodialysis system. The sign of the fixed charge is indicated above each membrane. Steady-state concentration profiles are shown for both directions of electric current flow, assuming the compartments between the membranes to be closed. After Naparstek *et al.* [18].

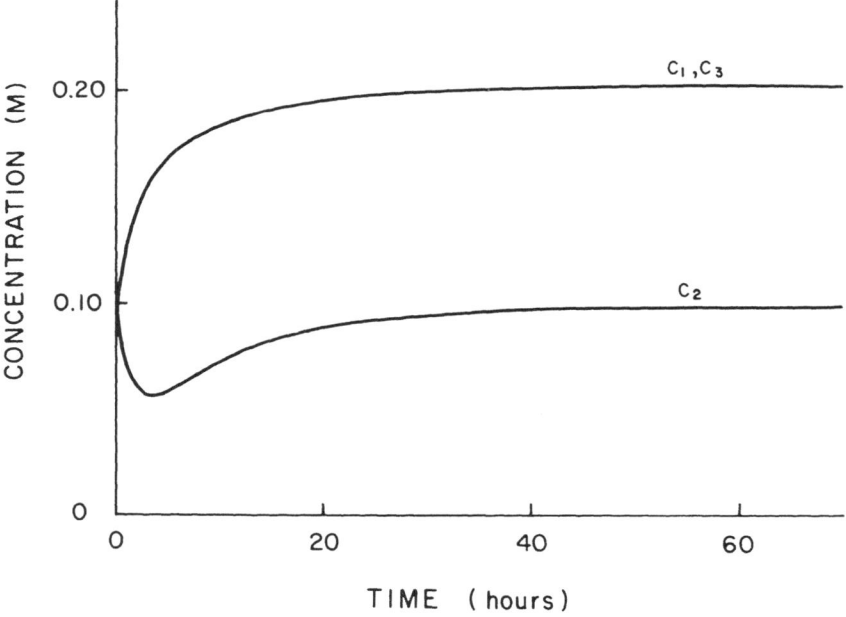

Fig. 9.   Simulation of the approach to stationarity of a model four-membrane electrodialysis system.
After Naparstek *et al.* [18].

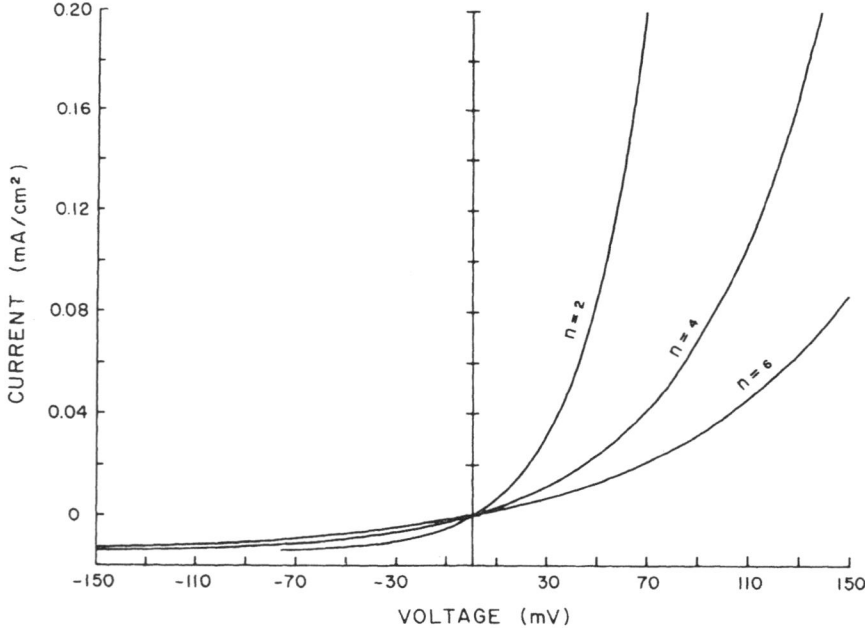

Fig. 10.   Simulated voltage-current curves for series arrays of bipolar membrane pairs held in pressure-contact with each other. The total number of membranes is indicated by *n*. After Naparstek
et al. [18].

intermembrane compartments are open rather than closed. Feed water flows past the membranes at a steady rate, usually in a parallel mode, product water and brine being obtained from alternate compartments. In such a system the transient concentration values shown in Figure 9 become stationary values, dependent on the rates of flow as well as the current. This suggests a method for optimizing flow rates at a given current or voltage in order to maximize desalination. For example, the flow rate corresponding to the minimum of the c_2 curve in Figure 9 maximizes desalination in that system – faster flows would cause the system to operate to the left of the minimum and slower flows to the right.

An important stationary-state property of the system is that of current rectification. This is most easily understood by referring to Figure 8: while the even-numbered intermembrane concentrations are independent of the current, the odd-numbered intermembrane concentrations are highly dependent on the direction of the current as well as its magnitude. The system is analogous to a chain of separate junction diodes connected in series. As a rectifier a multimembrane array of this kind appears to offer distinct advantages over a single bipolar pair. For example, a much larger voltage can be applied across the system without exceeding, within each compartment, the limiting voltage gradient in which water-splitting (i.e. the Bethe-Toropoff effect [19]) commences. Typical simulated characteristics are shown in Figure 10 for 1, 2, or 3 bipolar membrane pairs in series, held together in pressure-contact with each other. The results of an experimental study are shown in Figure 11.

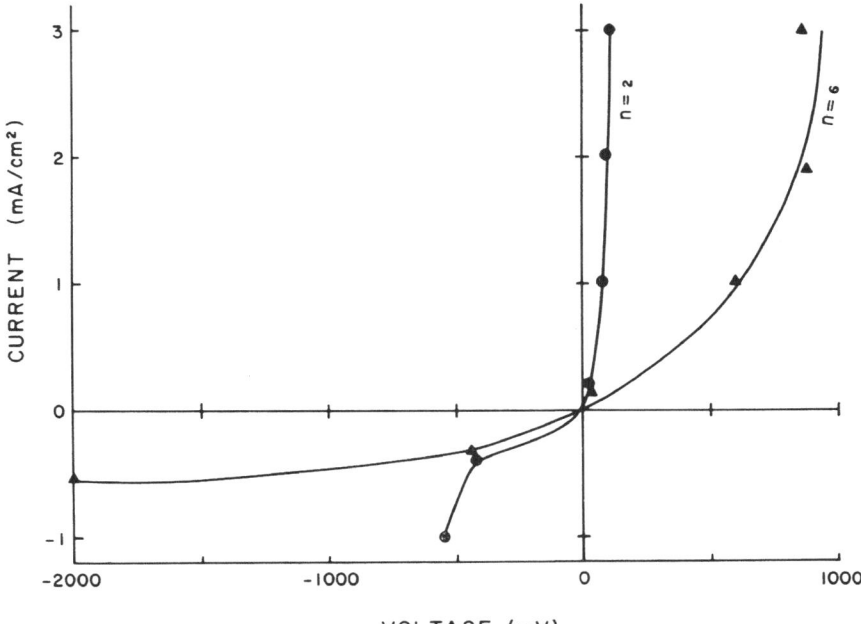

Fig. 11.   Experimental voltage-current curves for series arrays of bipolar membrane pairs held in pressure-contact with each other. The total number of membranes is indicated by n.
After Naparstek *et al.* [18].

4. Conclusions

The phenomena outlined above show that membrane arrays are capable of manifesting a rich variety of phenomena, not to be found in simple membranes. Cooperative effects between the elements lead to the emergence of totally new properties. If sufficient structure is introduced into the 'mixture' of elements, the whole becomes considerably more than the sum of its parts. This conclusion is, of course, well understood in the study of biological membranes, but it is still largely unexplored in engineering applications.

Acknowledgment

Supported by U.S.P.H.S. grant no. HL 14322 to the Harvard-M.I.T. Program in Health Sciences and Technology.

References

1. Sollner, K.: *Biochem. Z.* **244**, 370 (1932).
2. Neihof, R. and Sollner, K.: *J. Phys. and Colloid Chem.* **54**, 157 (1950).
3. Neihof, R. and Sollner, K.: *J. Gen. Physiol.* **38**, 613 (1955).
4. Grim, E. and Sollner, K.: *J. Gen. Physiol.* **40**, 887 (1957).
5. Weinstein, J. N., Bunow, B., Yanowitz, I., and Caplan, S. R.: *Polymer Preprints* **9**, 1564 (1968).
6. Weinstein, J. N., Bunow, B. J., and Caplan, S. R.: *Desalination* **11**, 341 (1972).
7. Kedem, O. and Katchalsky, A.: *Trans. Faraday Soc.* **59**, 1931 (1963).
8. Kedem, O. and Katchalsky, A.: *Trans. Faraday Soc.* **59**, 1918 (1963).
9. Weinstein, J. N., Misra, B. N., Kalif, D., and Caplan, S. R.: *Desalination* **12**, 1 (1973).
10. Weinstein, J. N. and Caplan, S. R.: *Science* **169**, 296 (1970).
11. Moore, M. R. and Caplan, S. R.: *Org. Coatings and Plastics Chem. Preprints* **30**, 226 (1970).
12. Brown, D. E. and Kramer, N. C.: *Trans. Amer. Soc. Artif. Int. Organs* **14**, 36 (1968).
13. Gardner, C. R., Weinstein, J. N., and Caplan, S. R.: *Desalination* **12**, 19 (1973).
14. Meares, P. and Ussing, H. H.: *Trans. Faraday Soc.* **55**, 142 (1959).
15. Meares, P. and Sutton, A. H.: *J. Colloid and Interface Sci.* **28**, 118 (1968).
16. McHardy, W. J., Meares, P., Sutton, A. H., and Thain, J. F.: *J. Colloid and Interface Sci.* **29**, 116 (1969).
17. Meares, P. and Foley, T.: *Office of Saline Water, Res. Develop. Progr. Rept. No. 584* (1969).
18. Naparstek, A., Caplan, S. R., and Katzir-Katchalsky, A.: *Israel J. Chem.* **11**, 255 (1973).
19. Bethe, A. and Toropoff, T.: *Z. Phys. Chem.* **88**, 686 (1914); **89**, 597 (1914).

IV

CARRIERS AND CHARGES

THE ION SELECTIVITY OF CARRIER MOLECULES,
MEMBRANES AND ENZYMES

SALLY KRASNE and GEORGE EISENMAN

Dept. of Physiology, UCLA School of Medicine, Los Angeles, Calif. 90024, U.S.A.

Abstract. The ion-selective conductances and permeabilities of bilayers in the presence of macrocyclic ion-carriers can in many instances be understood simply from a knowledge of the equilibrium selectivity of these carriers without having to consider their kinetic behavior. We have demonstrated elsewhere [1] how the selectivity of such carriers can be comprehended in terms of the equilibrium interactions between the ions and the individual ligand groups composing the ion-binding structure; and we review this work here, demonstrating that model solvents manifest selectivities comparable to those seen for these ion carriers and showing the striking similarities in selectivity among small cations seen between two carrier molecules of identical ligand type but very differing cavity sizes (the dodecadepsipeptide, valinomycin, and its hexadecadepsipeptide analogue). These results emphasize the importance of asymmetry (at the molecular level) in the interactions between ions and individual ligands vs. individual water molecules as a central factor in the origin of ionic selectivity. We also review how knowledge of the ionic selectivity patterns of molecules of known structure can be used to infer the types of ligands and their coordination for ion-specific systems which are not yet amenable to direct structural analysis such as the ion selective 'channels' in the cell membrane and the ion-activated 'sites' in enzymes.

1. Introduction

This chapter reviews the equilibrium basis for ionic selectivity, a subject which has been of considerable interest to biologists, biological chemists, and biophysicists and for which a considerable literature already exists. Since certain aspects of selectivity appear to have been explorered more thoroughly in the biological literature than in the chemical, we paraphrase here our recent review [1] of the present status of this problem.

One of the salient characteristics of all biological membranes and many enzymes [2] is their sensitivity to, and ability to distinguish among, such chemically similar cations as Li^+, Na^+, K^+, Rb^+, and Cs^+, as well as others such as NH_4^+ and Tl^+. Although this ion selectivity is dependent on ion size, even for the alkali metal cations it is not a simple monotonic function of the radius of the ions being selected, as has been discussed elsewhere [3–6]. For example, the permeability of the 'K$^+$ channel' of frog nerve [7] as well as the resting permeability of the squid axon [8], to the alkali metal cations is characterized by the selectivity sequence $(K > Rb > Cs > Na > Li)$ while the activation of crab nerve adenosine triphosphatase [9, 10] can occur either in the sequence $Na > Li > K > Cs > Rb$ when activated by magnesium alone or in the sequence $K > Rb > Cs > Na > Li$ when activated by sodium as well as magnesium [4].

Recently, a variety of cyclic molecules have been isolated from natural sources [11–14] as well as prepared synthetically [15] which exhibit selective ion complexation [16, 17] and act as ion carriers in artificial lipid bilayer membranes [18–24]. Molecular models of the naturally occurring carriers, nonactin, 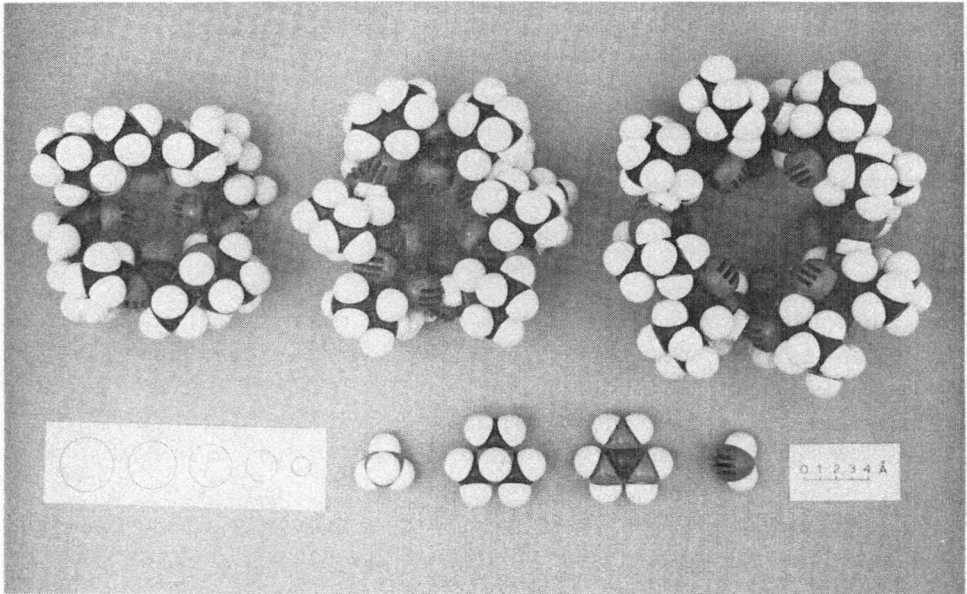 and valinomycin,

$$(D-Val-L-Lac-L-Val-D-HyIv)_3,$$

as well as the synthetic carrier hexadecavalinomycin,

$$(D-Val-L-Lac-L-Val-D-HyIv)_4,$$

synthesized by the Shemyakin Institute [25] are illustrated in Figure 1. These molecules share the common property of combining a lipophilic exterior with a polar interior consisting of oxygen ligands (e.g. ethers, ester carbonyls, amide carbonyls) energetically suitable for replacing all or part of the normal hydration shell of cations.

Fig. 1.   CPK models of three typical carriers and representative ions which they carry. From left to right above, the carriers are nonactin,   valinomycin, (D-Val-L-Lac-L-Val-D-HyIV)₃, and hexadecavalinomycin, (D-Val-L-Lac-L-Val-D-HyIV)₄. All of these are in the configurations inferred for their cation complexes [17, 23, 25–31]. From left to right below are presented the circumferences of Cs, Rb, K, Na, Li; CPK models of NH_4^+, $(CH_3)_3NH^+$, Hydrazinium$^+$; and a CPK model of the water molecule. Notice that in nonactin and valinomycin the cavity size is about appropriate to the middle-sized alkali cations K^+ or Rb^+ while it is considerably larger for hexadecavalinomycin. Also notice that nonactin is characterized by an approximately cubic coordination, with the four tetrahedrally oriented carbonyl ligands appearing more prominent and the four tetrahedrally oriented ether ligands less clearly visible in this view. The six octahedrally arrayed ester carbonyls of valinomycin are easily seen, as are the eight ester carbonyl ligands of hexadecavalinomycin which are arrayed at the corners of a square antiprism. Comparing the geometric array of the eight ligands in nonactin with that in hexadecavalinomycin, it can be seen that there is no tetrahedral array in the latter, in contrast to the situation in nonactin.

The molecular structures of the ion selective 'sites' of membranes and enzymes are of great interest but presently unknown; whereas for these cyclic molecules not only are the primary structures completely known but so are the conformations of their ion complexes in appropriate solvents [17, 23, 25, 26] or crystals [27–31]. These molecules contain the same ligands as those presumably present in enzymes or membranes and utilize natural molecular components (e.g. amino acids); they therefore provide unique model systems in which to study at the molecular level the elementary factors giving rise to ion selectivity.

As in the biological examples cited above, the alkali cation selectivity of these ion complexers (or carriers) is not a simple monotonic function of the sizes of the ions being selected [32–38]. For example, for nonactin the sequence of permeability ratios (K > Rb > Cs > Na > Li) has a maximum [24] at the intermediate sized K^+ as is illustrated in Figure 2A, where the logarithm of the experimentally observed permeability ratio has been plotted as a function of the reciprocal of the naked (Pauling [39]) cation radius. Other non-monotonic sequences are characteristic of different molecules; for example, a sequence (Rb > K > Cs > Na > Li) in which the maximum occurs at the Rb^+ ion is observed for valinomycin [18, 19] and hexadecavalinomycin [40], while for cyclic polyethers a variety of futher sequences are seen [41–44]. The detailed mechanisms by which these complexers act as carriers to enable ions to permeate bilayers has been discussed elsewhere, and it has been shown that these non-monotonic selectivities reflect well-defined chemical equilibria over a wide experimental range [24, 35, 45, 46]. We can therefore safely restrict present considerations to equilibrium interactions without becoming involved in the greater complications attendant when kinetic factors become important, as they certainly do in the 'kinetic domain' [35, 45] of experimental conditions even for the present carriers [46–51] or in the kinetic aspects of membrane permeation through channels.

The lack of a monotonic dependence on ion size for complexer selectivity might seem paradoxical since the binding energy between an ion and the ligands of the binding molecule is expected to increase monotonically with decreasing ion size (and therefore decreasing distance of closest approach).* However, the equilibrium selectivity of a complexer reflects not just the 'binding energy' of the complexer to each of the ions but represents the balance of the energy involved in the process of taking one ion out of water and putting it inside of the complexer versus that for the other ion (this is because the aqueous solution is generally chosen as the reference state in defining selectivity).** Thus, the selectivity reflects the difference between the 'binding energy' of the ions to the complexer and the 'binding energy' of the ions to water (i.e. the ion-hydration energy).

Although each of these binding energies is a monotonic function of ion size (see Figure 2B), it has been shown [1, 37, 38] that the *difference* between these energies can have a non-monotonic dependence on ion size if the binding energies of the ions

---

*This holds not only for ion: monopole interactions but also for more complex attractions (e.g. ion: multipole, induction forces, dispersion forces, etc.).

**In this paper we will use 'selectivity' in this narrow sense, that is, referred to water.

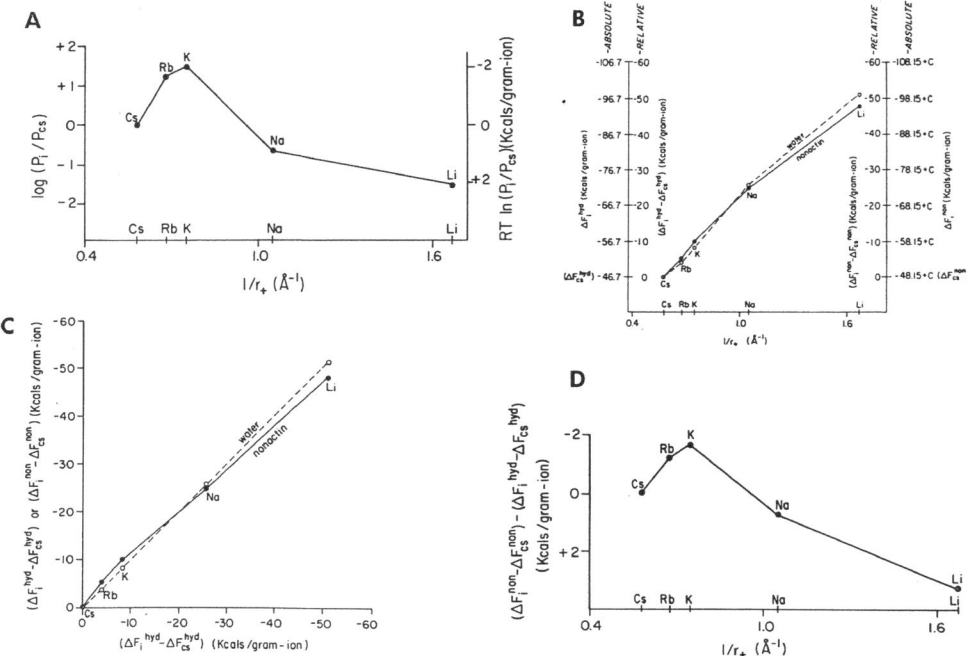

Fig. 2. (A) Non-monotonic dependence on ion size of the selectivity of nonactin. The logarithm of the permeability ratios relative to Cs+ determined from zero-current potential measurements of a bilayer membrane [24] in the presence of nonactin is plotted vs. the reciprocal of the cation radius [39]. The righthand ordinate gives the equivalent in energy units. (B) Monotonic dependence of the ion-binding energy to nonactin and of the ion-hydration energy on the reciprocal cation radius. The lefthand ordinate pertains to the ion-hydration energy (open circles, dashed line), and the right-hand ordinate pertains to the ion-binding energy of nonactin (closed circles, solid line). Each ordinate has two scales. The external scales, labelled 'absolute', are absolute values of free energy (relative to vacuum, see [52]) for which we have plotted Salomon's values [53] calculated from thermochemical data, and the internal scales, labelled 'relative', are values of free energy relative to those for Cs+. (C) The asymmetry between nonactin and water. The ordinate is the relative free energy of binding to nonactin (closed circles, solid lines) or the relative free energy of hydration (open circles, dashed lines) of the alkali cations, and the abscissa is the relative free energy of hydration for the alkali cations. (D) The selectivity of nonactin. The ordinate is the difference between the relative free energy of binding of the alkali cations to nonactin and the relative free energy of hydration of the alkali cations, and the abscissa is the reciprocal ion radius.

to the complexer are not a linear function of the hydration energies of the ions (thus producing a curved line when these binding energies are plotted against the hydration energies, as in Figure 2C). Such energy dependences which are not linear functions of each other have been termed 'asymmetrical' [1, 37, 38]. (By contrast, linearly related energy functions have been termed 'symmetrical').

The curvature in Figure 2C for nonactin vs. water reflects a difference in the relative ion-binding energies of nonactin and water which is maximal for K+, and of course, it is this energy difference which corresponds to the selectivity of nonactin. This energy difference is plotted in Figure 2D vs. the reciprocal cation radius and can be almost identically superimposed on the curve of the nonactin selectivity deduced from bilayer permeability ratios (cf. Figure 2A). The similarity between these two curves is no surprise since both the bilayer permeability ratios and the salt extraction ratios reflect the same underlying equilibrium selectivity of nonactin (as has been discussed elsewhere [24, 35, 45, 54]).

The pattern of selectivity of Figure 2D is one of the non-monotonic types of selectivity sequences (i.e. sequence IV of Eisenman [3, 55]) which result from asymmetrical interaction energies. It is worth noting that for carriers showing symmetrical interactions relative to water the selectivity will be a monotonic function of ion size (either yielding the sequences Li > Na > K > Rb > Cs, Cs > Rb > K > Na > Li, or the degenerate sequence Li = Na = K = Rb = Cs).

2. Theories on the Physical Origin of Asymmetries between Ion-Carriers and Water

There are two principal classes of theories as to the factors which produce the asymmetries observed for carriers or ion-binding 'sites', compared to water. One class, due to Eisenman [3, 55] and Ling [56], emphasizes the asymmetries expected in the interactions between cations and the individual ligand groups of the carrier or site (cf, 32, 35, 37). The other, 'steric', class of theories, propounded by Eigen and his colleagues [33, 57] as well as by Simon and Morf [34, 36], starts with the implicit assumption that the interactions of the individual ligand groups with ions are symmetrical to those of water molecules and then relegates any observed asymmetries of the carriers or sites to constraints on the packing of the ligands inside the carrier or site.

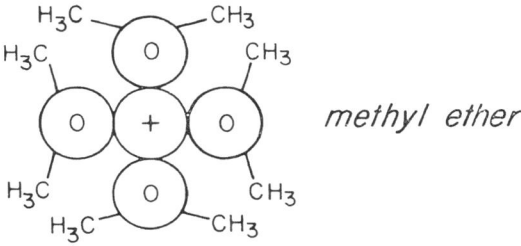

methyl ether

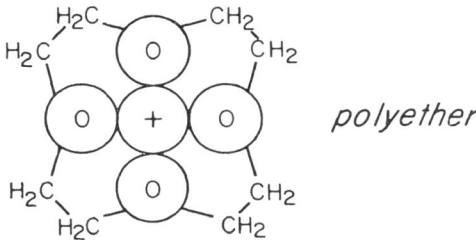

polyether

Fig. 3.   Ion solvation by methyl ether compared to 'ion solvation' by a polyether.

Although it seems likely that the observed selectivity among 'noble gas' cations of group Ia for a particular carrier molecule containing multidentate ligands should involve a contribution from both an asymmetry inherent in the individual ligands and an asymmetry arising from steric constraints in the packing of these ligands, we shall fine our considerations here to those properties of individual ligand groups which produce asymmetries. Indeed, we have shown [1, 37, 38] that solvation of group Ia cations by the ligands of appropriate solvents can mimic the selectivity seen in the 'solvation' of group Ia cations by various ion carriers, which implies that the special structural features of the macrocyclic carriers are not necessarily required for their selectivity among the alkali metal cations. Furthermore, the anomalous selectivities in carriers, membranes and enzymes for Tl^+ compared to the like-sized Rb^+ ion cannot be accounted for if one assumes that the ligands are themselves symmetrical to water and that selectivity is being determined only on a 'best fit' basis. Moreover, we will present evidence favoring the 'individual ligand' basis of selectivity over the steric 'fit' basis even for valinomycin from the similarity in the selectivity of valinomycin with that of hexadecavalinomycin which has a much less sterically restricted cavity.

3. Energies of Interactions with Model Solvents as Prototypes for Ion-Binding to Carriers

The solvation of ions by appropriate solvents can serve as a useful prototype for the 'solvation' of ions by macrocyclic carriers. In particular, solvents exist which seem likely to solvate ions through the same ligand groups as those in ion carriers (e.g. ester carbonyls, amide carbonyls, and ether oxygens); and the intrinsc selectivities of these ligand groups can serve as excellent prototypes for the selectivity of the same groups inside of a carrier molecule. Even the constraints upon the packing of solvent molecules might provide useful prototypes for the packing of such groups within certain carriers. The analogy we are suggesting is illustrated schematically in Figure 3 where an ion solvated by four oxygens of a cyclic polyther is compared with an ion solvated by four methyl ether molecules. In this representation the principle difference between the carriers and the solvent is the presence of backbone constraints on the carrier.

3.1. CATION SELECTIVITY OF AMIDE SOLVENTS

Somsen and Weeda's studies [58, 59] on ion solvation in formamide, *n*-methyl formamide, and dimethyl formamide provide the essential data for such a comparison. Figure 4A plots the enthalpies of ion solvation in amide solvents as a function of the reciprocal cation radius. For comparison, the hydration enthalpies are also presented. Although the enthalpy of solvation is comparable to the enthalpy of hydration and is also a monotonically increasing function of reciprocal cation radius for each of these solvents, there is an 'asymmetry' compared to water, as revealed by the curvatures in Figure 4B (cf. Figure 2C). The selectivity of these solvents (Figure 4C) can be compared with that for a typical carrier (Figure 2A). Notice: (1) the existence of a maximum in selectivity for K^+ for each of these amide solvents, (2) that the selectivity

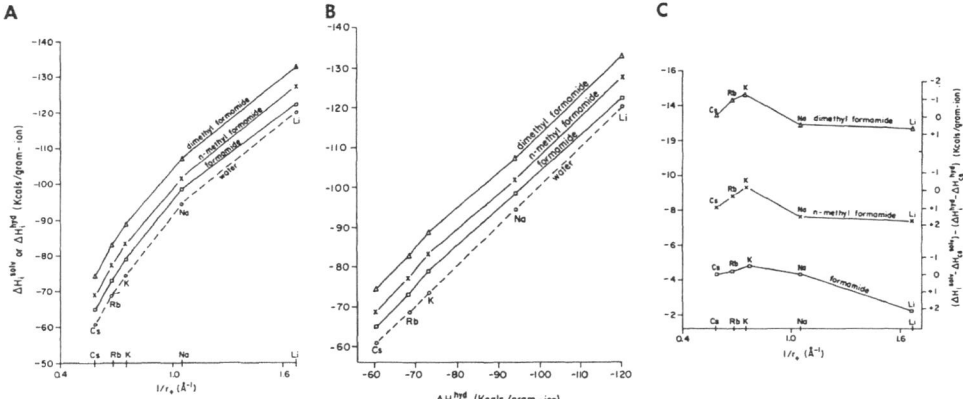

Fig. 4. (A) The monotonic dependence of solvation enthalpy on reciprocal ion radius. The ordinate is the solvation enthalpy in amide solvents and in water of the alkali cations as measured by Somsen and Weeda [58, 59], and the abscissa is the reciprocal ion radius. (B) The asymmetry between amide solvents and water. The ordinate is the absolute enthalpy of ion solvation in the amide solvents and in water and the abscissa is the absolute enthalpy of ion hydration. The curvatures of the lines for formamide; n-methyl formamide, and dimethyl formamide in this plot illustrate that the interactions of these solvents with the alkali cations are asymmetrical to the interactions of water with these ions. (C) The selectivity of amide solvents. The lefthand ordinate is the difference between the enthalpy of solvation in each amide solvent and in water of the alkali cations, and the righthand ordinate is this difference for each cation relative to that for Cs^+. The abscissa is the reciprocal ion radius. Note that the selectivity sequence for each of the amide solvents is the same as that observed for nonactin in Figures 2A and 2D (i.e. $K > Rb > Cs > Na > Li$).

sequence $(K > Rb > Cs > Na > Li)$ is the same as that for nonactin, and (3) the comparable magnitude of selectivity. It is apparent that the carbonyl ligands alone suffice to produce ion selectivity and that the special structural features of the macrocyclic carriers are not required for their selectivity.

Interestingly, the effect of methylation on the selectivity of the amide solvents (see [1, 37, 38]) is similar to that observed for the ion carriers nonactin, monactin, dinactin and trinactin (each of which has one more methyl group than the preceding one), and this effect of methylation on selectivity has been suggested to be due to an inductive [60] increase in the 'field strength' of the ligand oxygens [22].* Indeed, the 'pattern' of ionic selectivity derived for this series of structurally similar molecules [35] follows the expectations of the 'field strength' theory [1, 3, 32].

4. Calculated Selectivity for an Individual Carbonyl Ligand vs. an Individual Water Molecule

We have also calculated the selectivity expected for a crude model of a carbonyl ligand [1], using the discrete charge array representations of model water and carbonyl

---

*Note that formamide, n-methyl formamide and dimethyl formamide have dipole moments of 3.71, 3.84, and 3.86 debyes, respectively [59].

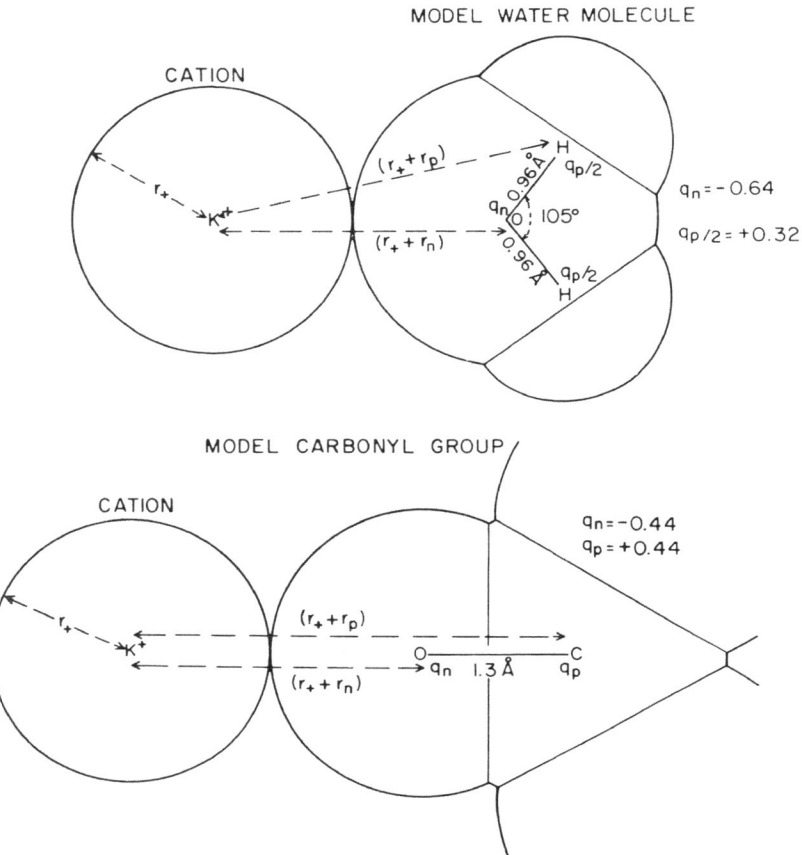

Fig. 5.   The charge distribution of a model water molecule and of a model carbonyl group. The top figure represents a cation (in this illustration, K^+) in contact with a single water molecule and the bottom figure represents a cation in contact with a single carbonyl group. The outlines of these ligands are drawn to scale using CPK models (the van der Waal's radii therefore being determined by these models). The van der Waal's radius of the oxygen is 1.4 Å, this being the value of r_n in this figure. The values of the partial charges, the bond lengths and the bond angle of the model water molecule are those proposed by Rowlinson [62], with the value of partial charge at the oxygen atom, q_n, being -0.64 electron units and that at each of the two hydrogen atoms, $q_p/2$, being $+0.32$ electron units, with the O−H distance being 0.96 Å and with the H−O−H angle being 105°. The C=O bond length of the model carbonyl group, 1.3 Å, is that proposed by Pauling [39]. The value of partial charge at the oxygen atom, q_n, is -0.44 electron units and the value of partial charge on the carbon atom, q_p, is $+0.44$ electron units. This value of partial charge was chosen so that the bond length times the partial charge would produce a dipole moment (2.75 D) consistent with those observed [63] for typical carbonyl containing molecules (e.g. acetaldehyde, 2.72 D, and acetone, 2.89 D). However, the precise value of partial charge was chosen to produce the particular selectivity sequence $K > Rb > Cs > Na > Li$ so that Figure 6 could be appropriately compared to the analogous figures in this section. (Note that values of partial charge which are larger in magnitude will produce 'higher field strength' selectivity sequences and values lower in magnitude will produce 'lower field strength' [3] selectivity sequences.) The radius of the cation is designated r_+. The distances from the center of charge of the cation to the centers of negative charge in the model water and carbonyl ligands are designated $(r_+ + r_n)$. The distances from the center of charge of the cation to the centers of positive charge in the model water and carbonyl ligands are designated $(r_+ + r_p)$.

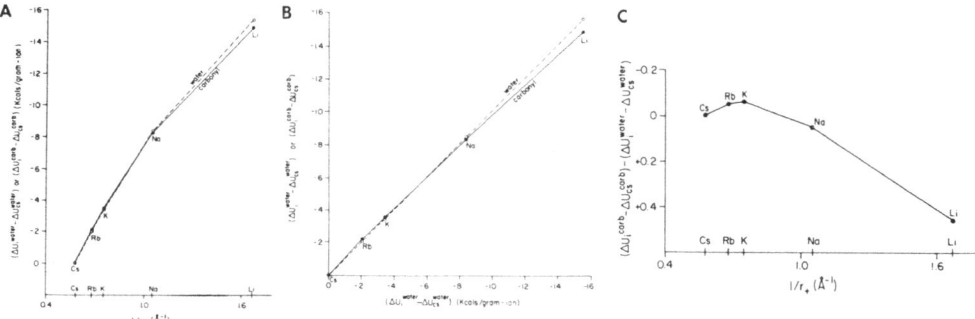

Fig. 6. (A) Monotonic dependence on the reciprocal ion radius of the calculated internal energy of ion binding to a model water molecule and to a model carbonyl group. The ordinate is the potential energy of interaction between the model carbonyl group of Figure 5 and the alkali cations (filled circles, solid line) and between the model water molecule of Figure 5 and the alkali cations (open circles, dashed line) both relative to that with Cs^+ and calculated using Equation (1). The abscissa is the reciprocal ion radius. (B). The asymmetry between the calculated internal energy of ion binding to a model carbonyl group and that to a model water molecule. The ordinate is the calculated internal energy of binding to the alkali cations (relative to that to Cs^+) of the model carbonyl group (filled circles, solid lines) and of the model water molecule (open circles, dashed lines), and the abscissa is the calculated internal energy of binging to the alkali cations (relative to that to Cs^+) of the model water molecule. The asymmetry between the model carbonyl group and the model water molecule is reflected by the curvature of the line for the model carbonyl group in this plot. (C) The calculated selectivity of a single model carbonyl group (referred to a single water molecule). The ordinate is the difference between the calculated internal energy of binding to the alkali cations (relative to that to Cs^+) of the model carbonyl group and of the model water molecule. The abscissa is the reciprocal ion radius.

groups pictured in Figure 5, in order to illustrate the selectivity implicit in replacing a single water molecule with a single ligand group having a different charge distribution than the water.* The asymmetry implied by the differences in charge arrays of the model ligand vs. the model water is illustrated in Figure 6 where Figure 6A (to be compared to Figures 2B and 4A)** is a plot of the internal energy of ion binding (relative to Cs^+) to the model water molecule and to the model carbonyl molecule as a function of the reciprocal ion radius, calculated using Coulomb's law for the internal energy, U (in Kcal/mole), between an ion of charge q_+ and radius r_+ and the molecules illustrated in Figure 5,

$$U = 332q_+ \left(\frac{q_n}{r_+ + r_n} + \frac{q_p}{r_+ + r_p} \right) \tag{1}$$

the parameters being defined in the figure legend.

*Talekar [64] has based a calculation of the K^+-binding energy to valinomycin on a discrete charge array representation by assaying the partial charges on the backbone atoms of valinomycin. This 'discrete charge array' representation, is not to be confused with 'point multipolar' representations [61] in which the charge distribution on a molecular assemblage is represented as an sum of point multipoles. Certain significant differences between these two types of representations (and their misuse (see for example [65]) are discussed elsewhere [37].
**The differences in the magnitudes of the energies in Figure 6, as compared to Figures 2 and 4, arise to a large degree from the fact that we are considering interactions with single ligands in Figure 6 whereas the empirical values in Figures 2 and 4 reflect multiple coordination.

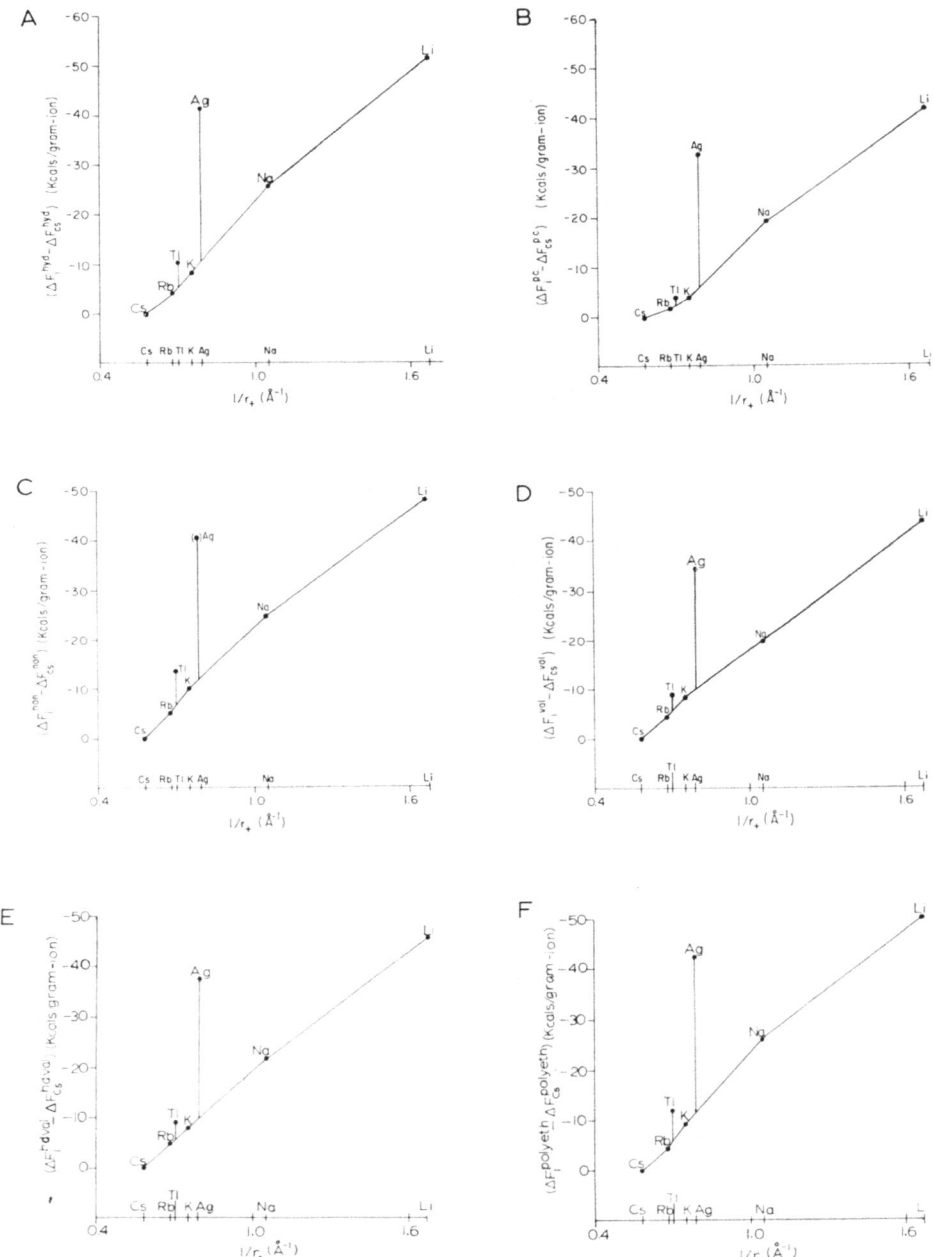

Fig. 7. Extra stabilization energies for the interactions of Tl$^+$ and Ag$^+$ with various ligands. (A) Excess hydration energy of Tl$^+$ and Ag$^+$. The ordinate is free energy of ion hydration from Salomon [53]. In all subfigures, the abscissa is the reciprocal of the crystallographic radii of the ions [39]. (B) Excess solvation energy of Tl$^+$ and Ag$^+$ in propylene carbonate. The ordinate is the relative free energy of ion solvation in propylene carbonate [53]. (C) Excess free energy of binding of Tl$^+$ and Ag$^+$ to nonactin. The ordinate is the relative free energy of ion binding to nonactin calculated from

Since the particular functional dependence of U on cation size depends upon the relationship between r_+, r_n and r_p, a ligand with different values of r_n and r_p than water (i.e. different 'discrete charge arrays') will have a different functional dependence of U on r_+ than is characteristic of the water molecule. The interaction energies of ions with the ligand and with the water molecule will therefore be asymmetrical. Comparing the distances $(r_+ + r_p)$ and $(r_+ + r_n)$ for the model water and model carbonyl in Figure 5, we see that the difference between $((r_+ + r_p)$ and $r_+ + r_n)$ is much greater for the carbonyl than for water (indeed, water is a ligand with an especially small distance between its positive and negative centers of charge compared to the other ligands (e.g. carbonyls) with which we are concerned). We suppose that this geometrical difference in charge distribution is the basis for the asymmetry between these ligands and water.

5. Non-'Noble Gas' Monoatomic and Polyatomic Cations

We have also examined [1, 37] non-'noble gas' monoatomic ions such as Ag^+ and Cu^+, whose outer orbitals contain ten d electrons, and Tl^+ and In^+, whose outermost orbitals contain two s electrons as well [66], and have also considered polyatomic ions (e.g. NH_4^+) which have an intrinsic, non-spherical 'shape' (as of course do those monoatomic ions with outer d^{10} and $d^{10}s^2$ orbitals). Such species have interactions with ligands which are above and beyond those of the group Ia cations, and we demonstrate below how these excess interactions can vary for different ions according to the particular ligand and thus might be 'diagnostic' of the ligand type.*

The excess ion-ligand interaction energies, as compared to the 'noble gas' cations of group Ia, are illustrated in Figure 7 for Tl^+ and Ag^+. Figure 7A demonstrates that the energies of hydration for Tl^+ and Ag^+ ions are much higher than those of alkali cations of comparable crystallographic radii. Figure 7B illustrates that this is also the case with the energy of the solvation of these ions by the carbonyl ligands of propylene carbonate. Indeed, excess energies for Tl^+ and Ag^+ are the general rule, being apparent also in their interactions with carriers having variously oriented carbonyl and ether ligands (tetrahedral carbonyl plus tetrahedral ether is typified by nonactin in Figure 7C, octahedral carbonyls by valinomycin in Figure 7D, carbonyls in a square antiprism [25] by hexadecavalinomycin in Figure 7E, and planar six-fold ethers by the cyclic polyether dicyclohexyl-18-crown-6 [42, 68] in Figure 7F).

---

bilayer permeability ratios [24]. The parenthesized point for Ag^+ has been extrapolated for nonactin from the value measured by Laprade [67] for the trinactin-Ag^+ complex. (D) Excess free energy of binding of Tl^+ and Ag^+ to valinomycin. The ordinate is the relative free energy of ion binding to valinomycin calculated from bilayer permeability ratios [35]. (E) Excess free energy of binding of Tl^+ and Ag^+ to hexadecavalinomycin. The ordinate is the relative free energy of ion binding to hexadeca-valinomycin calculated from bilayer permeability ratios [40]. (F) Excess free energy of binding of Tl^+ and Ag^+ to the polyether, dicyclohexyl-18-crown-6. The ordinate is the relative free energy of ion binding to dicyclohexyl-18-crown-6 calculated from aqueous binding constant measurements [42, 43].

*The physical basis for such excess interactions have been discussed elsewhere [1, 37].

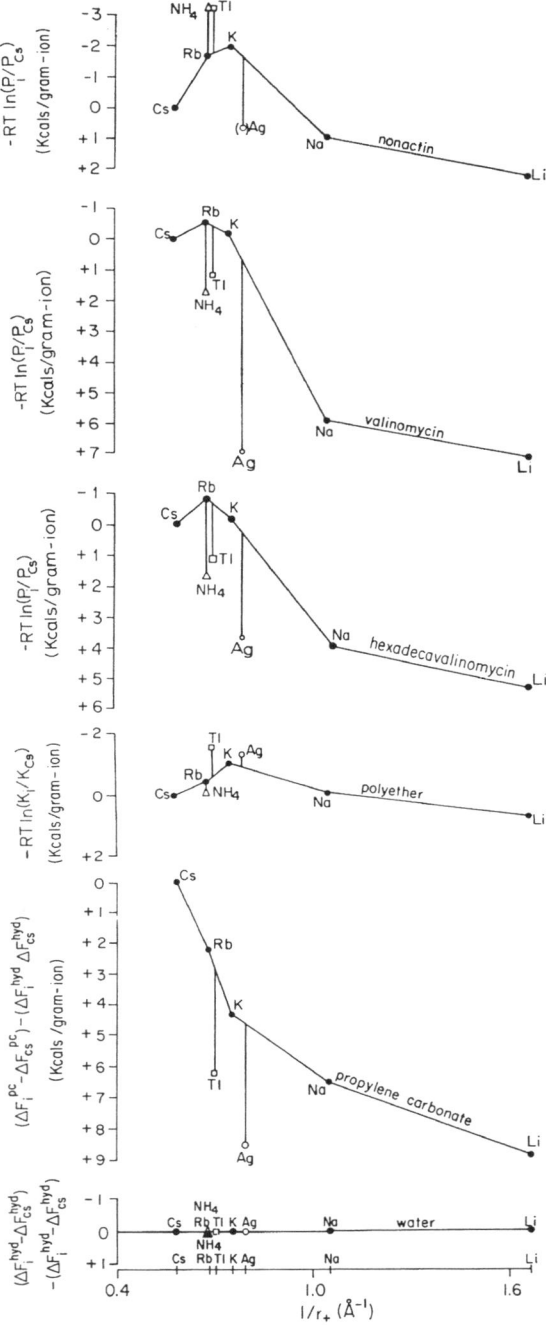

Fig. 8. Selectivity fingerprints of model ion-binding systems of known chemical composition. For each sub-figure the selectivities are plotted in the same manner as in Figures 2D, 4C and 6C but also including data for Tl+, Ag+, and NH4+. The ordinate is the energy equivalent of the selectivity for each ion compared to Cs+. The abscissa is the reciprocal of the crystallographic ion radius [39]. The nonactin [24], valinomycin [35], and hexadecavalinomycin [40] data are calculated from bilayer permeability ratios whereas the polyether data is calculated from aqueous binding constant measurements [42, 43] and the propylene carbonate data is taken directly from Salomon [53].

The excess energies for the interactions of Tl^+ and Ag^+ with these ligands when compared with those for their interactions with water manifest themselves as a selectivity for these ions which, depending upon the ligand (e.g. ether vs. ester), is either greater than or less than that for group Ia cations of comparable size. We have called selectivities less than or greater than those of comparably sized group Ia cations, 'sub-Ia' and 'supra-Ia', respectively [13, 37].

In Figure 8 are plotted 'selectivity fingerprints' [1, 37] for the group Ia cations as well as Tl^+ and Ag^+ (and NH_4^+, see below) of one solvent and of a variety of carriers of known ligand types. In this figure it is seen that Tl^+ is 'supra-Ia' in nonactin and in the polyether but is 'sub-Ia' in valinomycin, in hexadecavalinomycin, and in propylene carbonate; whereas Ag^+ is 'supra-Ia' only in the polyether, being 'sub-Ia' in nonactin, valinomycin, and propylene carbonate.* Knowing that the polyether has ether ligands, that valinomycin and hexadecavalinomycin have ester carbonyl ligands, that nonactin has both ether ligands and ester carbonyl ligands, and that propylene carbonate has carbonate carbonyl ligands allows us to infer the relationship between ligand types and the 'sub-Ia' or 'supra-Ia' positions of Tl^+ and Ag^+. In particular, all of the selectivity fingerprints in Figure 8 are consistent with the postulates that ether ligands produce 'supra-Ia' Tl^+ and Ag^+ selectivities whereas ester and carbonate carbonyl ligands produce 'sub-Ia' Tl^+ and Ag^+ selectivities. The only molecular fingerprint for which the consistency of this postulate is not immediately obvious is that of nonactin. In this molecule Tl^+ is 'supra-Ia' whereas Ag^+ is 'sub-Ia' .Since nonactin has four tetrahedrally arrayed ester carbonyls and four tetrahedrally arrayed ether groups we can see that consistency with the postulate demands that Ag^+ can contact only the carbonyl ligands wheareas Tl^+ is able to inertact with the ether groups as well. In fact this is precisely the ion-ligand coordination one would expect both from inspection of CPK models (cf. Figure 1) and by analogy to the X-ray crystal data on the Na^+-nonactin [29] and K^+-nonactin [27, 28] complexes.**

5.1. Polyatomic Ion Shape and the Spatial Array of Ligands

Polyatomic ions with particular 'shapes'. such as NH_4^+, may be valuable in assessing the coordination geometry of the ligands in sites of unknown structure. Thus, the tetrahedral NH_4^+ ion might be expected to have a more favorable interaction energy

---

*This means that, compared to the interaction energies with water, Tl^+ and Ag^+ are gaining relatively more (supra-Ia) or less (sub-Ia) energy with these ligand groups than would comparable sized 'noble gas' ions. To make this completely explicit, the curve for the selectivity of a water-like ligand (for which water is of course the appropriate model) is plotted at the bottom of Figure 8, where all of the points for these ions are shown to lie exactly on the horizontal line labelled 'water', corresponding to zero selectivity.

**The crystallographic radius of Ag^+ lies between that of Na^+ and that of K^+. In the K^+-nonactin complex [27, 28] the carbonyl oxygens and the ether oxygens are almost equidistant from the K^+ ion whereas in the smaller cavity of the Na^+-nonactin complex [29], the carbonyl oxygens are significantly closer to the Na^+ than are the ether oxygens (cf. Figure 1); so contact between Ag^+ and the carbonyl oxygens but not the ether oxygens is to be expected. In contrast to this situation, Tl^+, having a larger crystallographic radius than K^+ could easily contact all eight oxygens (as is approximately the case for K^+), and thus the expected 'supra-Ia' selectivity of ether groups for Tl^+ could manifest itself.

with tetrahedrally arrayed oxygens than with octahedrally arrayed oxygens even if the electric potential produced by the ligands at the center of the NH_4^+ ion is the same for both distributions of ligands [69]. Although we shall only cite NH_4^+ here to demonstrate how the shape of an ion can be used to infer ligand coordination, other monoatomic ions and molecular ions may be just as useful. One example is Hille's [7, 70] ingenious use of NH_4^+ derivatives with different sizes, shapes, and charge distributions (e.g. $CH_3NH_3^+$, $HONH_3^+$, $H_2NNH_3^+$, $(NH_2)_2\ C{=}NH_3^+$) to infer the spatial array and type of ligands in the Na^+ channel of frog nerve.

Included in the selectivity fingerprints of Figure 8 are the selectivities of the ion carriers for NH_4^+. Whereas nonactin has a 'supra-Ia' selectivity for NH_4^+, valinomycin, hexadecavalinomycin, and the polyether all have 'sub-Ia' selectivities for NH_4^+. Nonactin has four tetrahedrally directed carbonyl ligands and four tetrahedrally directed ether ligands, whereas neither valinomycin, hexadecavalinomycin, nor the polyether has tetrahedrally directed ligands, valinomycin having six octahedrally directed carbonyls, hexadecavalinomycin having eight carbonyls oriented in a square antiprism [25]. (see the molecular models of Figure 1), and the polyether having six planar ether ligands [42, 68, 72]. Since nonactin is the only carrier in Figure 8 with a tetrahedral array of ligands and also is the only carrier with a 'supra-Ia' selectivity for NH_4^+, it seems that a 'supra-Ia' selectivity for NH_4^+ can be taken to imply a tetrahedral orientation of the ligands, at least in these molecules. We have used this postulate elsewhere ([11], cf. Sections 6.1 and 6.2) to infer the array of ligands in ion-selective biological systems of unknown geometry.

6. Ion Selectivity in Biological Phenomena

Certain of the ion selectivities encountered in biological systems are also likely to reflect the equilibrium selectivities of the ion-binding sites, for example enzyme activation and ion binding to biological ion exchange sites. Certain other phenomena (e.g. ion permeation of cell membranes) presumably reflect non-equilibrium factors as well. We have considered examples of these phenomena elsewhere (cf. [1]), using the above-described selectivity considerations to infer the kinds of ligands likely to be present in the biological systems as well as their spatial orientation, but this subject goes beyond the scope of the present chapter.

7. A Comparison of Valinomycin and Hexadecavalinomycin

Present theories for the origin of ion selectivity fall into two classes, one attributing selectivity to steric factors (e.g. cavity size), the other attributing it to properties of the individual ligands (e.g. field strength). One approach to evaluating the relative roles of each of these factors in ion selectivity has been made possible through the pioneering syntheses and characterization by Ovchinnikov, Ivanov, and their colleagues of a number of important analogues of the dodecadepsipeptide, valinomycin [17, 23, 25].

TABLE I

A comparison of the permeability ratios for valinomycin and hexadecavalinomycin (glyceryl oleate-decane membranes) [40]

	Valinomycin P_i/P_K	Hexadecavalinomycin P_i/P_K
Li$^+$	0.0000042	0.00011
Na$^+$	0.000036	0.001
K$^+$	1.0	1.0
Rb$^+$	1.8	3.0
Cs$^+$	0.76	0.74
Ag$^+$	0.0000056	0.0016
Tl$^+$	0.105	0.11
NH$_4$$^+$	0.039	0.045
CH$_3$NH$_3^+$	0.00039	0.21
(CH$_2$)$_3$NH$^+$	≤ 0.0000046	0.32
Choline$^+$	≤ 0.000031	1.3
Acetylcholine$^+$	0.000016	4.5
Formamidinium$^+$	≤ 0.0059	0.052
Guanidinium$^+$	≤ 0.00072	0.09

Table I compares the permeability ratios for valinomycin and hexadecavalinomycin for various monoatomic, as well as larger polyatomic, cations [40] (these two molecules differing only in the number of repeating sub-units, and thus in their cavity sizes). For the smaller cations these permeability ratios have been plotted as selectivity 'fingerprints' in Figure 8. The most noteworthy feature of these fingerprints is their striking similarity for the small cations, in contrast to the large differences seen with the larger cations. Indeed, the group Ia selectivity sequence (Rb > K > Cs > Na > Li) is virtually identical for valinomycin and for hexadecavalinomycin, the only significant difference in the selectivities of these two molecules being a 2 kcal decrease in the spread of the selectivity for hexadecavalinomycin as compared with valinomycin; in addition, the 'sub-Ia' behavior for Tl$^+$ and NH$_4^+$ are the same for these two molecules.

If the cavity size were the determining factor in the selectivity of valinomycin among smaller monovalent cations [33, 36, 71], it is difficult to see why hexadecavalinomycin should show an identical selectivity sequence among group Ia cations and such a similar quantitative 'fingerprint'. Thus, this result is consistent with the view that the 'field strength' [3, 32, 55] of the individual ligand determines the selectivity for smaller ions since the ligands in these two molecules are chemically identical.

In contrast, the favoring of the larger polyatomic cations by hexadecavalinomycin relative to valinomycin seen in Table I appears to be a particularly clear example of how steric factors can act to exclude such larger species from a molecule once the ionic species is indeed large compared to the energetically feasible cavity size of the molecule.

Acknowledgement

This work has been made possible by the support of the National Science Foundation (Grant GB 30835) and the National Institutes of Health (Grant NS 09931-04).

References

1. Eisenman, G. and Krasne, S.: *MTP International Review of Science, Biochemistry Series*, Vol. 2, (ed. by C. F. Fox), Butterworths, London, 27–59 (1975).
2. Suelter, C. H.: *Science*, **168**, 1789 (1970).
3. Eisenman, G.: in A. Kleinzeller and A. Kotyk (ed.), *Symposium on Membrane Transport and Metabolism*, Academic Press, New York, 1961, p. 163.
4. Eisenman, G.: *Proc. 23rd Int. Congr. of Physiol. Sci.: Exerpta Medica* **87**, 489 (1965).
5. Diamond, J. M. and Wright, E.: *Ann. Rev. Physiol.* **31**, 581 (1969).
6. Suelter, C. H.: *Metal Ions in Biology*, Vol. 3 (ed. by H. Segel), Marcel Dekker, New York, 1972.
7. Hille, B.: *J. Gen. Physiol.* **61**, 669 (1973).
8. Hagiwara, S., Eaton, D. C., Stuart, A. E., and Rosenthal, N. P.: *J. Memb. Biol.* **9**, 373 (1972).
9. Skou, J. C.: *Biochim. Biophys. Acta* **23**, 394 (1957).
10. Skou, J. C.: *Biochim. Biophys. Acta* **42**, 6 (1960).
11. Kinsky, S. C.: *J. Bacteriol.* **82**, 889 (1961).
12. Lampen, J. O., Arnow, P. M., and Safferman, R. S.: *J. Bacteriol.* **80**, 200 (1961).
13. Moore, C. and Pressman, B. C.: *Biochem. Biophys. Res. Commun.* **15**, 562 (1964).
14. Lardy, H. A., Graven, S. N. and Estrada-0., S.: *Fed. Proc.* **26**, 1355 (1967).
15. Pedersen, C. J.: *J. Amer. Chem. Soc.* **89**, 7017 (1967).
16. Stefanac, Z. and Simon, W.: *Microchem. J.* **12**, 125 (1967).
17. Shemyakin, M. M., Ovchinnikov, Y. A., Ivanov, V. I., Antonov, V. K., Vinogradova, E. I., Shkrob, A. M., Malenkov, G. G., Evstratov, A. V., Laine, I. A., Melnik, E. I., and Ryabova, 1. D.: *J. Memb. Biol.* **1**, 402 (1969).
18. Mueller, P. and Rudin, D. O.: *Biochem. Biophys. Res. Commun.* **26**, 398 (1967).
19. Lev, A. A. and Buzhinsky, E. P.: *Tsitologiya* **9**, 102 (1967).
20. Pressman, B. C., Harris, E. J., Jagger, W. S., and Johnson, J. H.: *Proc. Nat. Acad. Sci.* **58**, 1949 (1967).
21. Andreoli, T. E., Tieffenberg, M., and Tosteson, D. C.: *J. Gen. Physiol.* **50**, 2527 (1967).
22. Eisenman, G., Ciani, S. M., and Szabo, G.: *Fed. Proc.* **27**, 6 (1968).
23. Ovchinnikov, Y. A., Ivanov, V. T., Evstratov, A. V., Bystrov, V. F., Abdullaev, N. D., Popov, E. M., Lipkind, G. M., Arkhipora, S. F., Efremov, E. S., and Shemyakin, M. M.: *Biochem. Biophys. Res. Commun.* **37**, 668 (1969).
24. Szabo, G., Eisenman, G., and Ciani, S.: *J. Memb. Biol.* **3**, 346 (1969).
25. Ovchinnikov, Yu. A.: *23rd Int. Congr. of Pure and Appl. Chem.* **2**, 211 (1971).
26. Prestegard, J. H. and Chan, S. I.: *J. Amer. Chem. Soc.* **92**, 4440 (1970).
27. Kilbourn, B. T., Dunitz, J. D., Pioda, L. A. R., and Simon, W.: *J. Mol. Biol.* **30**, 559 (1967).
28. Dobler, M., Dunitz, J. D., and Kilbourn, B. T.: *Helv. Chim. Acta* **52**, 2573 (1969).
29. Dunitz, J. D.: personal communication.
30. Pinkerton, M., Steinrauf, L. K., and Dawkins, P.: *Biochem. Biophys. Res. Commun.* **35**, 512 (1969).
31. Dobler, M., Dunitz, J. D., Krajewski, J.: *J. Mol. Biol.* **42**, 603 (1969).
32. Eisenman, G.: in R. A. Durst (ed.), *Ion-Selective Electrodes*, National Bureau of Standards Special Publication **314**, 1969.
33. Diebler, H., Eigen, M., Ilgenfritz, G., Maas, G., and Winkler, R.: *Pure Appl. Chem.* **20**, 93 (1969).
34. Simon, W. and Morf, W. E.: in E. Munoz, F. Garcia-Ferrardiz and D. Vasquez (eds.), *Symposium on Molecular Mechanisms of Antibiotic Action on Protein Biosynthesis and Membranes*, Elsevier Publishing Co., Amsterdam, 1971.
35. Eisenman, G., Szabo, G., Ciani, S., McLaughlin, S. G. A., and Krasne, S.: in *Progress in Surface and Membrane Science*, Vol. 6, (ed. by J. F. Danielli, M. Rosenberg, and D. Cadenhead), Academic Press, New York, 1973, p. 139.

36. Simon, W. and Morf, W. E.: in *Membranes – A Series of Advances*, Vol. 2 (ed. by G. Eisenman), Marcel Dekker, New York 1973, Chapter 4.
37. Krasne, S. and Eisenman, G.: in *Membranes – A Series of Advances*, Vol. 2 (ed. by G. Eisenman), Marcel Dekker, New York, 1973, Section V of Chapter 3.
38. Eisenman, G. and Krasne, S.: Lecture to Symposium on Membrane Structure and Function. *IVth Int. Biophys. Cong., Moscow, U.S.S.R.,* August, 1972.
39. Pauling, L.: *The Nature of the Chemical Bond*, Cornel Univ. Press, Ithaca, New York, 1960.
40. Krasne, S. and Eisenman, G.: unpublished data, 1973.
41. Chock, P. B.: *Proc. Nat. Acad. Sci., U.S.A.,* **69**, 1939 (1972).
42. Frensdorff, H. K.: *J. Am. Chem. Soc.* **93**, 600 (1971).
43. Izatt, R. M., Nelson, D. P., Rytting, J. H., Haymore, B. L., and Christensen, J. J.: *J. Am. Chem. Soc.* **93**, 1619 (1971).
44. McLaughlin, S. G. A., Szabo, G., Ciani, S., and Eisenman, G.: *J. Memb. Biol.* **9**, 3 (1972).
45. Eisenman, G., Ciani, S. M., and Szabo, G.: *J. Memb. Biol.* **1**, 249 (1969).
46. Szabo, G., Eisenman, G., Laprade, R., Ciani, S., and Krasne, S.: *Membranes – A Series of Advances*, Vol. 2 (ed. by G. Eisenman), Marcel Dekker, New York, 1973, Chapter 3, p. 179.
47. Ciani, S. M., Eisenman, G., Laprade, R., and Szabo, G.: *Membranes – A Series of Advances*, Vol. 2. (ed. by G. Eisenman, Marcel Dekker, New York, 1973, Chapter 2, p. 61.
48. Markin, V. S., Kristalik, L. I., Liberman, E. A., and Topaly, V. P.: *Biofizika* **14**, 256 (1969).
49. Läuger, P. and Stark, G.: *Biochim. Biophys. Acta* **211**, 458 (1970).
50. Stark, G. and Benz, R.: *J. Memb. Biol.* **5**, 133 (1971).
51. Grell, E., Funck, Th., and Eggers, F.: in E. Munoz, F. Garcia-Ferrandiz, and D. Vasquez (eds.), *Symposium on Molecular Mechanisms of Antibiotic Action on Protein Biosynthesis and Membranes*, Elsevier Publ. Co., Amsterdam, 1972, p. 646.
52. Bockris, J. and Reddy, A.: *Modern Electrochemisty*, Vol. 1, 1970, Chapter 2.
53. Salomon, M.: *J. Phys. Chem.* **74**, 2519 (1970).
54. Eisenman, G., Szabo, G., MacLughlin, S. G. A., and Ciani, S. M.: *Bioenergetics* **4**, 93 (1973).
55. Eisenman, G.: *Biophys. J.* **2**, Part 2, 259 (1962).
56. Ling, G. N.: *A Physical Theory of the Living State: The Association-Induction Hypothesis*, Blaisdell, New York, 1962.
57. Eigen, M. and Winkler, R.: *Neurosci Res. Prog. Bull.* **9**, 330 (1971).
58. Somsen, G. and Weeda, L.: *J. Electroanal. Chem.* **29**, 375 (1971).
59. Somsen, G. and Weeda, L.: *Rec. Trav. Chim.* **90**, 81 (1971).
60. Gould, E. S.: *Mechanism and Structure in Organic Chemistry*, Holt, Rinehart and Winston, New York, 1959.
61. Hirschfelder, J. O., Curtiss, C. F., and Bird, R. B.: *Molecular Theory of Gases and Liquids*, John Wiley and Sons, Inc., 1954.
62. Rowlinson, J. S.: *Trans. Faraday Soc.* **47**, 120 (1951).
63. Hodgkin, C. D., Weast, R. C., and Selby, S. M.: *Handbook of Chemistry and Physics*, The Chemical Rubber Company, Cleveland, Ohio, 1960, p. 2536.
64. Talekar, S. V.: *Quantum Chemical Studies of the Electronic Structure of Macrocyclic Antibiotic Molecules: Valinomycin and its Potassium Complex*, Ph. D. Diss., All-India Institute of Medical Sciences, New Delhi, India, 1970.
65. Rein, R., Rabinowitz, J. R., and Swissler, T. J.: *J. Theor. Biol.* **34**, 215 (1972).
66. Cotton, F. A. and Wilkinson, G.: *Advanced Inorganic Chemistry*, 2nd ed., Interscience, New York, 1967.
67. Laprade, R.: personal communication, 1972.
68. Fenton, D. E., Mercer, M., and Truter, M. R.: *Biochem. Biophys. Res. Commun.* **48**, 10 (1972).
69. Eisenman, G., Szabo, G., McLaughlin, S., and Ciani, S.: in E. Munoz, F. Garciz-Ferrandiz, D. Vasquez (eds.), *Symposium on Molecular Mechanisms of Antibiotic Action on Protein Biosynthesis and Membranes*, Elsevier Publ. Co., Amsterdam 1972, p. 545.
70. Hille, B.: *J. Gen. Physiol.* **58**, 599 (1971).
71. Christensen, J. J., Hill, J. O. and Izatt, R. M.: *Science* **174**, 459 (1971).

VALINOMYCIN-INDUCED POTASSIUM SPECIFICITY
IN CHARGED MEMBRANES

ORA KEDEM

Laboratory of Membranes and Bioregulation, Weizmann Institute of Science, Rehovot

and

M. PERRY and RENÉ BLOCH

Rehovot Research Products, Kyriat Weizmann, Rehovot, Israel

Abstract. The mechanism of the high potassium specificity of membranes containing valinomycin was studied. It was shown that cellulose acetate and cellulose acetate nitrate membranes become highly permselective if they contain a small amount of fixed charges. Addition of valinomycin to the membranes induced high ion specificity, provided that permselectivity was already present as a consequence of the fixed charges. A method was developed for the preparation of modified cellulose acetate membranes carrying a controlled fixed-charge density.

In contact with mixed NaCl/KCl solutions the charged membranes contained both potassium and sodium, with some preference for potassium 1 mM Na, 3 mM K+ for 5 mM KCl+150 mM NaCl in outer solution). Addition of valinomycin to the membrane changed the ion composition only slightly (from 1.4 to 1.1 mM sodium). It is well known that valinomycin forms practically no complex with sodium, and the formation constant of the potassium complex depends on the medium. Apparently in our membrane only a fraction of the potassium counter-ions were complexed with valinomycin.

Charged membranes showed practically ideal perm-selectivity to cations, giving a potentiometric slope of 59 mV for a tenfold change in cation activity. Thus the cation transport number was found to be unity within the experimental error. Potentiometric measurements with mixed KCl/NaCl solutions gave large 'sodium errors': both potassium and sodium could contribute to ion transport.

Upon addition of valinomycin the transport number of potassium became practically unity even in the presence of a large excess of sodium. Ion uptake measurements showed that sodium is not excluded from the membrane. The highly specific potassium permeability must therefore be due to a relatively high average mobility of potassium; ion-pair formation with the fixed charges is decreased by the complexation between potassium and valinomycin.

LIST OF SYMBOLS

a_+', a_+''	activities of monovalent cations on the two sides of the membrane
a_K', a_K''	activities of potassium on the two sides of the membrane
$\bar{c}_+^*, \bar{c}_-^*$	membrane concentrations of free dissociated cations and anions respectively
$\bar{c}_t$	total salt concentration in the membrane
f	error factor defined in Equation (12)
F	Faraday constant
I	electric current
J_i	flow of species i
K_d^s	dissociation constant of a salt in a membrane
K_d^f	dissociation constant of an ionizable fixed group in a membrane
R	gas constant
T	absolute temperature
t_i	reduced transport number of species i in the membrane
u_+, u_-	mobilities of the dissociated cations and anions respectively in the membrane
z	valency of the permanent ion.
χ_t	concentration of total fixed ionizable groups in the membrane
χ	concentration of the associated part of fixed ionizable groups
χ_-^*	concentration of the free part of fixed ionizable groups

Eric Sélégny (ed.), Charged Gels and Membranes II, 125–135. All rights reserved.

composition dependent part of the chemical potential of component i
electric potentials on side one and two respectively of the membrane
membrane potential

1. Introduction

As was described by Professor Sollner [1], the charged membranes which are now
widely used in technology, have a long history in physiological research. Many interesting analogies to ion transport in living systems were observed. In two important
respects, however, charged membranes clearly failed as models: high osmotic selectivity (reflection coefficient near unity) combined with substantial permeability to
water, and ion-specificity. Attempts to introduce specificity, as model systems or for
technologically interesting separations, failed for a rather deep reason: specific binding
sites attached to the matrix increased the uptake of the desired ion by the membrane,
but decreased its mobility, compensating or overcompensating the effect of preferred
uptake.

Very high specificity, comparable to that of the cell membrane, was achieved by the
addition of macrocyclic antibiotics to lipid bilayer membranes. Extensive studies
showed that this was due to highly specific complex formation and solubility of the
complex in the lipid membrane [2].

Comparable specificity was achieved with the macrocyclic compounds also in relatively thick synthetic membranes. Recent results reported and discussed in the following lead to some unexpected conclusions about the mechanism of this selectivity.

2. 'Thick' versus 'Thin' Membranes

The thickness of the bilayer membrane is comparable to the dimensions of single
macromolecules, while even so-called ultrathin polymeric membranes are a few microns thick – two orders of magnitude above the thickness of the bilayer membrane.
This quantitative difference in dimensions gives rise to several qualitative differences
in the possible mechanisms of ion transport in both types of systems, in particular:

Electroneutrality: polymer membranes, including plasticised and liquid 'membranes'
held by a porous support, comprise a separate, macroscopic, electroneutral subsystem
[3]. Even in the presence of arbitrary potential gradients, cations and anions present
in the membrane balance each other; more accurately, the imbalance related to any
space charge is stoichiometrically negligible [4]. This is not at all true for the bilayer
membrane: the total quantity of ions contained in it is very small. A concentration of
e.g. potassium ions, large enough to cause a significant increase in conductance,
constitutes a very small space charge [2]. Hence, in the bilayer membrane the permeability of a single ion can be measured and interpreted, irrespective of other ion distribution coefficients and fluxes. In a thick membrane all charges, mobile and fixed, have
to be considered simultaneously.

Fick's law: Eyring described diffusion as a series of single jumps, overcoming at
each step an activation energy barrier. This approach has been applied to membrane

transport [5] and it has been shown that when a large number of consecutive jumps is necessary to pass the membrane, the single-jump treatment leads to linear diffusion flow, following Fick's law. Again, this does not necessarily apply to the bilayer membranes. Considering the passage of an ion through the hydrophobic barrier as a single jump, non-linear dependence on the driving force may be expected [6].

Surface equilibrium: finally, in a 'thick' membrane, migration through the membrane itself is almost always rate-determining and thus any processes at the membrane surface are very close to equilibrium. (We are excluding here high electric currents which might disturb ion-equilibria at the surfaces [7].) On the other hand, in the bilayer membranes the rate of complexation-decomplexation may be of the same order, or even slower, than the permeation itself [8].

All these different features have to be borne in mind when a model developed and tested for one class of membranes is applied to the other.

3. Permselectivity and Fixed Charges in Non-Aqueous Membranes

In accord with customary terminology, 'permselectivity' is defined as the discrimination between anions and cations, irrespective of their chemical nature. If a membrane separates two solutions of the same electrolyte, at different concentrations, the electric potential in the absence of a current is a measure of the permselectivity. The equilibrium potential across an ideally cation selective membrane is

$$\psi' - \psi'' = -\frac{-RT}{zF} \ln \frac{a'_+}{a''_+} \tag{1}$$

ψ is the potential measured with standard calomel electrodes, contacting the solution through salt bridges. Superscripts $'$ and $''$ refer to the two solutions. (For other notations cf. list of symbols.) Holding the concentration of one of the solutions (denoted by $'$) constant and plotting the measured potentials as a function of the logarithm of the mean ion activity in the other solutions, a straight line with a slope of 59 mV per decade is obtained for an ideally permselective membrane. In the following, the potentiometric slope, 'the slope' for short, refers to such a series of measurements.

From the general considerations outlined in the last section we were led to expect that fixed negative charges would be an indispensable element for ion-specific membranes based on uncharged macrocyclic compounds. This was confirmed by a comparison of porous cellulose-acetate membranes and Millipore filters, both soaked with a mixture of plasticizers, dimethyl- and dibutylsebacate. Cellulose-acetate (CA) is essentially uncharged, though it may carry a very small amount of carboxyl groups formed by oxidation (cf. papers by Spiegler *et al.* and Pusch *et al.* in this volume). Millipore is known [9] to carry negatively charged groups, apparently introduced by the detergents used in the preparation. Figure 1 shows the results of potentiometric measurements carried out with these two kinds of membranes.

The slope of 59 mV per decade obtained with Millipore indicates complete cation selectivity, while cellulose-acetate gives a slope of only 29 mV, far from ideal cation

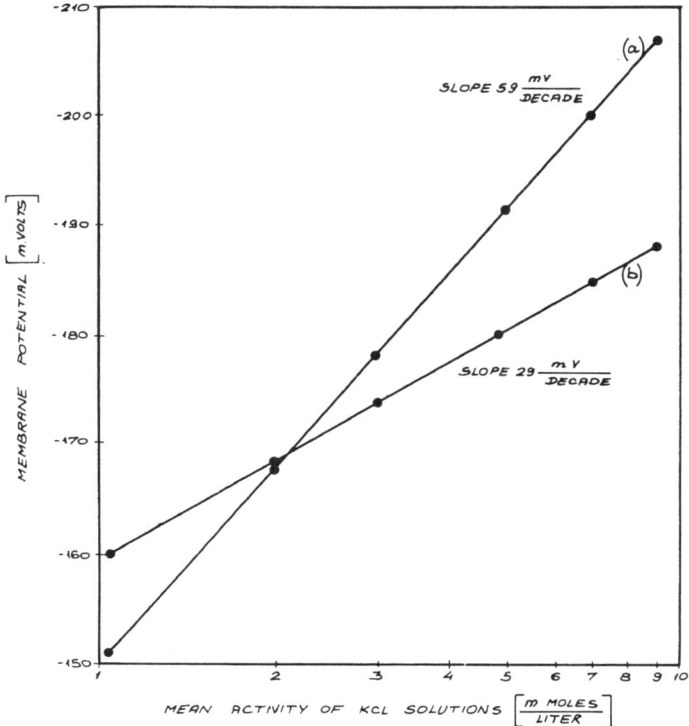

Fig. 1.   Potentiometric responses of plasticized polymeric membranes in potassium chloride solution
of various concentrations.
(a) charged, porous Millipore filter soaked with a mixture of dimethyl and dibutyl sebacates.
(b) porous cellulose acetate membrane soaked with a mixture of dimethyl and dibutyl sebacates.
The concentration of the fixed potassium-chloride solution was 1 mMoles/l, the concentration
range of the outer potassium chloride solutions, 1–10 mMoles/l. The membrane potentials were
measured by means of a silver-silver chloride electrode immersed in a fixed 1 mMolar KCl solution
and a standard calomel electrode contacting the outer solution through
lithium-trichloroacetate salt bridge.

selectivity. This partial permselectivity of CA may be due to the small amount of
fixed charges, to differences in mobility or to selective ion-adsorption.

For a systematic investigation of the effect of fixed charges on selectivity, plasticized
membranes were prepared from CA, and pretreated to contain negatively charged
groups. Thus, upon oxidation of CA before membrane-casting, an increase in this
potentiometric slope was observed, but it was difficult to prepare well defined membranes by this method. Better control of charge density was achieved by covalently
binding to CA an azinyl dye (Procion Brilliant Orange dye) carrying sulphonic groups.
This charged cellulose acetate (CABO) could then be dissolved together with unmodified CA and a plasticizer in acetone and cast into homogeneous plasticized
membranes with graded charge density. Table I summarizes the potentiometric slopes
obtained in KCl solutions. It is clear that a quite small charge density, 0.3 mMoles/l

TABLE I

Potentiometric slopes of membranes with different charge density

Membrane	% CABO w/w	Charge density mMoles/l	Slope mV/decade
Cellulose acetate	very small[a]	–	25–30
Cellulose acetate-CABO	15	0.3	59.0
Cellulose acetate-CABO	31	0.6	58.5
Cellulose acetate-CABO	50	1	56.5
Cellulose acetate-CABO	100	2	56.0
Millipore	–	4	59.0

The potentials across the membrane were determined by means of two standard calomel electrodes. The concentration of potassium chloride on the fixed side was 1 mMoles/l while the concentration of potassium chloride in the second solution varied from 0.1 mMoles/l to 0.1 Moles/l.

[a] Very small charge density that may originally exist in pure cellulose acetate.

of membrane phase, suffices to impart to the membrane high permselectivity, indistinguishable from ideal, even in the presence of 0.1 Moles/l KCl. A slight decrease in permselectivity with increasing charge density was observed, probably due to water uptake and swelling – unfortunately the plasticized cellulose acetate is not as mechanically stable as the Millipore.

The relation between charged groups on the membrane matrix and permselectivity is of course well known in aqueous systems. By the 'fixed charge' model it was possible to show that Donnan equilibria between membrane and solution phases can lead to pronounced permselectivity, without any other mechanism of ion exclusion, provided the fixed charge density is much larger than the electrolyte concentration in the solution [10]. However, as already pointed out, charge densities on the CABO membranes are very small compared to the salt concentration in the solutions. The strong influence of a rather small charge density on the permselectivity, can be understood qualitatively from a simple consideration of the equilibria between the organic membrane and the aqueous phases.

The membrane potential in a multicomponent system is given by [11]

$$\Delta\psi = -\frac{1}{F}\sum_i t_i \Delta\mu_i^c \tag{2}$$

where μ_i^c is the composition-dependent part of the thermodynamic potential of component i, $\Delta\mu_i^c$ is the difference of each μ_i^c between the two compartments separated by the membrane, and t_i the reduced transport number:

$$t_i = \frac{J_i F}{I} \quad \text{when} \quad \Delta\mu_{i,n}^c = 0 \tag{3}$$

J_i = flow of species i, I = electric current. The transport number t_i gives thus the flow of species i per Faraday of current passed through the membrane, in the absence of concentration gradients. Hence, in principle, all permeating species contribute to the potential, including the uncharged ones, if their flows are coupled to the flows of ions.

In an organic solvent, in the absence of mobile carriers and complexing agents, most of the ions will be present as neutral ion pairs. These do not migrate in an electric field and therefore the transport numbers are determined by the dissociated ionic fraction only. For a single uni-univalent electrolyte, assuming that the influence of water flow is negligible:

$$t_+ = \frac{u_+ \bar{c}_+^*}{u_+ \bar{c}_+^* + u_- \bar{c}_-^*}$$

$$t_- = \frac{u_- \bar{c}_-^*}{u_+ \bar{c}_+^* + u_- \bar{c}_-^*}$$

(4)

where $u_\pm$ are the ionic mobilities in the membrane and c_+^* the concentrations of the free ions.

It is instructive to rewrite Equation (4) as:

$$t_+ = \frac{1}{1 + \dfrac{u_-}{u_+} \dfrac{\bar{c}_-^*}{\bar{c}_+^*}}$$

(4a)

If the mobilities are of the same order of magnitude, permselectivity is determined by the ratio between free co-ions and counter-ions; in the absence of fixed charges this is of course unity.

Introduction of fixed charges can influence the total ion concentration in the organic phase only slightly, since the unionized part remains practically unaffected [12], while the ratio between free co-ions and counter-ions may be changed drastically.

Electroneutrality in the organic membrane requires:

$$\bar{c}_+^* = \chi_-^* + \bar{c}_-^*$$

(5)

where χ_-^* is the concentration of the dissociated fixed charge groups. Hence the ratio between the free co-ions and the free counter-ions becomes:

$$\frac{\bar{c}_-^*}{\bar{c}_+^*} = \frac{\bar{c}_-^*/\chi_-^*}{1 + \bar{c}_-^*/\chi_-^*}$$

(6)

χ_-^* is related to the concentration of the total fixed ionizable groups; and χ_t and the concentration of their non-dissociated part χ by the relation:

$$\chi_t = \chi + \chi_-^*$$

(7)

For a membrane containing fixed negatively charged groups, cations will form ion pairs both with the charged groups and the co-ions. The dissociation equilibria in the organic phase are:

$$\bar{c}_-^* \times \bar{c}_+^* = K_d^s (\bar{c}_t - \bar{c}_-^*) \tag{8}$$

$$\chi_-^* \times \bar{c}_+^* = K_d^f (\chi_t - \chi_-^*) \tag{9}$$

where $\bar{c}_t$ is the total co-ion concentration in the membrane.

Hence the ratio between the concentrations of free mobile anions and dissociated charged groups is:

$$\frac{\bar{c}_-^*}{\chi_-^*} = \frac{K_d^s}{K_d^f} \frac{\bar{c}_t - \bar{c}_-^*}{\chi_t - \chi_-^*} \tag{10}$$

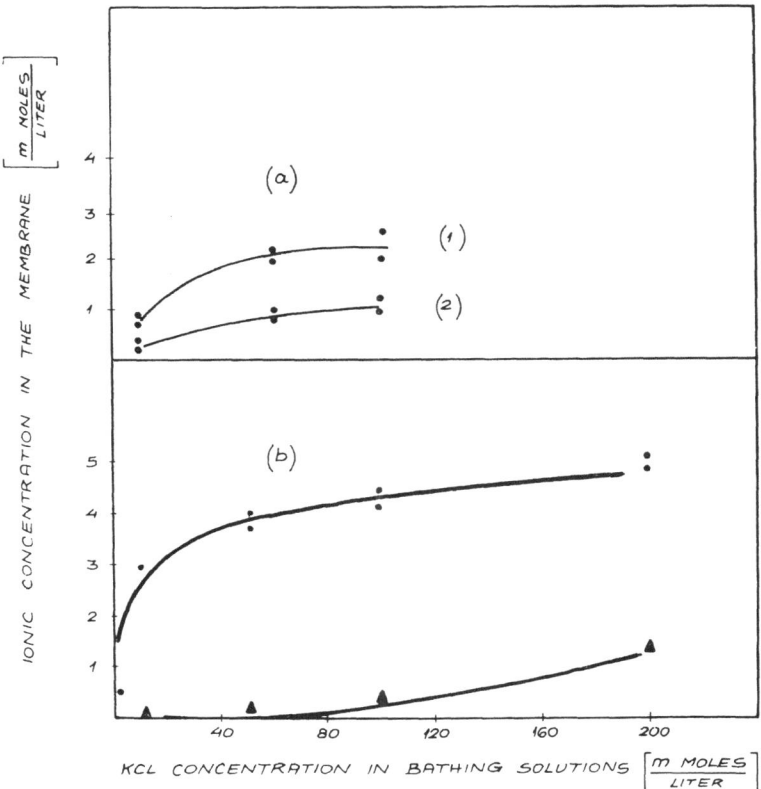

Fig. 2. Potassium – O and chloride – Δ ion concentrations in charged polymeric membranes as a function of potassium chloride concentration in the bathing solutions. Determined with the aid of radioisotopes K[42] and Cl[36].

(a) homogeneous cellulose-acetate (CABO) plasticized membranes with various concentrations of charged groups: (1) 100% w/w CABO and (2) 50% w/w CABO; equilibrated with potassium chloride solutions of different concentrations.

(b) charged Millipore filters soaked in a mixed dimethyl and dibutyl sebacate plasticizer equilibrated with buffered solutions of potassium chloride at pH 6.55.

Assuming small dissociation, the Equation (10) may be rewritten to a good approximation as:

$$\frac{\bar{c}_-^*}{\bar{\chi}_-^*} = \frac{K_d^s \bar{c}_t}{K_d^f \bar{\chi}_t} \tag{10a}$$

Equation (10a) shows that the free co-ion concentration, $\bar{c}_-^*$, may be a small fraction of $\bar{\chi}_-^*$, and therefore of the free counter-ion concentration, $\bar{c}_+^*$, even if the salt concentration in the aqueous solution is larger than χ_t. Thus high permselectivity can be achieved with a relatively small fixed charge density. From Equations (10) and (10a) it is clear that the ratio between the two dissociation constants is an important factor, and the permselectivity i.e., the potentiometric slope may be dependent on the nature of the anion, as this determines the value of K_d^s and $\bar{c}_t$. In fact experimental evidence for the dependence of permselectivity on the nature of the anion has been shown by Boles and Buck [13].

Ion content in the charged cellulosic membrane can be determined with the aid of the radioisotopes K^{42} and Cl^{36}. Potassium content of homogeneous CA/CABO plasticized membranes equilibrated with KCl solutions of different concentrations, and potassium and chloride content of Millipore equilibrated with similar buffered solutions, are shown in Figure 2. Potassium content increases with increasing charge density in CA/CABO. There is a substantial difference between potassium and chloride concentration in the Millipore, reflecting the negative fixed charge density. However, as expected, chloride is far from totally excluded, and from this total ion-composition alone one could not predict the ideal permselectivity observed. In fact, ideal permselectivity of Millipore membranes is not spoilt if the chloride content is about a third of the potassium content. This evidently indicates that the dissociation constant of the salt is smaller than that of fixed ionizable groups, i.e.

$$K_d^s < K_d^f$$

4. The Influence of Valinomycin on Permselectivity and Ion-Selectivity

Table II shows that in uncharged as well as charged CA membranes, the permselectivity as measured by the membrane potential in potassium chloride solutions is unchanged, within the experimental error, by the addition of valinomycin.

As expected, charges have a drastic influence on the potassium specificity of the membrane. While addition of valinomycin makes uncharged CA membranes only slightly more selective towards potassium, plasticized Millipore membranes containing valinomycin show ideal selectivity towards potassium. The membrane potential is uniquely determined by the potassium activities in both solutions, and an ideal potassium slope is found even in the presence of 100 fold excess of sodium.

A test of interference, in our case by Na^+, adapted to the analytical uses of specific membranes is the so-called 'sodium-error'. For ideal potassium specificity

$$\Delta\psi = 59 \log \frac{a_K''}{a_K'} \tag{11}$$

TABLE II

K/Na selectivity in charged and no-charged cellulosic membranes

	Charge density mMoles/l	Valinomycin concentration mMoles/l	Slope mV/decade	Sodium-error %
1	2	3	4	5
Cellulose acetate[a]	very small[c]	0	25–30	500
Cellulose acetate[a]	very small[c]	6	25–30	350
Cellulose acetate[b]	very small[c]	10	25–30	300
Cellulose acetate-CABO	0.3	9	59.0	0.5
Cellulose acetate-CABO	0.6	15	58.5	3
Cellulose acetate-CABO	1	24	56.5	7
Cellulose acetate-CABO	2	38	56.0	9
Millipore filter	4	0	59.0	70
Millipore filter	4	6	59.0	0.5

The composition of the solutions being in contact with the two sides of the tested membranes was unchanged for the whole set. One solution contained 1 mMole/l of potassium chloride and the other consisted of a mixture of 1 mMole/l potassium chloride and 10 mMole/l sodium chloride.

[a]Porous filter swollen with a plasticizer.
[b]Homogeneous plasticized membrane.
[c]Very small charge density that may originally exist in pure cellulose acetate.

where a' is the potassium activity in the constant standard solution and a'' the potassium-activity in the examined solution, even if NaCl is added to it. If sodium interferes, one may define the error by a factor f

$$\log f = \frac{\Delta \psi}{59} - \log \frac{a''_K}{a'_K} \qquad (12)$$

In the absence of interference f is unity. Its deviation from unity, in percent, is the 'sodium-error'.

Table II also shows the influence of valinomycin on the series of charged membranes discussed in the previous section. Columns 4 and 5 show permselectivity and sodium-error. Only when the membrane is permselective, as a consequence of fixed charges, the addition of valinomycin into the membrane imparts to it high potassium/sodium selectivity.

The influence of valinomycin concentration on potassium/soduim selectivity was examined with two sets of Millipore filters (Table III). One set of membranes was equilibrated in a solution containing KCl and a great excess of NaCl in order to allow the interfering sodium ion to equilibrate with the membrane phase, while the other was equilibrated in pure KCl solutions for reference purpose.

Surprisingly, a concentration of 0.7 mMoles/l of valinomycin in the membrane is enough to reduce the sodium-error to less than 1%, while the total charge density is about six times larger. Following extraction data and interpretation of selectivity in lipid bilayer membranes, one might expect that in the presence of valinomycin, sodium

TABLE III

Permselectivity and K/Na specificity of a plasticized Millipore filter as a function of valinomycin concentration

Valinomycin concentration mMoles/l	Membranes equilibrated in pure KCl solution 1 mMole/l		Membranes equilibrated in KCl–NaCl solution (5mMoles/1KCl+150mMoles/1NaCl)	
	Slope (mV/decade)	Sodium error %	Slope (mV/decade)	Sodium error %
0	59.0	70	–	–
0.3	59.5	5	57.5	8
0.7	59.0	1	59.0	1
1.2	59.5	0.5	59.5	0.5
5.8	59.5	0.5	59.5	0.5

The charged Millipore filters having 4 mMoles/l of fixed ionizable groups, were equilibrated prior to potentiometric determinations in two different bathing solutions: one set was immersed in a solution of pure potassium chloride and the other in a mixture of potassium chloride and sodium chloride. Concentration of an inner KCl solution – 1 mMole/l, composition of the second solution used in sodium-error measurements – 1 mMole/l KCl+0.15 Mole/l NaCl).

TABLE IV

Sodium uptake in charged Millipore filters as a function of the valinomycin concentration

Valinomycin concentration mMoles/l	Sodium concentration Mmoles l/
0	1.4 ± 0.3
0.7	1.2 ± 0.3
5.8	1.1 ± 0.3

The equilibration of the membranes was performed in a mixed solution containing 5 mMoles/l potassium chloride and 0.15 Moles/l sodium chloride. The concentration of fixed ionizable groups is 4 mMoles/l.

would be completely displaced from the membrane leaving the charged potassium-valinomycin complex as the only counter-ion [14]. If selectivity can be achieved with an amount of valinomycin equivalent to only a small fraction of the negative fixed charges, selective extraction of potassium cannot be the mechanism. This conclusion is confirmed and an alternative explanation is obtained by an analysis of sodium content in plasticized Millipore membranes.

After immersing membranes without valinomycin in a solution containing KCl and a large excess of NaCl, only about a third of the counter-ions in the membrane were found to be sodium ions (Table IV). It is thus clear that potassium is considerably preferred over sodium by the membrane itself, even without valinomycin. Moreover, addition of valinomycin decreases the sodium content only slightly, though the same membrane shows ideal potassium-selective potentiometric behavior.

The high transport number of potassium in the charged membrane containing valinomycin can be explained on the basis of the much lower degree of ion pair formtion of the potassium-valinomycin complex in the organic membranes as compared to the dissociation of the uncomplexed interfering ions. This higher dissociation constant is due to much larger radius of closest approach of the potassium-valinomycin complex. As a consequence, the selective complex formation between the potassium ion and valinomycin imparts selective permeability to the potassium ion by decreasing the association with the fixed negative charge.

Table IV shows furthermore that even an excess of valinomycin affects only slightly the total ion uptake. This is not surprising in view of a very small concentration of the unassociated ions in the membrane. Valinomycin complexes mainly with unassociated potassium ions and therefore the concentration of the potassium-valinomycin complex formed will be small as compared to the total concentration of the ionic species. This concentration of potassium-valinomycin complex, however, is still much higher than the concentrations of the unassociated non-complexed ions, thus determining the membrane potential.

In conclusion, our experiments contribute the following elements to an explanation of the mechanism of permeability and potassium specificity of hydrophobic membranes.

(1) In thick membranes the permselective conductivity due to potassium-valinomycin complexation requires the presence of fixed negative charges.

(2) Exclusion of interfering ions is not a necessary condition for ion specificity.

(3) The high transport number of potassium is due to the decreased ion-pair formation with the fixed charges due to the complexation between potassium and valinomycin.

References

1. Sollner, K.: in E. Sélégny (ed.), *Charged Gels and Membranes I*, D. Reidel Publ. Co., Dordrecht–Holland, p. 3.
2. Krasne, S. and Eisenman, G.: this volume, p. 107.
3. Lev, A. A., Malev V. and Osipov, V.: in *Membranes*, Vol. 2, 'Lipid Bilayers and Antibiotics' (ed. by G. Eisenman), Marcel Dekker, New York, 1973, p. 481.
4. Läuger, P. and Neumcke, B.: *ibid.*, p. 13.
5. Zvolinski, B. J., Eyring, H., and Reese, G. E.: *J. Phys. Chem.* **53**, 1426 (1949).
6. Stark, G. and Benz, R.: *J. Memb. Biol.* **5**, 133 (1971).
7. Oren, Y. and Litan, A.: submitted to *J. Phys. Chem.* (1974).
8. Eigen, M.: *Pure Appl. Chem.* **6**, 105 (1963).
9. Ilani, A.: *J. Gen. Physiol.* **46**, 839 (1963).
10. Teorell, T.: in *Progress in Biophysics and Biophysical Chemistry*, Vol. 3 (ed. by J. A. V. Butler and J. T. Randall, F. R. S.), Pergamon Press Ltd., London, 1953, p. 331.
11. Staverman, A. J.: *Trans. Farad. Soc.* **48**, 176 (1952).
12. Helfferich, F.: *Ion Exchange*, McGraw-Hill Book Co. Inc., New York, N.Y., 1962, p. 507.
13. Boles, J. H., and Buck, R. P.: *Anal Chem.* **45**, 2057 (1973).
14. Szabo, G., Eisenman, G., Laprade, R., Ciani, S. M., and Krasne, S.: in *Membranes*, Vol. 2, Lipid Bilayers and Antibiotics" (ed. by G. Eisenman), Marcel Dekker, New York, 1973, p. 191.

ION EXCHANGE AND STRUCTURAL PROPERTIES OF ALKALI ION MACROMOLECULAR CARRIERS IN LIQUID MEMBRANES

R. VAROQUI and E. PEFFERKORN

Centre de Recherches sur les Macromolecules, C.N.R.S. 67083 – Strasbourg CEDEX, France

Abstract. A synthetic amphiphilic polyacid (the copolymer of maleic acid and cetylvinylether) is used to develop metal ion exchange between N-octanol and water phases. The organic polymeric solution in contact with the aqueous phase behaves as a weak acid ion exchanger with selective properties. A thorough thermodynamic analysis of the two phase ion-exchange equilibria between alkaline and alkaline earth metals and acids is given. The mechanism by which specific association of ions on polymeric sites occurs is discussed using the data of coion and water uptake and by considering solvent-ion and ion-ion interactions in both phases. The conformational properties of the polymeric chain as a function of hydrogen-metal substitution are analysed, and an attempt is made to correlate the selective binding of metal ions with the conformational and configurational properties of the polymer. In particular, the average chain dimensions are explained using theoretical arguments and considering the complexing of metal ions on particular polymeric sites. Some typical experimental results of facilitated ion transport through a liquid hydrocarbon barrier and of coupling effects between flows of H^+, Na^+ and Ca^{2+} resulting in countertransport are given in the second part of the paper.

1. Introduction

In liquid ion exchange, a solution of an ionogenic compound dissolved in a water immiscible solvent is brought into contact with an aqueous electrolyte solution and ion exchange proceeds through the liquid/liquid interface. The exchange is made possible by specific complexing of the permeant ion with the ion-exchanger molecule in the organic phase. The selectivity coefficients are usually much greater for liquid-exchangers than for the more common ion-exchange resins, and also faster exchange and greater permselectivity are obtained. Increased attention has therefore been directed in recent years to the use of liquid ion-exchangers of various compositions in separation processes or for electrode sensing devices [1–4].

Moreover, the properties of non-aqueous liquid membranes when the liquid ion-exchanger separates two aqueous electrolyte solutions have been the subject of extensive research, the stimuli for this development originating largely from biological phenomena; their permeability is based on complex formation of the permeants with membrane molecules, and the elucidation of the mechanism of carrier mediated ion transport through artificial liquid membranes may help the understanding of basic principles of transport in biological membranes [5–7].

Since most of the water immiscible organic solvents have a low dielectric constant, it is desirable that the organic solute contains lipophilic residues of sufficient size to constrain the molecule completely to the organic medium. Typical liquid-ion-exchangers with a high solubility in a water immiscible organic phase are low molecular weight components such as quaternary ammonium derivatives or substituted phosphonic or sulphonic acids. In a recent work we investigated the behaviour of a particular class of ionogenic polymers which are soluble in organic media and insoluble in water and therefore fulfil the requirements for liquid ion-exchangers. These com-

Eric Sélégny (ed.), Charged Gels and Membranes II, 137–170. All rights reserved.

ponents are synthetic cationic or anionic amphiphilic polyelectrolytes of molecular weights ranging from 10^5 to 10^6. We shall describe here the properties of a polyacid with the following chemical structure:

$$-(CH - CH - CH_2 - CH-)_n$$
$$| \quad | \qquad \qquad | \qquad \qquad (1)$$
$$COOH \; COOH \qquad OC_{16}H_{33}$$

Owing to its dual hydrophobic-hydrophilic character, this polymer displays some particular properties: for instance, X-ray diffraction studies on aqueous concentrated gels have shown the existence of lamellar and cylindrical mesomorphous phases with different conductivities under definite conditions of hydration and neutralization [8–9]; intramolecular phase separations in water and in a binary solvent mixture have also been reported [10].

In the present study the liquid ion exchange properties of the polymer are investigated. The liquid exchangers or liquid membranes were made from a solution of the polymer in its acid or partially alkali metal form in N-octanol. N-octanol was chosen as solvent because of its very low solubility in water, and also because a large number of partition coefficients of ions or neutral components have already been determined for the water/octanol system in order to predict binding with respect to a reference component [11].

In the first part of this contribution we discuss hydrogen-metal and metal-metal ion-exchange equilibria for alkaline and alkaline earth metals, simultaneously analyzing the polymer structure and conformational changes which occur when a metal is substituted for hydrogen. Emphasis will be placed upon the mechanism through which the metal affinities, their complexing on particular polymer sites and the conformation and dimension of the macromolecular carrier are inter-related. This point, which concerns the important structural features of polyelectrolytes in a low dielectric constant medium, should be of importance when discussing the concept of carriers and carrier processes in more elaborate systems. In fact, natural membranes have been considered as macromolecular complexes embedded in a hydrophobic medium, and their permeability to cations has been described in terms of effects underlying conformational changes of membrane macromolecules as a result of the competitive binding of mono- and divalent counterions at polymeric sites [12].

In the second part we report some typical experimental results to illustrate facilitated ion transport through a hydrocarbon barrier and the coupling effects arising between H^+, Na^+ and Ca^{2+} fluxes.

2. Ion Exchange in Relation to Structural Properties

2.1. *Experimental section*

2.1.1. *Polymers*

The anionic polyelectrolyte (1) was obtained by copolymerizing maleic anhydride with vinyl hexadecyl ether followed by hydrolysis according to standard procedures [10], [13]. Radical polymerization yielded a 1–1 copolymer of weight average molecular weight $M_w \simeq 3 \times 10^5$. The polymer was fractionated with a $\frac{2}{3}$ acetone and $\frac{1}{3}$ benzene solvent mixture with methanol as precipitant. After purification of the resulting fractions, the following relationships between intrinsic viscosity $[\eta]$ and molecular-weight in tetrahydrofuran and water saturated N-octanol solutions were found:

$$[\eta] = 1.32 \times 10^{-2} \, M_w^{0.68}$$
$$[\eta] = 8.05 \times 10^{-3} \, M_w^{0.70} \tag{2}$$

In the present work a fractionated sample of 2×10^5 molecular weight was used throughout.

2.1.2. Partition Coefficient Measurements

N-octanol is about 2.15 molar with respect to water at saturation ($\simeq 1$ mole of water for 3 moles of octanol); on the other hand water dissolves only a small amount of N-octanol $\simeq 10^{-2}$ wt %. In order to saturate octanol with water, and conversely water with octanol, octanol was shaken for several days with water and decantation was allowed to take place over one week. Deionized water of 18 MΩ after passage through successive Millipore filters and NaCl, KCl, CsCl, CaCl$_2$, HCl electrolytes of analytical grade were used. The apparatus for partition coefficient measurements consisted of a covered glass jar in which a 1 wt % polymer solution in water saturated N-octanol was spread over a water phase containing a mixture of simple electrolytes. Since the polysoap molecules are surface active, the volume of the oil layer with respect to the interfacial surface area was chosen large enough to avoid any appreciable decrease of bulk polymer content by interfacial adsorption. Equilibrium between the two phases was accelerated by stirring near the interface by means of a double bladed stirrer, or by gentle rocking of the cell in such a way as to avoid micro-emulsion formation. Equilibrium was reached when the ionic composition vs. time reached in each phase a plateau value. The vessel was fitted with glass and counterelectrodes extending into the aqueous solution and the pH was monitored and recorded with an automated titrator (Tacussel U-5). Quantitative analysis of the ion concentration was made using small amounts of radiotracers Na[22], K[42], Cs[137], Ca[45], Cl[36] which distribute themselves between the two phases in the ratio of the non-labelled species. In a few instances, the composition of the organic phase was also analysed by atomic absorption spectrophotometry (Perkin Elmer 290 B). Electrical conductances were determined with an autobalance Wayne Kerr bridge model 13641 at a frequency of 1.6 kcps. The temperature was always 25 °C.

2.1.3. Water Content Determination of the Organic Phase

The water content of the octanol phase was found to be strongly dependent on the ionic composition of the system. The differential amount of water absorbed or desorbed during the ion-exchange process was obtained by measuring the specific volume of the octanol phase as a function of polymer concentration. The significance of partial specific volumes in multicomponent systems has been discussed by Casassa and Eisenberg, and for a detailed account the reader is referred to their papers [14–15]. Equilibrium experiments were performed at constant chemical potential of the solvents: the macromolecular solution at varying polymer concentration was equilibrated with the water phase at constant composition. It has been shown that in this case it is possible to define from the partial specific volume data an excess or deficiency of the diffusible solvent components which can be attributed to 'preferential solvation' or 'hydration' of the macromolecular species. More precisely, if we denote by Γ the weight of excess water in grams to be removed from the organic solution per gram of polymer added at constant chemical potential of water and octanol μ_w, μ_o, then

$$\Gamma = - \left(\frac{\partial c_w}{\partial c_p}\right)_{\mu_o, \, \mu_w} \tag{2}$$

The 'adsorption' or 'membrane distribution coefficient' Γ is obtained from density measurements according to the expression:

$$\Gamma = \frac{\left(\dfrac{\partial \rho_s}{\partial c_p}\right)_{\mu_o, \, \mu_w} - \left(\dfrac{\partial \rho_s}{\partial c_p}\right)_{w, \, o}}{1 - \overline{V}_w \, \rho_s^0} \tag{3}$$

$(\partial \rho_s / \partial c_p)_{\mu_o, \, \mu_w}$ is the density increment of the organic phase with respect to the polymer concentration c_p at equilibrium with the water phase

$(\partial \rho_s / \partial c_p)_{w, \, o}$ is the density increment of the organic phase at constant solvent composition, ie., when the amount of water with respect to octanol in the organic phase is kept constant at varying polymer concentration.

ρ_s^0 is the density of the organic phase extrapolated to zero polymer concentration; for $\overline{V}_w$ value, the partial specific volume of water in the oil phase 0.997, was taken. The densities of organic solutions were measured at $o < c_p < 0.02$ g cm^{-3} by means of a digital precision densitometer [16].

2.2. Results

2.2.1. Hydrogen-Metal Titration Data

The pH of the aqueous phase in two-phase potentiometric acid-base titrations for Na^+, K^+, Cs^+, Ca^{2+} hydroxides and in the presence of corresponding chlorides is given in Figure 1. The aqueous concentration m of the electrolyte was 10^{-2} molar and α the degree of neutralization of the polyacid was obtained from

$$\alpha = z_A n_A - n_{Cl} \tag{5}$$

where z_A is the valence of the ion considered, n_A and n_{Cl} are the concentrations of metal A and chloride in the organic phase expressed in moles per unit equivalent of exchange capacity of the polymer. (One monomole of polymer, which is made up of a residue of maleic acid and of hexadecyl vinyl ether, has two exchange sites.) Chloride

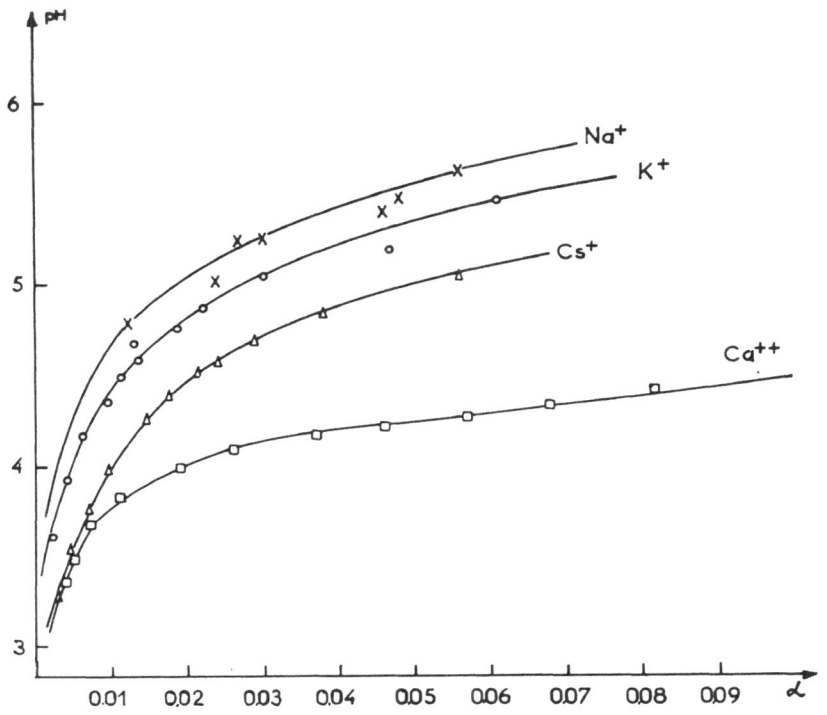

Fig. 1.   Two-phase potentiometric acid-base titration data for $Na^+(X)$, $K^+(O)$, $Cs^+(\triangle)$, $Ca^{2+}(\square)$
in the presence of 10^{-2} M NaCl, KCl, CsCl, CaCl$_2$ respectively.
Full curves according to Equation (5)
$$pH = 6.60 + (1 - 5.5\alpha + 40\alpha^2) \log(\alpha/1-\alpha) \ (X)$$
$$pH = 6.20 + (1 - 12.8\alpha + 146\alpha^2) \log(\alpha/1-\alpha) \ (O)$$
$$pH = 5.70 + (1 - 17.6\alpha + 168\alpha^2) \log(\alpha/1-\alpha) \ (\triangle)$$
$$pH = 4.79 + (1 - 3.0\alpha) \log(\alpha^{1/2}/1-\alpha) \ (\square)$$

co-ions have been used througout this work.

These titration curves shift to lower pH values on going from Na^+ to Cs^+, that is with increasing size of the crystal radius of the monovalent cations. A modified Henderson-Hasselbach equation of the form

$$pH = k_H^A + (1 + a\alpha + b\alpha^2) \log \frac{\alpha}{1-\alpha} \tag{6}$$

where a, b and k_H^A are constants for a given ion was fitted to the experimental points. The values of the a and b parameters given in Figure 1 indicate that the titration curves depart somewhat from an ideal one.

Table I gives the values of the n_{Cl}/n_{Na} ratio at $\alpha = 0.06$ as a function of salt concentration m and polymer concentration c_p. Electrolyte sorption by pure octanol is given in the second column of Table I.

Most of our results were obtained at $m = 10^{-2}$ M for which n_{Cl} is seen to be negligibly small as compared to n_A (first line of Table I).

TABLE I

Electrolyte sorption of the organic phase as a function of polymer ($\alpha = .06$) and
external salt concentrations, $\Delta\% = \pm 5$

		c_p Wt %		
	0	1	5	10
(water)		(organic)		
m(molar)	n_{Cl}(molar)		n_{Cl}/n_{Na}	
10^{-2}	1.7×10^{-6}	9.2×10^{-4}	3.9×10^{-4}	3.3×10^{-4}
10^{-1}	1.8×10^{-5}	9.7×10^{-3}	4.2×10^{-3}	3.5×10^{-3}
5×10^{-1}	1.1×10^{-4}	5.8×10^{-2}	2.6×10^{-2}	2.2×10^{-2}
1	2.9×10^{-4}	1.3×10^{-1}	6.3×10^{-2}	5.3×10^{-2}
2	8.2×10^{-4}	3.1×10^{-1}	1.6×10^{-1}	1.4×10^{-1}

The polymer in its acid form is readily soluble in water saturated octanol, which is a good solvent as evidenced by the high exponent 0.70 in the Mark Houwink relationship (2). This results from the affinity of the large hydrocarbon residues attached to the polymer backbone for the aliphatic part of the fatty alcohol, and from the fact that water molecules in the octanol phase act as small effective dipolar solvating agents for the carboxyl acid groups.

Octanol becomes, however, a poor solvent at high external pH when more and more metal ions are brought onto the polymer coil, and phase separation occurs in the organic phase at a limiting α value of approximately 0.1 for Na^+. The solubility domain increases in the order $Na^+ < K^+ < Cs^+ < Ca^{2+}$ and for Ca^{2+} the limiting α value is about 0.40. Since the polymer is insoluble in water for all α values of the solubility domain in octanol, it is completely prevented from moving from the organic to the water phase. The octanol solubility of the polymer increases with the salt concentration in the water phase; at higher external salt concentrations, octanol absorbs an increasing amount of electrolyte and water (cf. Table I), and preferential solvation phenomena then probably account for the greater solubility of the partially neutralized polyacid.

At salt concentrations below 5×10^{-3} M, surface forces promote association of the surface active polymer at the interface; for instance, when Na^+ is the counterion, a very tiny visible film develops at the junction of both phases for m equal to 10^{-3} M. However, light scattering measurements indicate that colloidal aggregates of large molecules are not formed in the polymer solution.

2.2.2. Metal-Metal Exchange

Figures 2 and 3 summarize the data refering to metal-metal ion-exchange. β_A and β'_A the ionic fractions of a metal A in octanol and water respectively for $Na^+ - Cs^+$, $Na^+ - K^+$ and $Na^+ - Ca^{2+}$ couples are defined in relation (7):

$$\beta_A = z_A n_A/(z_A n_A + z_B n_B), \qquad \beta'_A = z_A c_A/(z_A c_A + z_B c_B) \qquad (7)$$

For each isotherm α is constant. α is taken as the total number of equivalents of metal ions A plus B per unit equivalent of polymer, neglecting co-ion sorption i.e.,

$$\alpha = z_A n_A + z_B n_B \tag{8}$$

Non-linear isotherms are observed; the selectivity of the oil phase increases in the order $Na < K < Cs < Ca$. Curve II in Figure 2 shows data for $(Na^+ - Cs^+)$ at different α values: all points, whatever the value of α, fall on the same line. The variation of β_{Na} for the couple $(Na^+ - Ca^{2+})$ are plotted against β'_{Na} in Figure 3. An overwhelming preference of the organic phase for Ca^{2+} is observed, for instance the selectivity coefficient $\beta_{Na} \cdot \beta'_{Ca}/\beta'_{Na} \cdot \beta_{Ca}$ of the two counter ions is about 3.4×10^{-3} at $\alpha = 0.14$ and increases further with α.

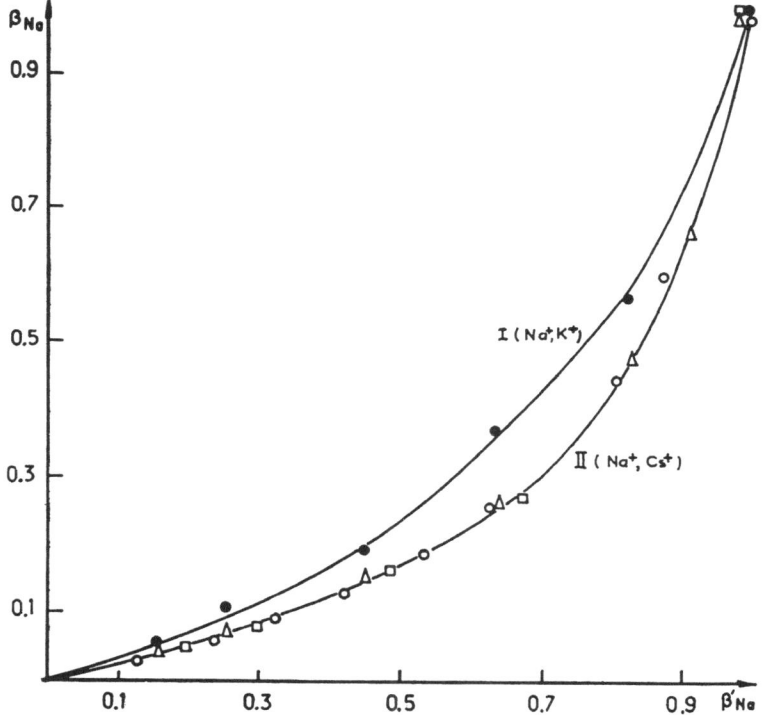

Fig. 2. Ion-exchange isotherm for $(Na^+ - K^+)$, curve I, and for $(Na^+ - Cs^+)$, curve II, at $10^{-2}\ M$ total external electrolyte concentration. The ionic fraction of polymer sites occupied by sodium is plotted against the ionic fraction of sodium in the aqueous solution: $\alpha = 6\%\ (\bullet)$: $\alpha = 3\%\ (\bigcirc)$; $\alpha = 6\%\ (\triangle)$; $\alpha = 9\%\ (\square)$.

For mono-monovalent exchange, regular isotherms of the type

$$\ln\left[(1-\beta)/\beta\right]\left[\beta'/(1-\beta')\right] = k_{Na}^A \tag{9}$$

are obtained, whereas for mono-divalent exchange the isotherms depart from a regular shape according to:

$$\ln\left[(1-\beta)/\beta^2\right]\left[\beta'^2/(1-\beta')\right]^{0.87} = k_{Na}^{Ca} + \ln\alpha \tag{10}$$

k_{Na}^A and k_{Na}^{Ca} being constants.

2.3. *Quantitative analysis of ion exchange results*

2.3.1. *Thermodynamic Treatment*
In the present section we discuss the underlying causes of the selective properties.

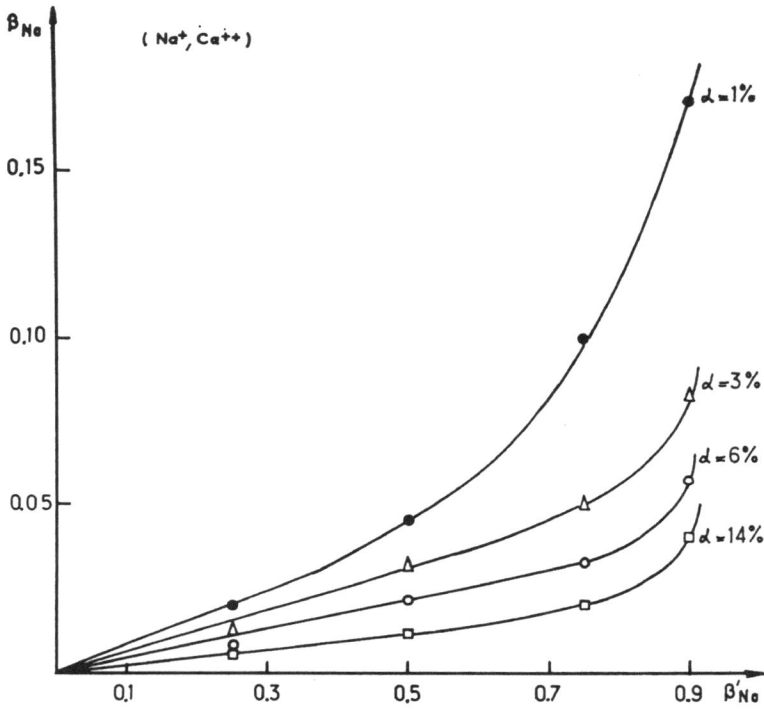

Fig. 3.　Ion-exchange isotherm for $Na^+ - Ca^{2+}$ exchange at 10^{-2} M total external salt concentration and different total amounts of polymer sites occupied by $Na^+ + Ca^{2+}$.

An equation for the ion-exchange isotherms is derived by expressing the free energy change dG of the organic phase relative to one equivalent of polymer exchange capacity:

$$dG = \sum_i \mu_i \, dn_i, (i = A, B, R, O, W) \tag{11}$$

μ_i is the chemical potential of component i in the organic phase. Suitable components of the system are exchange ions A and B, the polymer, water and octanol, the last three being denoted by subscripts R, W and O in the text. The choice of ionic species as components is permissible provided the final equations are expressed in determinate combinations of individual activities [17].

A suitable expression for dG is obtained by writing the chemical potential in terms of specific amounts,

$$\mu_i = \mu_i^0 + RT \ln (z_i n_i f_i / \sum z_i n_i) \tag{12}$$

and expressing the exchange reaction of two ions A and B according to Equations (13) in conjunction with the definition of the thermodynamic equilibrium constant K_A^B according to Equations (14) and (15)

$$z_B A^{z_A+} + z_A B_R^{z_B+} \rightleftarrows z_A B^{z_B+} + z_B A_R^{z_A+} \tag{13}$$

$$K_B^A = (z_A n_A f_A)^{z_B} (a_B')^{z_A}/(z_B n_B f_B)^{z_A} (a_A')^{z_B} = K_{cB}^A (f_A)^{z_B}/(f_B)^{z_A} \tag{14}$$

$$RT\ln K_B^A = (z_B \mu_A'^0 - z_A \mu_B'^0) - (z_B \mu_{RA}^0 - z_A \mu_{RB}^0) \tag{15}$$

f_i, z_i are the rational activity coefficient and the valence of ion i respectively, a_i is the activity of i in the water phase. $\mu_A'^0$, $\mu_B'^0$ are the standard chemical potentials of A and B in the aqueous solution and μ_{Ri}^0 is given by

$$\mu_{Ri}^0 = z_i \mu_R^0 + \mu_i^0 \tag{16}$$

μ_R^0 being the standard chemical potential of one equivalent of the exchanging group on the polymer. K_{cB}^A defined by Equation (14) is the 'corrected' selectivity coefficient for A and B.

Equating dG to zero and using Equation (14), a simple expression for the ion-exchange isotherm is obtained after some elementary rearrangements and neglecting co-ion sorption:

$$\ln \left(\frac{K_B^A}{K_{cB}^A}\right)^{1/z_B} = \frac{1}{RT} \frac{d\,G_m^E}{dn_A} + n_w \frac{d\ln a_w}{dn_A} + n_0 \frac{d\ln a_0}{dn_A} + 1 - \frac{z_A}{z_B} \tag{17}$$

G_m^E is the excess free energy of the electroneutral polymer per unit equivalent of exchange capacity, i.e;,

$$G_m^E = RT(n_A \ln f_A + n_B \ln f_B + \ln f_R) \tag{18}$$

Equation (17) may be written for hydrogen-metal exchange in the conventional form with $\alpha = z_A n_A$, $n_B = n_H = 1-\alpha$, $pH = -\log a_B'$

$$2.3\,pH - \ln \frac{\alpha^{1/z_A}}{1-\alpha} = \frac{1}{z_A}(\ln K_A^H + 1 - z_A - \ln a_A') +$$
$$+ \frac{1}{RT} \frac{d\,G_m^E}{d\alpha} + n_w \frac{d\ln a_w}{d\alpha} + n_0 \frac{d\ln a_0}{d\alpha} \tag{19}$$

z_A is equal to 1 and 2 for mono and divalent ions respectively. The two phase pH metric equation is formally similar to the one derived and discussed by Sélégny and coworkers for water swollen cross-linked ion-exchangers (17). It differs from the usual single phase potentiometric titration equation in the additional term $\ln a_A$ and the differential increment of the logarithm of the water and octanol activities, multiplied by the mole number.

If K_{cB}^A is expressed in terms of the ionic fractions β_A and β_B as defined by Equation (7), Equation (17) transforms for mono-monovalent and mono-divalent metal exchange into:

$$\ln \frac{a'_A \beta_B}{a'_B \beta_A} = \ln K^B_A + \frac{1}{\alpha} \left(\frac{1}{RT} \frac{d G^E_m}{d\beta} + n_w \frac{d \ln a_w}{d\beta} + n_0 \frac{d \ln a_0}{d\beta} \right) \tag{20}$$

$$\ln \left(\frac{a'_A}{\beta_A} \right) \left(\frac{\beta_B}{a'_B} \right)^{\frac{1}{2}} = \frac{1}{2} (\ln K^B_A + \ln \alpha + 1) + \frac{1}{\alpha} \left(\frac{1}{RT} \frac{d G^E_m}{d\beta} + n_w \frac{d \ln a_w}{d\beta} + \right.$$
$$\left. + n_0 \frac{d \ln a_0}{d\beta} \right) \tag{21}$$

Equations (17) to (21) are based on general principles and require no assumptions, save Equation (14) in which ion exchange equilibrium is treated as a reversible bimolecular equilibrium reaction, [18]. The effect of the external salt concentration on the two phase titration curves is expressed by the term $\ln a'_A$ on the r.h.s. of Equation (19). In Figure 4 the three upper curves are titration curves obtained with NaOH for three different NaCl concentrations 10^{-2} M, 10^{-1} M and 1 M. A large decrease of pH is observed for increasing salt concentrations. However, if pH $+ \ln a'_{Na}$ is plotted

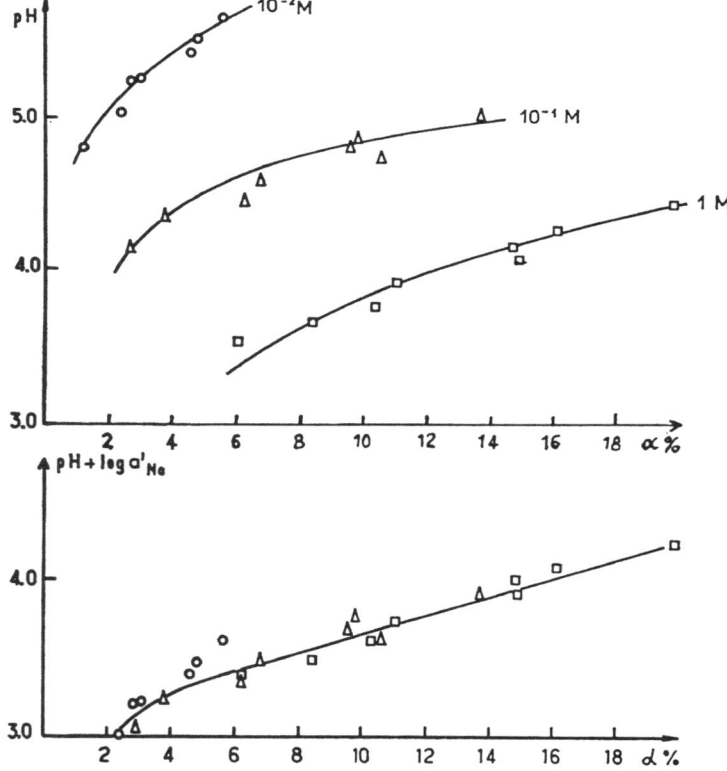

Fig. 4.   Two-phase potentiometric titration data for Na$^+$ in the presence of 10^{-2} M ($\circ$), 10^{-1} M
($\triangle$) and 1 M ($\square$)
NaCl solutions (upper curves)
pH $= 6.60 + (1-5.5\alpha + 40\alpha^2) \log (\alpha/1-\alpha)$ ($\circ$)
pH $= 5.50 + (1-5.5\alpha + 22\alpha^2) \log (\alpha/1-\alpha)$ ($\triangle$)
pH $= 4.00 + (1-7.1\alpha-18.5\alpha^2) \log (\alpha/1-\alpha)$ ($\square$)
lower curve gives pH $+ \log a'_{Na}$ according to Equation (19).

against α, the result is seen to fall within experimental error on the same curve and therefore application of the mass action law (14) correctly predicts the effect of ionic strength. From this result it may also be deduced that the water and octanol sorption, as well as the excess free energy change contribution to the free energy balance are about the same at the three electrolyte concentrations.

2.3.2. Thermodynamic Equilibrium Constant and Hydration Number of Metal Ions in Organic Media

The thermodynamic equilibrium constant K_B^A is obtained by integrating Equation (17):

$$n_A \ln K_B^A = \int_0^{n_A} \ln K_{cB}^A \, dn_A + z_B \int_0^{n_A} \sum_i n_i \, d \ln a_i +$$
$$+ n_A(z_B - z_A) + z_B \left[\frac{G_m^E(n_A) - G_m^E(0)}{RT} \right], (i = O, W) \tag{28}$$

The second integral on the right hand side expresses the effects of changes in water and octanol activities on the exchange equilibria. Allowance for this term should only slightly affect the value of K_A^B. This may be shown in the following way.

Water dissolves only a very small amount of octanol, less than 10^{-4} parts, and n_0 can safely be assumed constant, so that.

$$z_B \int_0^{n_A} n_o \, d \ln a_o \simeq z_B n_o \ln \frac{f_o(n_A)}{f_o(0)} \tag{23}$$

On the other hand, water sorption or desorption is not negligible. Density measurements made according to Equation (3) and (4) to determine water sorption during hydrogen-metal exchanges show that for all couples $Na^+ - H^+$, $K^+ - H^+$, $Cs^+ - H^+$, $Ca^{2+} - H^+$, parameter Γ of Equation (3) is a linear function of α:

$$\Gamma \,(Na^+ \,-H^+) \; 0.159 + 0.526\alpha$$
$$\Gamma \,(K^+ \;\; -H^+) = 0.159 - 0.088\alpha$$
$$\Gamma \,(Cs^+ \;\, -H^+) = 0.159 - 0.266\alpha$$
$$\Gamma \,(Ca^{2+} -H^+) = 0.159 + 0.050\alpha$$

The number of moles of water which accompanies the substitution of one hydrogen by a metal ion is derived from these data using the following relationship:

$$\frac{dn_w}{d\alpha} = \frac{d}{d\alpha}\{\Gamma \times [10.7 + 5.55 \times 10^{-2}\alpha \,(M_i - 1)]\} \tag{24}$$

where M_i is the atomic weight of the ion considered. The values of $dn_w/d\alpha$ are given in Table II. The transference from the aqueous phase to the organic phase of a polar ionic solute is accompanied by that amount of water which acts as solvating agent for that species in the organic phase, and the differential increment of water is obviously related to the number of moles of water adsorbed in the basic hydration shell around the metal ions. It can be seen that the exchange of Na^+, K^+, Cs^+, Ca^{2+} with hydrogen involves respectively about $+5$, -1, -3, $+1$ moles of water; while Na^+ and Ca^{2+}

give rise to a positive absorption of water from the aqueous to the organic phase, the hydration number of the acid group COOH exceeds that of K^+ and Cs^+ by about one and three water molecules respectively.

For mixtures of electrolytes AX, BX, HX at pH > 4 and total ionic concentration $AX + BX = 10^{-2}$ M, $\ln a_w$ computed from the osmotic coefficient ϕ listed by Robinson and Stokes, [19], assuming the additivity rule, is found to be almost constant, so that despite water sorption the water correction term in Equation (22) can be neglected in computing K_B^A.

Since no independent experimental determination of the activity coefficient f_0 was performed in this work, the K_B^A values cannot be rigorously computed by Equation (22) in conjunction with Equation (23) from the available experimental data. The experimentally available parameter is the average apparent thermodynamic equilibrium constant K_{aB}^A defined as

$$n_A \ln K_{aB}^A = \int_0^{n_A} \ln K_{cB}^A \, dn_A + z_B \int_0^{n_A} n_w \, d \ln a_w + n_A(z_B - z_A) \qquad (25)$$

From Equations (22), (23) and (25) it is readily seen that K_{aB}^A is related to the true thermodynamic constant K_B^A by:

$$\ln K_{aB}^A = \ln K_B^A + [G_{m, o}^E (\beta_A = 0) - G_{m, o}^E (\beta_A = 1)]/RT$$

$$\ln K_{aH}^A = \ln K_H^A + [G_{m, o}^E (\alpha = 0) - G_{m, o}^E (\alpha)]/RT\alpha \qquad (26)$$

where $G_{m, o}^E$ is the excess free energy of the electroneutral polymer + octanol molecules.

In the present context, there is little point in attempting to determine K_A^B with great accuracy and a comparison of the different K_{aB}^A will prove sufficient in discussing the selective properties of the polymer, inasmuch as the variation of the excess free, energy $G_{m, o}^E$ is small owing to the absence of long range electrostatic forces in the low dielectric constant octanol medium* (cf. the discussion of the next paragraph).

Table II gives the values of $\ln K_{aH}^H$ and $\ln K_{aNa}^A$ for sodium exchange with K^+, Cs^+ and Ca^{++}. The differences of the standard chemical potentials $\mu_A^{\prime 0} - \mu_B^{\prime 0}$ of ions in water (fifth and eighth line of Table II) were taken equal to the differences in the free

TABLE II

Thermodynamic data on water sorption and computed chemical potential differences of ions in both phases

	$H^+ - Na^+$	$H^+ - K^+$	$H^+ - Cs^+$	$H^+ - Ca^{2+}$
$dn_w/d\alpha$(moles)	$+ 5.6 \pm 1.2$	-0.94 ± 0.2	-2.84 ± 0.20	$+ 0.66 \pm 0.20$
$\ln K_{aA}^H$	10.94	10.32	9.56	18.50
$\mu_{RA}^0 - z_A \mu_{RH}^0$(kcal g^{-1}ion^{-1})	168.6	186.1	198.4	152.1
$\mu_A^{\prime 0} - z_A \mu_H^{\prime 0}$(kcal g^{-1}ion^{-1})	162.3	179.9	192.7	141.0
	$Na^+ - K^+$	$Na^+ - Cs^+$	$Na^+ - Ca^{2+}$	
$\ln K_{aNa}^A$	1.10	1.56	6.02^2	
$z_A \mu_{RNa}^0 - \mu_{RA}^0$(kcal g^{-1}ion^{-1})	-17.0	-29.5	187.4	
$z_A \mu_{Na}^{\prime 0} - \mu_A^{\prime 0}$(kcal g^{-1}ion^{-1})	-17.6	-30.4	182.6	

^{2} In K_{aNa}^{Ca} is a function of α as shown in Figure 3. The quoted value in Table II refers to $\alpha = 0.14$.

energy of hydration of the corresponding ions in dilute solutions. Experimental values of the free energy of solvation in water are found from Rosseinsky's tabulation [20]. $(\mu_{RH}^0 - z_A \mu_{RH}^0)$ and $(z_A \mu_{RNa}^0 - \mu_{RA}^0)$ were computed according to Equation (15).

2.3.3. Interpretation of the Ionic Selectivities

From Equation (15), the selectivity order and its magnitude is given by the sign and value of $(\mu_{RA}^0 - \mu_{RB}^0) - (\mu_A'^0 - \mu_B'^0)$. Ions are forced to the phase which provides the largest free energy difference relative to that phase. Alkali metal ions with the smallest dimensions prefer water as the medium owing to solvation effects, and though the intrinsic selectivity of the organic phase still favours the smallest cation as indicated by the tabulated $\mu_{RA}^0 - \mu_{RB}^0$ values, these ions are forced out of the organic phase into the aqueous phase. On the other hand, by comparing the $\mu_{RA}^0 - \mu_{RH}^0$ and $\mu_A'^0 - \mu_H'^0$ values it can be deduced that both phases are largely favourable to H^+, but carboxylic acid groups are preferentially formed to the detriment of acid in the water phase.

In order to explain from a molecular point of view the selectivity sequence according to the listed $\Delta\mu'^0$ and $\Delta\mu^0$ values, it is necessary to account for ion-ion and ion-solvent interactions in both phases. Diamond and Whitney have emphasized the role of ion-solvent interactions in their extensive studies of ion selectivities of highly swollen polyelectrolyte gels [21]. In the present case, while effects of ion-water interactions at small electrolyte concentrations still play the major role in describing the competitive behaviour of ions for the water phase, both ion-solvent and ion-ion interactions need to be considered in the organic phase, and it is believed that the stronger ion-ion interaction in the latter phase is the special feature which provides the selectivity order according to the listed $(\mu_{RA}^0 - \mu_{RB}^0)$ values. For metal ions in the octanol medium, electrostatic short range ion pairing, with different intrinsic free energies for different kinds of ions pairs RA, is probable. The energy of ion pair formation is expected to be inversely proportional to the ionic radius, and should also be related to the differences between the solvent-ion and solvent-ion pair interaction energies [22, 23].

An experimental verification of the extent of ion association is readily obtained from conductance data. In Figure 5 the equivalent conductance Λ_p of the polymer in the organic phase is plotted against the square root of the polyion concentration for 9% neutralization with different hydroxydes. The conductances of NaCl, KCl, CsCl and $CaCl_2$ in octanol + water were also determined at different salt concentrations. By applying to these results the method of Fuoss-Shedlovsky [24], the degree of dissociation α_i of the metal ions on the polymer was computed for different α's (the dielectric constant of water saturated octanol was found to be 8.92 [25]). Table III

---

* The K_{aB}^A, K_{aH}^A parameters would be identical to K_B^A and K_H^A respectively if the reference state for the components were chosen in accord with $G_{m,o}^E = 0$ in the state specified by both limits of integration in Equation (25). Clearly, this has a precise meaning only if the reference states (i.e., $f_A \cdot f_R = 1$, $f_B \cdot f_R = 1$) correspond to the homoionic form of the polymer in equilibrium with an infinitely diluted solution of the corresponding salt. For hydrogen-metal exchange, the availability of experimental data is, however, limited to a small domain of α (as already reported, phase separation occurs at values of α greater than 0.1), so that integration from one homoionic form to the other homoionic form is not possible in the present case.

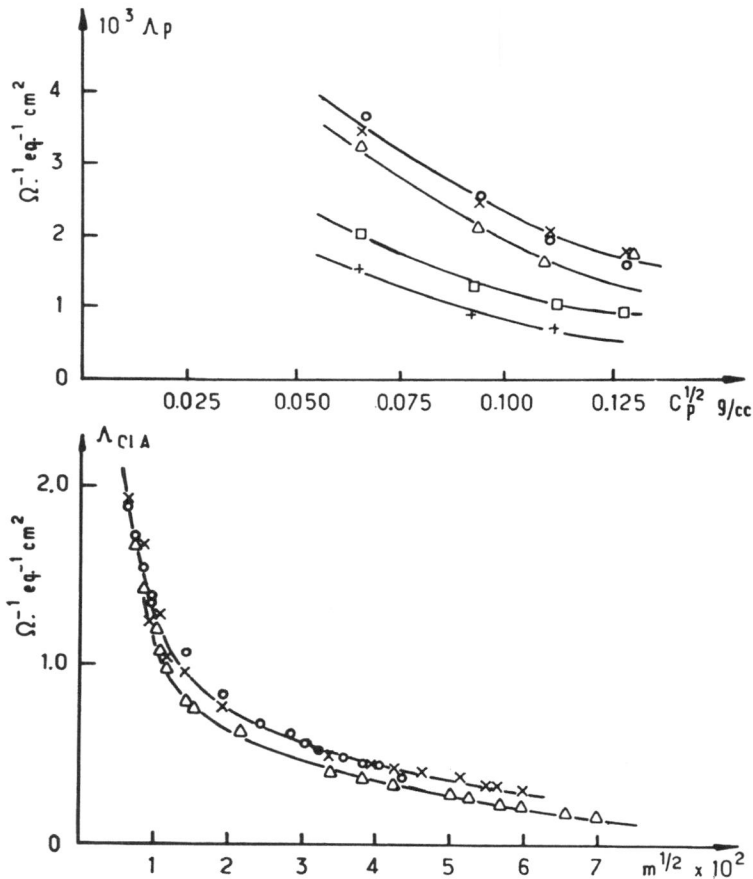

Fig. 5. Upper curves – Equivalent conductances of the polymer in the organic phase against the root square of polymer concentration at $\alpha = 0.09$, Na$^+$ (X), K$^+$(O), Cs$^+$($\triangle$), Ca^{2+} ($\square$), polymer in the acid form ($\alpha = \circ$) (+)

Bottom curves – Equivalent conductances of NaCl (X), KCl ($\circ$), CsCl ($\triangle$) in water saturated octanol solutions against the root square of electrolyte molarity.

TABLE III

Degree of dissociation of alkali metal ions on the partially neutralized polycarboxylic acid in water saturated octanol solutions.

α	$\alpha_{i, Na}$	$\alpha_{i, K}$	$\alpha_{i, Cs}$	$\alpha_{i, Ca}$
.03	4.29×10^{-2}	4.20×10^{-3}	2.27×10^{-3}	9.7×10^{-4}
.06	3.50×10^{-2}	3.25×10^{-3}	1.55×10^{-3}	5.4×10^{-4}
.09	2.80×10^{-2}	2.80×10^{-3}	1.73×10^{-3}	6.9×10^{-4}

$\alpha_{i, H} (\alpha = 0) = 2.04 \times 10^{-5}$

summarizes these results (the derivation of α_i is outlined in Appendix I).

The degree of dissociation is extremely small and decreases, as expected, with increasing metal substitution.

The strong interaction of the divalent calcium ion also deserves some comment. In addition to the classical 'electro-selectivity' for ions of higher charge there is an intrinsic preference of the organic phase for calcium ions, which for the $Ca^{2+} - Na^+$ couple amounts to 187.4 kcal ion^{-1} g^{-1}. There is also ample evidence in dilute aqueous polyelectrolyte solutions, that bivalent counterions are strongly associated with weak carboxylic polyacids. This behaviour may be related to a change in the electronic structure of these ions when, as for calcium, polydentate ligands may form with one maleic acid residue. In fact, our hydratation data show conclusively that calcium binding involves fewer water molecules than does sodium, therefore the calcium ion may be in close contact with the COO^- group, while the ions are probably separated by solvent molecules in the RNa ion pair.

2.3.4. Excess Free Energy and Conformational Properties of the Polymer

From the experimental results (6), (9) and (10) and the theoretical Equations (19) to (21), the following relations for the excess free energy increments are obtained:

$$
\frac{dG^E_{m,o}}{d\alpha} = RT(a\alpha + b\alpha^2) \ln \frac{\alpha^{1/z_A}}{1-\alpha}
$$

$$
\frac{dG^E_{m,o}}{d\beta} \simeq 0 \tag{27}
$$

$$
\frac{dG^E_{m,o}}{d\beta} = 0.076 \, \alpha \, RT \ln \frac{\alpha\beta^2}{1-\beta}
$$

By integration of (27) over the maximum α or β range, it is found that the average excess free energy differences

$$
[G^E_{m,o}(\alpha) - G^E_{m,o}(\alpha = 0)] / \alpha \, RT \text{ and } [G^E_{m,o}(\beta = 0) - G^E_{m,o}(\beta = 1)] / RT
$$

amounts to 0.7, 0.2 and 0.1 for CsOH, KOH, NaOH acid-base titrations, and to 3×10^{-2} ($\alpha = 0.14$) for $Na^+ - Ca^{2+}$ exchange.

The non-ideality of the isotherms can be attributed to the fact that the probability of contacts between metals located at different sites of the chain increases with α; the modification of intramolecular contacts as metal substitution proceeds yields different pair contact energies, and a rearrangement of the polymer conformation and configuration. Viscosity measurements provide an easy and precise determination of the polymer dimensions, and from experimental viscosity data we shall try to obtain a more precise representation of the mechanism of complex formation in relation with the structure of the ion-exchange molecule.

2.3.5. Thermodynamics of Polymer Conformational Changes

Viscosity measurements were performed to provide the average polymer dimensions according to the classical Flory-Fox formula (28), [26]:

$$
[\eta]_\text{J} = \phi \, (\overline{R^2})^{3/2}/M \tag{28}
$$

$[\eta]$ is the intrinsic viscosity (ml g^1), $\phi = 3.3 \, 10^{24}$ cgs, M the molecular weight and

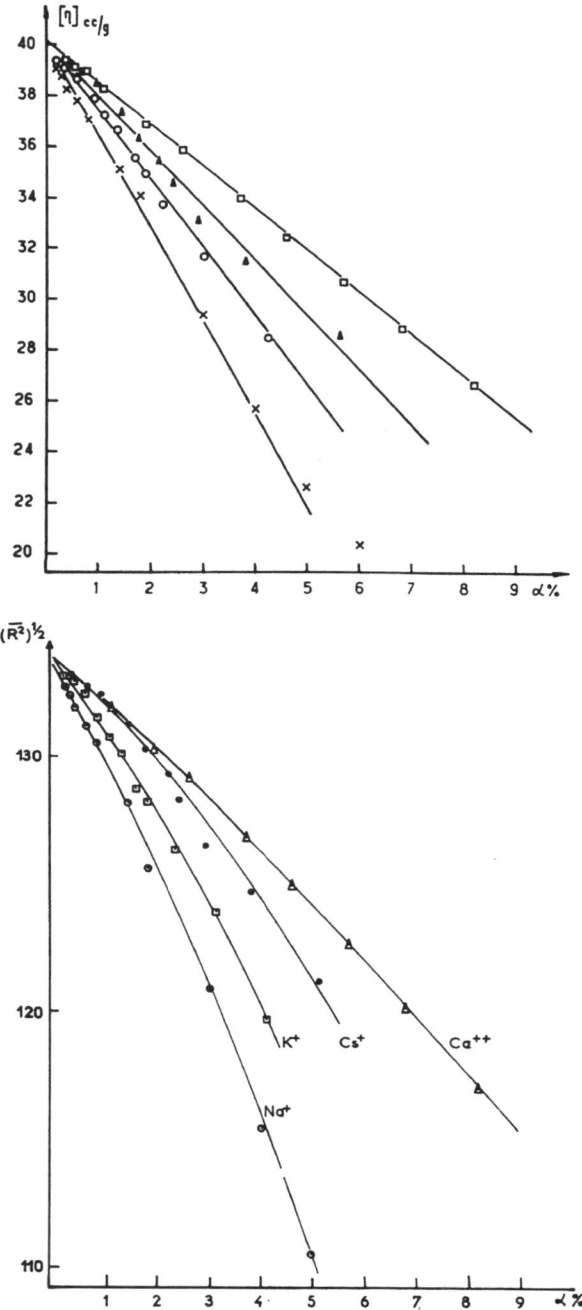

Fig. 6. Limiting intrinsic viscosities (upper curves) of polymer $M_w = 2 \times 10^5$ in water saturated octanol solutions as a function of fractional metal occupation. Experimental points: $Na^+(X)$; $K^+(\bigcirc)$; $Cs^+(\triangle)$; $Ca^{2+}(\square)$. Lower curves: $(R^2)^{\frac{1}{2}}$ against α computed by Equation (28). Full lines: according to Equations (33), (34) and (36) with the χ's values given in the text.

$(\overline{R^2})^{\frac{1}{2}}$, the root mean square distance of monomeric unit from the centre of gravity of the molecule, is the radius of gyration of the coil. The intrinsic viscosity of the organic phase at equilibrium with the water phase was measured and used together with the pH (α) data to determine $(\overline{R^2})^{\frac{1}{2}}$ from Equation (28) as a function of α. The results are shown in Figure 6 from which it can be seen that $(\overline{R^2})^{\frac{1}{2}}$ decreases rapidly as α increases, for example for a 5% Na$^+$ occupation, the chain dimension is decreased by a factor of about 17%. A marked dependence of $(\overline{R^2})^{\frac{1}{2}}$ on the nature of the ions is observed. The trend is the same as that reported for the affinities of ions to bind on polymeric sites (cf. Figure 1 and Table II), sodium ions resulting in the smallest dimension.

The ability to undergo configurational changes by interactions with cations has also been reported for antibiotic carriers in organic media [27, 28]. In the present case, since the polymer is a multi-site carrier and as the conformation of the chain is that of a coil the correlation between conformational changes and the reactivity of the monomeric sites involves some fundamental differences.

Is was shown that the metal ions of small ionic radius are the most hydrated. Since the organic solvent is a 'poor' solvating medium for hydrated bound metal ions, it is likely that the metal ions seek maximum internal polymer-polymer contacts with the oxygen of a carboxylic or ether group belonging to a closely located monomer according to the following scheme*:

$$
\begin{array}{l}
| \quad \swarrow^{\displaystyle 0\text{---Na----}0}\diagdown \\
C-C \qquad\qquad\qquad\qquad \diagdown C-C \\
| \quad \diagdown_{\displaystyle 0\text{---H----}0}\diagup \quad |
\end{array}
\qquad
\begin{array}{l}
| \quad \swarrow^{\displaystyle 0} \quad R\diagdown \\
C-C \qquad\qquad\quad 0-C \\
| \quad \diagdown_{\displaystyle 0\text{----Na}}\diagup \quad |
\end{array}
\qquad (29)
$$

Intramolecular polymer-polymer bonds result in conformational changes with a larger dimensional change for the polymer with the more hydrated ion.

Moreover, since monomeric units are distributed on the average about the molecular centre of gravity according to an approximately Gaussian distribution, the cations find a better environment in the central region of the coil by virtue of the high monomer density in that region. One may therefore also expect that the concentration of metal ions about the molecular centre greatly exceeds that calculated on a stoichiometric basis assuming a constant linear density α along the polymer chain.

Using an approch which parallels the derivation of the polymer dimensions for non-charged polymers in dilute solutions [26], we have derived the thermodynamic interaction parameters and metal distribution function quantitatively. In the model, long range electrostatic effects are neglected; it is furthermore assumed that the perturbations of the dimensions of the chain may be considered as determined by the equilibrium between the conformational free energy and the free energy of internal polymer-polymer and polymer-solvent contact energies, intermolecular polymer-polymer interactions being neglected as a first approximation.

* It was shown that the binding of cations on small lipophilic carrier molecules involves similar interactions, e.g., in macrocyclic carriers, the molecule is folded around the ion and the cation is surrounded by the ether and carbonyl oxygens which provids a favourable energetic environment [29, 30].

Let us then imagine an idealized polymer coil composed of αX monomeric units of kind RA and $(1-\alpha) X$ of kind RH. For the sake of simplicity, both units and solvent are assumed to have identical dimensions and the segments joining the centres of consecutive repetitive units are distributed on the average about the molecular centre of gravity in accordance with a Gaussian formula:

$$\rho = X \left(\frac{3}{2\gamma^2 \overline{R_0^2}}\right)^{3/2} \exp\left(-\frac{3r^2}{2\gamma\ \overline{R_0^2}}\right) \tag{30}$$

ρ is the number of segments per unit volume at distance r and γ is the expansion coefficient of the chain.

$\overline{R^2}$ is the mean square distance from the centre of gravity averaged over all segments in the unperturbed state. The free energy $\delta(\Delta G_M)$ of mixing polymer segments and solvent molecules in a volume element $d\tau$ at a distance r is given by the expression:

$$\delta(\Delta G_M) = kT\{[(1-\overline{V}\rho)(\chi-1)-\overline{V}\rho/2]$$

$$+ \alpha(r)\ln \alpha(r) + [1-\alpha(r)]\ln [1-\alpha(r)]\}\rho d\tau \tag{31}$$

$\overline{V}$ is the volume of the monomer. The first bracket on the r.h.s. of Equation (31) is Flory's expression, which is based on the classical lattice model where only nearest neighbour interactions are taken into account and the number of pair contacts of a eiven kind is assumed proportional to the volume fraction. The second term is the gntropy of mixing of groups RA and RH. Since counterions A and H may not be distributed at random along the chain, i.e., $\alpha(r) \neq \alpha_{stoich}$, the entropy of mixing RA and RH on the chain is included in the free energy, and in that respect Equation (31) differs from that for statistical copolymers. χ is a quadratic function of the local composition $\alpha(r)$

$$\chi = \chi_{S,RH} + [\chi_{S,RA}-(\chi_{S,RH} + \chi_{RH,RA})]\alpha(r) + \chi_{RH,RA}\,\alpha^2(r) \tag{32}$$

$\chi_{S,RA}$ and $\chi_{S,RH}$ are the classical dimensionless polymer-solvent interaction parameters and $\chi_{RH,RA}$ characterizes the RH, RA interactions, [31]. Equation (32) is readily derived by extending Flory's theory to copolymers. The total free energy ΔG_M of mixing polymer segments and solvent molecules is obtained by summing Equation (31) over the total space and adding the free energy ΔG_{el} of deformation of the polymer chain

$$\Delta G_M = \int_0^\infty \delta(\Delta G_M) 4\pi r^2\, dr + \Delta G_{el}$$

$$\Delta G_{el} = kT[3(\gamma^2-1)/2-3\ln \gamma] \tag{33}$$

The equilibrium values of $\alpha(r)$ and γ are obtained by minimizing the free energy with respect to these parameters, subject to the restriction of constant number of metal ions:

$$d(\Delta G_M) = \left[\frac{\partial(\Delta G_M)}{\partial\alpha(r)}\right]_\gamma d\alpha(r) + \left[\frac{\partial(\Delta G_M)}{\partial\gamma}\right]_{\alpha(r)} d\gamma = 0 \tag{34}$$

$$\int_0^\infty \rho\alpha(r)\,d\tau = \alpha X \tag{35}$$

A solution of Equation (34) and (35) is given in Appendix II. Computations of the equilibrium values of γ and $\alpha(r)$ have been performed with the following parameter values:

$$M = 2 \times 10^5; \quad X = 1040; \quad \mathcal{N}\overline{V} = 2.185; \quad b = 1.54 \times 10^{-8}$$

$$R_0^2 = (2/6)(2.6b)^2 X; \quad \chi_{S,RH} = 0.375 \tag{36}$$

In the absence of any data concerning the unperturbed dimension of our polymer in N-octanol, we have chosen the effective bond length $2.6b$ to be identical to that observed for polyisobutylene in benzene [32]. $\overline{V}$ was computed from the values of the partial specific volume of the polymer (see esp. part 2.1.3) according to

$$\overline{V} = \overline{V}_p M / X \tag{37}$$

Our polymer is composed of a diacid followed by a hexadecyl ether side chain, and hence the assumption of defining a 'segment' in the usual way as equal in volume to a solvent molecule is not strictly true. The calculations which have been performed using experimental data (36) apply rather to a polymer of identical contour length, molecular weight, volume and number of acid groups but of slightly different structure

$$\begin{array}{c}
(CH - CH - CH - CH)_n \\
| \quad\quad | \quad\quad | \quad\quad | \\
COOH \; C_8H_{17} \; COOH \; OC_8H_{17}
\end{array} \tag{38}$$

If one takes the view that the long vinyl ether chain overlaps the two acid groups in structure (1) then polymer (38) in which one repetititve unit is composed of one acid group and one half of the alkyl chain, should have comparable thermodynamic properties; the conclusion we shall reach here does not greatly depend on this extra assumption. It should also be mentioned that the water molecules which make up the hydration shells of an acid or a metal group must be considered as part of the monomers, the χ parameters refering then to the total contact energy between hydrated ion-paired complexés RA or RH and the binary octanol-water mixture. The parameter $\chi_{S,RH}$ which characterises the interaction energy between solvent molecules and the monomer in its acid form, was derived from the experimental limiting viscosity figure of the polyacid Figure 6 and applying Equation (28) and (39)*.

$$\gamma^5 - \gamma^3 = (3/4\,\overline{R_0^2})^{3/2}\, X^2 \overline{V}(0.5 - \chi_{S,RH}) \tag{39}$$

* A calculation of Kurata and Yamakawa (*J. Chem. Phys.* **29**, 311, 1958) led these authors to propose the relation $\gamma_{[\eta]}^3 = \gamma^{2.43}$ between the viscosity expansion coefficient and the geometrical expansion coefficient in Equation (30). In the present treatment the two coefficients were taken equal. It should also be mentioned that numerous attempts have been made to improve the original polymer theory, notably in the refinement of the mathematical treatment of the models which lead to Equations (31) and (32). In this text the theory is used in its original form, refinements appearing of trivial significance at the present stage of experimental work.

Good agreement between $(\overline{R^2})_{th}^{\frac{1}{2}}$ and $(\overline{R^2})_{exp}^{\frac{1}{2}}$ is obtained – see Figure 6 – for the following values of the unknown $\chi_{S,RA}$ and $\chi_{RH,RA}$ parameters (there is no alternative solution, as shown in Appendix II):

	Na	K	Cs
$\chi_{S,RA}$	2.65	1.87	1.60
$\chi_{RA,RH}$	-2.13	-1.80	-1.37

According to the lattice model the χ parameters account for the excess bond energy ω in a heterobond RA, RH over that of the average of the homobonds RA, RA and RH, RH, i.e.,

$$\chi_{RA,RH} \propto \omega_{RA,RH} - (\omega_{RA,RA} + \omega_{RH,RH})/2$$
$$\chi_{S,RA} \propto \omega_{S,RA} - (\omega_{RA,RA} + \omega_{S,S})/2$$

Both computed χ values are of a reasonable order of magnitude ($RT\chi \simeq 1.2$ kcal); while interactions S, RA are repulsive, interactions RA, RH are attractive. A large positive value for $\chi_{S,RA}$ was expected. On the other hand, the meaning from a molecular view point of the negative values of $\chi_{RH,RA}$ can be explained if one looks more closely at the chemical reaction which leads to the definition of this parameter:

$$[RH, RH] \quad + \quad [RA, RA] \quad \rightleftarrows \quad 2\,[RA, RH] \tag{40}$$

The reaction product RA, RH is expected to be the most stable, if according to the scheme the cation A in RA, RH faces the electronegative oxygen atom and leads to attractive interaction. On the other hand, repulsive forces between the alike positively charged cations prevail in RA, RA. Reaction (40) reproduces only interactions which are likely to occur between the hydrated polar heads of the monomers. In fact , in each monomeric unit a hydrocarbon tail is attached to the polar head and the χ's, as defined by the model, are relative to the total contact energy between hydrated ion paired complexes, so that additional interactions may also play a role.

The most striking result which emerges from the theory is shown in Figure 7. It can be seen that the linear metal ion density $\alpha(r)$, which was computed according to Equation (34) and (35) for Na^+, depends substantially on the spatial location of the monomeric unit. The location of a functional group on the chain is influenced by its environment and metal ions have a tendency to bury themselves in the centre of the macromolecular coil where, in view of the high segment density, contacts with the solvent molecules are minimized. The 'admission' of metal ions into the organic medium by means of the carrier molecule results therefore in a large rearrangement of the polymer conformation and configuration and the polymer may be visualized as a complex with an internal high metal density and a hydrophobic external region, the latter accounting for the solubility of the whole complex in organic media. In Figure 8 the ratio of $\alpha(r = 0)$ at the centre of the coil to that at infinite distance $\alpha(r = \infty)$ is

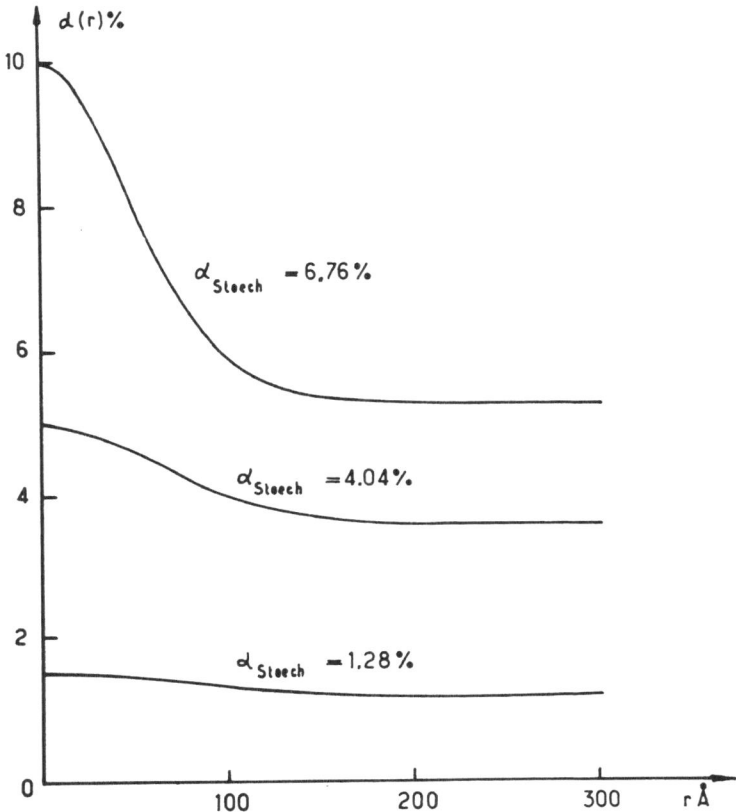

Fig. 7.   Linear distribution of metal ions along the polymer chain as a function of the distance from
the center of the polymer coil for α_{stoich}. (Na$^+$) equal to 6.76%, 4.04% and 1.28%.

plotted against the degree of polymerization X, and it can be seen that the effect is
strongly enhanced as the chain length decreases. This result is due to the fact that at
the centre of a macromolecular coil the segment density varies as the inverse 3/2
power of the total segment number for a Gaussian distribution. However, the foregoing considerations cannot be extrapolated to polymers of too small a size, for the
Gaussian distribution (30) would not then apply.

    Before concluding this section we would like to emphasize once again the argument
by which the properties have been derived. While polyelectrolyte chains in an aqueous
medium consist rather of long fully extended segments, chain polyelectrolytes in a low
dielectric constant medium can be envisaged as uncharged coils with intramolecular
contacts. Redistribution of functional groups along the chain then depends whether
specific intramolecular interactions are strong enough, and whether the polymer
volume fraction at the origin is large enough. This last condition is requisite since the
probability of intramolecular contacts is proportional to the polymer volume fraction.
Both conditions which are clearly expressed by Equation (A. IV) of Appendix II

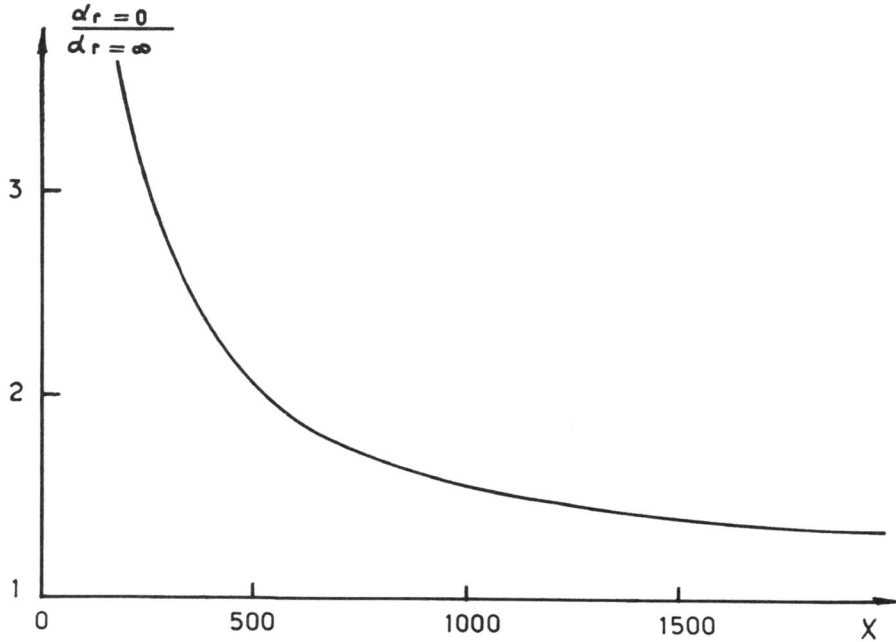

Fig. 8.   Redistribution of metal ions as a function of the degree of polymerization X.

namely that the functional $(1-\rho\overline{V})\Delta\chi$ must be substantially smaller than $\Delta\chi$ at the origin for $\alpha(r)$ to be a function of r, are satisfied by our polymer.

3. Carrier Mediated Ion Transport

During the last two decades a large amount of theoretical and experimental work has been devoted to studies of carrier mediated transport through membranes and the efforts in this direction have been reviewed in the recent literature on membranes. Some workers were more interested in the study of transport of ions or neutral species through thin lipid membranes; thus valinomycin, enniatin and nonactin have currently been used as biological analogues for carriers [6], [33–36]. These components are macrocyclic molecules with a non-polar exterior and an internal ring cavity where the ester or ether groups, through their ring oxygens, provide an environment similar to water for cations. Synthetic macrocyles such as polyethers or polymacrocycles have also been used as carriers [30], [37–38]. Ion exchange through lipidic barriers can also occur by means of an ion-exchange molecule, ion-ion pair formation being in this case the feature which provides association between the carrier and the small ion. Such systems have been used as models to elucidate the mechanism of ion transport through non aqueous 'thick' liquid membranes, and have found increased applications in specific areas [39–42]. Enzyme mediated facilitation was analysed in the first volume of this series [42 bis].

   In the preceding section we discussed the specificity, architecture and mechanism of alkali metal binding on polyacid carriers. The use of these amphiphilic polyelectrolytes as model molecules in liquid membranes may be interesting from a number of aspects, and in the present section we report some typical examples of facilitated diffusion of ions through a hydrocarbon barrier.

3.1. *Experimental membrane system*

The liquid membrane was a 1 wt % polymer solution floating over two aqueous solutions 1 and 2 separated by a diaphragm as depicted in Figure 9. The oil phase, about 1.5 cm thick, was stirred by means of a blade rotating at approximately 30 r.p.m. The cell was similar in most respects to the one devised by Schulman and coworkers [43].

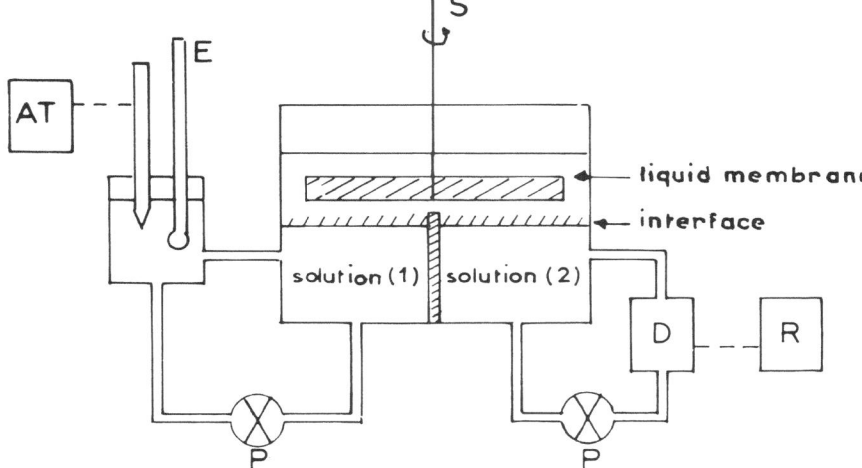

Fig. 9.   Diagrammatic representation of the membrane cell and recording devices. (S) stirring device, (P) circulating pump, (D) radiation, detector (flow cell), (R) ratemeter and recorder, (E) glass electrode and counter electrode, (AT) automatic titrator.

The homogenization of solutions 1 and 2 was accomplished by circulation of the aqueous solutions through reservoirs 1 and 2, each of one litre capacity, by means of centrifugal pumps. During the experiment the acid concentration in the left cell compartment was kept constant by a slow and controlled addition of a base by means of an automatic titrator (Tacussel U.5). Sodium fluxes through the oil layer from side 1 to 2 were analysed using Na^{22} isotope labelling. For this purpose, the flow circuit of the receiving right compartment included a flow cell into which a crystal scintillation counter was inserted. The net flux J_i of an ion species i through unit area of membrane is given by the difference between the unidirectional fluxes $J_i(f)$ and $J_i(b)$ in a steady state situation [44].

$$J_i = J_i(f) - J_i(b)$$

$$J_i(f) = V_2 \frac{m_i(1)}{m_{ir}(1)} \frac{d\ m_{ir}(2)}{dt}, \ m_{ir}(1) \gg m_{ir}(2) \tag{41}$$

$$J_i(b) = V_1 \frac{m_i(2)}{m_{ir}(2)} \frac{d\ m_{ir}(1)}{dt}, \ m_{ir}(2) \gg m_{ir}(1)$$

J_i is expressed in moles per unit area per unit time. V_1 and V_2 are the aqueous solution volumes, $m_i(1)/m_{i_r}(1)$ and $m_i(2)/m_{i_r}(2)$ the concentration ratios of non-labelled to labelled species i in compartments (1) and (2) and dm_{i_r}/dt is the time derivative of the radioactivity in a given compartment.

Diffusion of small ions in the water phase close to the oil/water interface is much faster than diffusion in the calm oil layers just ahead of the interface, and hence the latter process is the rate controlling step, therefore concentration gradients are set up in the unstirred Nernst oil layers and the effective membrane thickness is twice the thickness of the unstirred layers assuming that the flow in the absence of vigorous stirring is laminar near the interface and that each Nernst layer is of uniform thickness over the whole interfacial area. Isobaric and isothermal conditions at 25 °C were maintained during the experiments.

3.2. Membrane transport phenomena

3.2.1. Carrier Transport
Carrier transport is illustrated in Figure 10 for the following symmetrical arrangement.

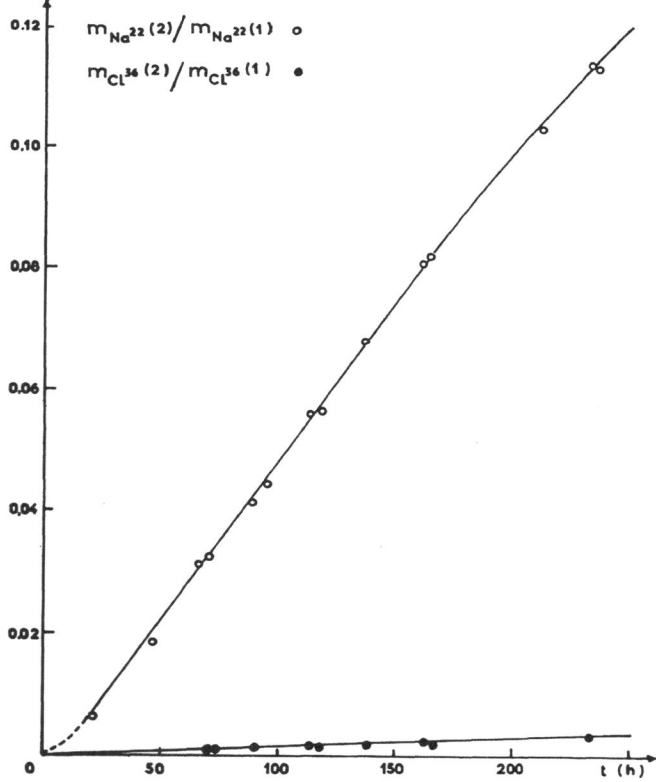

Fig. 10.   The kinetics of Na$^+$ ($\circ$) and Cl$^-$ ($\bullet$) transference across the membrane.

Solution (1)		Solution (2)
pH = 4.9	Membrane	pH = 4.9
10^{-1}M NaCl		10^{-1}M NaCl

System I

The membrane was a 1 wt % polyelectrolyte solution in N-octanol as described in the experimental section. Side (1) of the membrane bathing solution was radioactivated with traces of Na^{22} and Cl^{36} and the increase of activity in side (2) was recorded. The ordinate in Figure 10 represents the concentration ratio $m_{ir}(2)/m_{ir}(1)$ of labelled species in compartments (2) and (1) as a function of time (hours) for $i \equiv Na^{22}$ and $i \equiv Cl^{36}$. According to Equation (41), the ratio of the slopes of the two straight lines in Figure 10 is equal to the ratio of the unidirectional fluxes of Na^{+} and Cl^{-}. This ratio is equal to 2.8×10^{-2}. The flux of Cl^{-} ions was found to be identical to the salt flux recorded in the absence of the polymer, and the comparatively higher flux of Na^{+} is generated by association with the polymer carrier.

3.2.2. *Counter-Transport System*
Curves (I) and (II) if Figure 11 describe the net flux $J_{Na^{+}}$ of sodium ions according to Equation (41) through the membrane for the two following experimental arrangements:

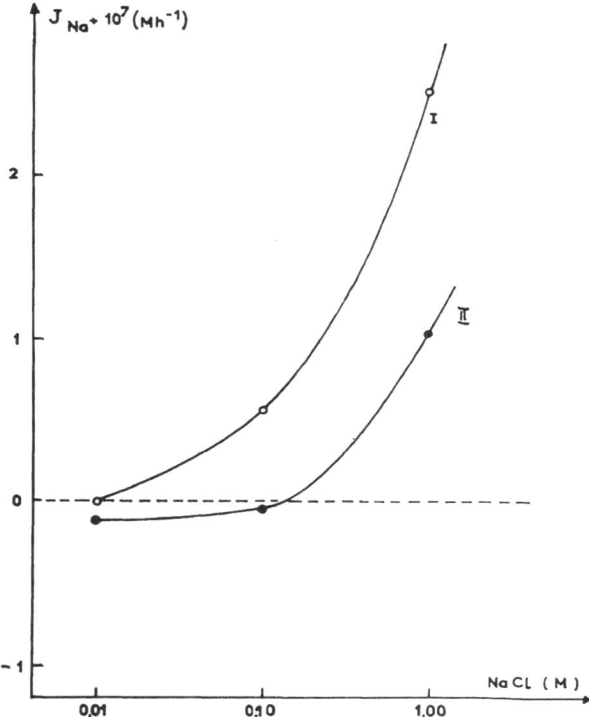

Fig. 11.   The kinetics of Na^{+} transference across the membrane as a function of NaCl concentration with the aqueous phases at identical pH ($\circ$, I) and with the aqueous phases at different pH ($\bullet$, II).

Solution (1)   Solution (2)
pH = 5   pH = 5

| $10^{-2} M$ | | Membrane
| $10^{-1} M$ | NaCl | $10^{-2} M$ NaCl
| $1 M$ | |

System II

Solution (1)   Membrane   Solution (2)
pH = 3   pH = 5

| $10^{-2} M$ | |
| $10^{-1} M$ | NaCl | $10^{-2} M$ NaCl
| $1 M$ | |

System III

The framed concentration was varied and is represented as the abscissa in Figure 11. In the presence of a pH difference – system III – the net sodium flux is negative for 10^{-2} M and 10^{-1} M NaCl in the left compartment. Under such circumstances, as hydrogen ions flow in the direction of decreasing hydrogen ion concentration, sodium ions are driven from the right to the left compartment, i.e., from the dilute to the concentrated side. A phenomenological description of coupled flows is readily obtained from the thermodynamics of non-equilibrium phenomena. The salt flux J_1 $(1 \equiv \text{ClNa})$ and the acid flux J_2 $(2 \equiv \text{ClH})$ across a unit area of membrane are related to the differences of the chemical potentials μ_1' and μ_2' of the electroneutral components in the aqueous solutions by linear relationships:

$$
\begin{aligned}
J_1 &= L_{11}\, \Delta\mu_1' + L_{12}\, \Delta\mu_2' \\
J_2 &= L_{21}\, \Delta\mu_1' + L_{22}\, \Delta\mu_2' \\
L_{11} &\geqslant 0,\ L_{22} \geqslant 0,\ L_{12} = L_{21}
\end{aligned}
\tag{42}
$$

The coupling parameter L_{12} is expected to be negative for carrier mediated flow [45], so that with suitable values of $\Delta\mu_1'$ and $\Delta\mu_2'$ it is possible for J_1 to have a direction opposite to that expected from the sign of $\Delta\mu_1'$. We shall admit that the rate of chemical reactions in the membrane interior is much faster than that of diffusion, and also that chemical equilibrium applies at both membranes interfaces*:

$$
(RNa)_x\, (RH)_y + NaCl \rightleftarrows (RNa)_{x+1}\, (RH)_{y-1} + HCl
$$

$$
\mu_{RNa} + \mu_{HCl} = \mu_{RH} + \mu_{NaCl}
\tag{43}
$$

$$
\mu_i(\delta) = \mu_i'(\delta)\,;\ \mu_i(0) = \mu_i'(0),\ (i = 1, 2)
\tag{44}
$$

---

* This *a priori* statement need not trouble us here, for diffusions through 'thick' membranes are much slower than ion-exchange reactions. A much more complicated situation would be encountered in transport through thin liquid or bilayer membranes. In such systems chemical equilibrium in the membrane may not be fulfilled, and the concentration profiles of the diffusing components will not then be linear as in the present case. Moreover, the membrane-liquid interfacial properties might also be associated with the transport of ions and the description given here would not be suitable [46, 47].

In equation (43) the chemical potentials μ_{RNa} and μ_{RH} refer to one equivalent of the polymer COONa or COOH groups respectively and μ_{NaCl}, μ_{HCl} are the chemical potentials of the free electrolytes at a given location. Equation (44) express the thermodynamic equilibrium condition for electrolytes NaCl and HCl at interfaces δ and 0, the primes referring to the aqueous phases.

The transport of Na^+ and H^+ in the membrane is due to carrier and free electrolyte transport contribution:

$$J_1 = \mathscr{L}_{11} \nabla\mu_1 + \mathscr{L}_{1^*1^*} \nabla\mu_{1^*}$$

$$J_2 = \mathscr{L}_{22} \nabla\mu_2 + \mathscr{L}_{2^*2^*} \nabla\mu_{2^*}$$
(45)

In Equation (45), the $\mathscr{L}_{ii}$ coefficients are locally defined in the membrane phase and $1^* \equiv RNa$, $2^* \equiv RH$. Cross coefficients and free ion transport is neglected in (45). System (45) merely states that the fluxes of each electroneutral species is proportional to the chemical potential gradient of the ion paired complexes 1, 1*, 2 and 2*.

Defining the average transport parameter $\overline{\mathscr{L}}_{ii}$ by

$$\overline{\mathscr{L}}_{ii} = \frac{1}{\delta\Delta\mu i} \int_0^\delta \mathscr{L}_{ii} \, d\mu_i, \ (i = 1, 2, 1^*, 2^*)$$
(46)

and noting that the carrier mechanism implies that

$$\mathscr{L}_{1^*1^*} \nabla\mu_{1^*} = -\mathscr{L}_{2^*2^*} \nabla\mu_{2^*}$$
(47)

it is easy to combine Equations (42–47) and to define the L_{11}, L_{22}, L_{12} coefficients in terms of the $\overline{\mathscr{L}}_{ii}$:

$$L_{11} + L_{12} = \overline{\mathscr{L}}_{11}, \ L_{22} + L_{21} = \overline{\mathscr{L}}_{22}$$

$$L_{12} = L_{21} = -\frac{\overline{\mathscr{L}}_{1^*1^*} \times \overline{\mathscr{L}}_{2^*2^*}}{\overline{\mathscr{L}}_{1^*1^*} + \overline{\mathscr{L}}_{2^*2^*}}$$
(48)

Since all $\overline{\mathscr{L}}_{ii}$ are positive quantities, L_{12}, L_{21} are negative by Equation (48) and the condition of incongruent transport $J_i < 0$ is:

$$\frac{\Delta\mu_2}{\Delta\mu_1} > 1 + \overline{\mathscr{L}}_{11}/\overline{\mathscr{L}}_{2^*2^*} + \overline{\mathscr{L}}_{11}/\overline{\mathscr{L}}_{1^*1^*}$$
(49)

If one makes the simplifying assumption that in the membrane phase the activities may be replaced by mole fractions, one may relate the transport parameter $\overline{\mathscr{L}}_{ii}$ to the concentrations and diffusion coefficients and (49) reduces then to the manifest condition:

$$\int_\delta^0 D_1 \, dn_1 < C_p X \int_0^\delta D_p \, d\alpha$$
(50)

D_1 and D_p being the diffusion coefficients of electrolyte 1 and of the polymer respectively at concentration n_1, C_p and composition α. X is the total number of monomeric units. Calculations of both terms in Equation (50) was performed for situation (II) using the following parameter values:

$$D_1 = 0.45 \times 10^{-6} \text{ cm}^2 \text{ s}^{-1}; \quad C_p = 5 \times 10^{-8} \text{ m g}^{-1}; \quad X = 1040$$
$$n_1(10^{-2} \text{ M}) = 2.9 \times 10^{-9} \text{ m g}^{-1}; \quad n_1(10^{-1} \text{ M}) = 3.0 \times 10^{-8} \text{ m g}^{-1}$$
$$n_1(1 \text{ M}) = 4.9 \times 10^{-7} \text{ m g}^{-1}$$

The n_1 values were taken from Table I, D_1 was measured in N-octanol by the capillary technique [48] and was found independent of the salt concentration. On the other hand, D_p is inversely proportional to the average polymer dimension according to formula (52):

$$D_p = \frac{kT}{5.1 \, \eta_0 (\overline{R^2})^{\frac{1}{2}}} \tag{52}$$

where η_0 is the viscosity of the solvent. The integral of the r.h.s. of Equation (50) can be evaluated numerically using (52) with the dependence of $(\overline{R^2})^{\frac{1}{2}}$ on α given in Figure 4. In our experiment III, we purposely used a rather small pH difference so that the mean value $D_p = [D_p(x=0) + D_p(x=\delta)]/2$ can be taken as a first approximation in evaluating the r.h.s. integral in Equation (50). In Table IV are compiled both the carrier and the free electrolyte diffusional flow contributions. The carrier contribution amounts to 3.1×10^{-14} approximately.

TABLE IV

Free electrolyte diffusion and carrier transport contribution

m_1 (aq. phase)	$D_1 [n_1(0) - n_1(\delta)]$	$C_p \times \int_0^{\delta} D_p \, d\alpha$
10^{-2} M	$\simeq 0$	3.1×10^{-14}
10^{-1} M	1.2×10^{-14}	3.1×10^{-14}
1 M	22×10^{-14}	3.1×10^{-14}

3.2.3. $Na^+ - Ca^{2+}$ Coupling Effects

In the preceding paragraph we described carrier mediated coupling effects for the H^+, Na^+ species. One may of course observe similar strong coupling effects arising between other ions coexisting in the aqueous solution, when these ions complete with different affinities to bind on the membrane macromolecule.

A typical example of coupled carrier flow effects for $Na^+ - Ca^{2+}$ ions is illustrated in Figure 12 for system IV

Solution (1)		Solution (2)
pH = 5	Membrane	pH = 5
10^{-2} M NaCl		10^{-2} M NaCl
5×10^{-4} M CaCl$_2$	System IV	

At the beginning of the experiment the system was symmetrical with 10^{-2} M NaCl and pH = 5.0 aqueous phases on both sides. Solution (1) was rendered radio-active with Na^{22} and the increase of activity with time in compartment (2) was re-

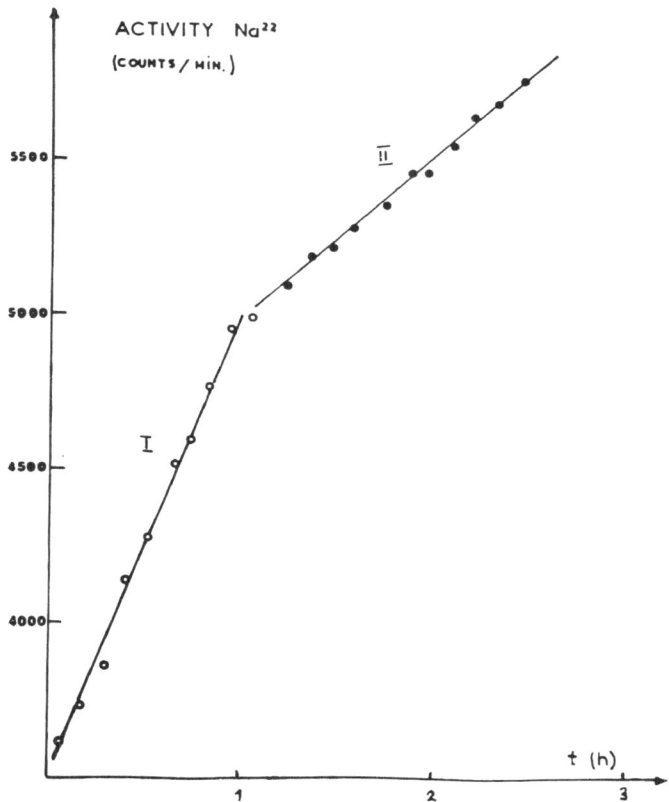

Fig. 12.   The kinetics of Na$^+$ transference across the membrane in the absence of Ca^{2+} ($\circ$, I), and
in the presence of 5×10^{-4} M, Ca^{2+} ($\bullet$, II).

corded and resulted in the straight line (I) in Figure 12. After the addition of the small
amount of 5×10^{-4} M CaCl$_2$ to solution (1), the Na^{22} flux decreased by a factor of
about three and changed to line (II), the unidirectional fluxes of Na$^+$ (from left to
right) being respectively 1.3×10^{-8} mole h^{-1} in (I) and 0.46×10^{-8} moles h^{-1} in (II).
During this experiment, the pH of both aqueous phases was constantly adjusted to 5.

In experiment IV, calcium ions, though at low concentration in water, occupy most
of the polymer sites and therefore impede the sodium ion flux. The possible biological
significance of sodium-calcium interferences in living system has been emphasized in
an number of papers [12], [49]. In particular, ion flux changes as produced by a
change in the ratio of univalent cations to divalent cations in the outer solution of the
physiological membrane have been reported. However, biological properties have been
explained on the basis of a conformational transition determined primarily by the
univalent/divalent cation ratio. In our dilute systems, the effects of the addition of
calcium on sodium transport is based upon ionic selectivities toward carrier molecules.
The conformational change of the macromolecular system, though important, is not
the most important factor in this case. However, the change of univalent-divalent

cation ratio in the outer solution may possibly produce phase transitions in more concentrated systems and could then considerably modify the transport properties of hydrophobic membranes. In fact, X-ray, diffraction studies of aqueous gels of the polyelectrolyte have shown the existence of different phases with different conductivity properties [8], [9]; hence, a structural and kinetic study of the properties of such membranes as a function of polymer concentration and ionic composition would also be of interest in the future.

Notation

List of the principal symbols used in the paper:

C_p — concentration of polymer in the organic solution g g^{-1}

m — concentration of electrolyte in moles per unit volume in the aqueous phase

M, M_w, molecular weight and weight average molecular weight

ρ_s — density of organic solution g cm^{-3}

$\bar{V}_\rho$ — partial specific volume of the polymer (including hydration water)

$\bar{V}$ — partial molar volume of one polymeric segment

W, O, R: subscripts for water, octanol and polyion components

RA, RB, RH: subscripts for association complexes of polymer and ions

α — stoichiometric degree of neutralization

$\alpha_{i,A}$ — degree of ionization relative to species A

z_i — positive valence of the ith ionic species

n_i — concentration of the ith component including ions and electro-neutral components in moles per one equivalent of exchange capacity of the polymer

β_i, β_i' — equivalent ionic fractions of the ith ion in octanol and water phases respectively

a_i, a_i' — activities of component i in octanol and water phases respectively

μ_i, μ_i' — chemical potentials of component i

f_i — activity coefficient

K_B^A — thermodynamic equilibrium constant relative to the exchange of ion A and B

K_{aB}^A — apparent thermodynamic constant defined by Equation (26)

K_{cB}^A — 'corrected' selectivity coefficient

G — Gibbs free energy of the organic phase per equivalent exchange capacity of the polymer

$G_m^E, G_{m,o}^E$ excess Gibbs free energy of the electroneutral polymer and electroneutral polymer + octanol molecules respectively

G_M — free energy change of mixing polymer and solvent molecules

χ — total numbers of segments in one polymer molecules

ρ — number of polymer segments per unit volume

γ — expansion coefficient used in Flory's theory

$\overline{R_0^2}$ — mean square distance from the center of gravity of the polymer coil in the unperturbed state

b — bond lengths between two successive segments in the polymer coil

χ_{ij} — Flory's interaction parameter in the lattice model theory

$[\eta]$ intrinsic viscosity in cm^3 g^{-1} of polymer solution
ϕ universal constant in the viscosity molecular weight relationship
$\mathscr{N}$ Avogadro number

Appendix I

The equivalent conductance Λ_p of the polymer is given by the following equation

$$\Lambda_p = \frac{\mathscr{F}^2}{F_p}[(1-\alpha)\alpha_{i,H} + \alpha\alpha_{i,A}] + (1-\alpha)\alpha_{i,H}\Lambda_H^0 + \alpha\alpha_{i,A}\Lambda_A^0 \tag{A.I}$$

The first bracket on the r.h.s. is the polyion contribution, the second bracket is the free metal and hydrogen ion conductance contribution.

F_p is the frictional coefficient of the polymer in octanol, which is related to the intrinsic viscosity $[\eta]$ by the classical expression (ref. [26], pp. 362 and 399):

$$F_p = \frac{6^{3/2}\pi\eta_0}{3.7}([\eta] M_w/\phi)^{1/3} \tag{A. II}$$

η_0 is the solvent viscosity, values of $[\eta]$ being given in part 2.3.5. The limiting conductances Λ_A^0 and Λ_H^0 were obtained from the conductances of the alkali-halides in water saturated octanol solutions by applying the method of Fuoss and Shedlovsky in ref. [24], Equation (9).

The two unknown parameters $\alpha_{i,H}$ and $\alpha_{i,A}$ are readily obtained by combining $\Lambda_p(\alpha)$ and $\Lambda_p(\alpha = 0)$.

Appendix II

The variation of $d(\Delta G_M)$ are subject to the restrictions of constant number of metal ions, Equation (35), and constant number of total monomers. Multiplying Equation (35) by a constant A

$$A \int_0^\infty \alpha(r)\delta\rho \, d\tau + A \int_0^\infty \rho \, \delta[\alpha(r)] \, d\tau = 0 \tag{A. III}$$

and adding (A. III) to (34), the equilibrium conditions at a specified thermodynamic state are defined by:

$$\left[\frac{\partial\delta(\Delta G_M)}{\partial\alpha(r)}\right]_\gamma + A\rho \, d\tau = 0$$

$$\int_0^\infty \left\{\left[\frac{\partial\delta(\Delta G_M)}{\partial\rho}\right]_{\alpha(r)} + A\alpha(r) \, d\tau\right\}\frac{\partial\rho}{\partial\gamma} = 3\left(\frac{1}{\gamma}-\gamma\right)$$

or if $\delta(\Delta G_M)$ is known

$$(1-\rho\bar{V})[\Delta\chi + 2\chi_{RA,RH} \, \alpha(r)] + \ln\frac{\alpha(r)}{1-\alpha(r)} + A = 0 \tag{A. IV}$$

$$\int_0^\infty \rho\alpha(r)\,d\tau = \alpha X \tag{A. V}$$

$$\int_0^\infty \{\rho\bar{V}[1-2\chi_{S,\,RH}-\Delta\chi\alpha(r)]-\chi_{RA,\,RH}\,\alpha^2(r)\}\,\frac{\partial\rho}{\partial\gamma}\,d\tau$$

$$+\int_0^\infty \ln\,[1-\alpha(r)]\,\frac{\partial\rho}{\partial\gamma}\,d\tau = 3\left(\frac{1}{\gamma}-\gamma\right) \tag{A. VI}$$

where $\Delta\chi$ stands for

$$\Delta\chi = \chi_{S,\,RH}-(\chi_{S,\,RH}+\chi_{RA,\,RH})$$

The constant A of Equation (A. IV) may be computed for any arbitrary value of $\alpha(r=0)$, subject to the condition $0\langle\alpha(r=0)\langle1$. Equation (A. IV) may than be solved numerically, writing

$$\alpha(r) = \frac{(\rho\bar{V}-1)\Delta\chi-A-\ln\dfrac{\alpha(r)}{1-\alpha(r)}}{2(1-\rho\bar{V})\chi_{RA,\,RH}}$$

If the r.h.s. is plotted agaiń́s $\alpha(r)$ the true value is obtained from the intersection of this plot with a plot of $\alpha(r)$ against $\alpha(r)$ on the same graph.

The $\Delta\chi$ value is obtained as follows. From Equation (A. IV) and (A. V) it is possible to show that if $\alpha \to 0$, than $\alpha(r) \to 0$. Noting furthermore that $\rho\bar{V} < 1$, $\alpha(r)$ for small α values is given by:

$$\alpha(r) = \exp\,(-A)\exp\,(-\Delta\chi)\left[1 + \rho\bar{V} + \frac{(\rho\bar{V})^2}{2} + ...\right]^{\Delta\chi} \tag{A. VII}$$

Combining (A. IV), (A. V) and (A. VII) the value of α for small values is obtained:

$$\exp\,(-A) = \frac{\alpha\,\exp\,\Delta\chi}{1 + \bar{V}\Delta\chi XC\gamma^{-3}},\quad C = \left[\frac{3}{4\pi R_0^2}\right]^{3/2} \tag{A. VIII}$$

If Equations (A. VII), (A. VIII) are combined with (A. VI) one obtains after integration:

$$\gamma^5 - \gamma^3 = C\bar{V}X^2\,[0.5-\chi_{S,\,RH}-\alpha\Delta\chi] \tag{A. IX}$$

Near $\alpha = 0$, γ may be expanded into a series in powers of α:

$$\gamma = \gamma_0 + \left(\frac{\partial\gamma}{\partial\alpha}\right)_{\alpha\,=\,0}\alpha + ...$$

If the expression for γ is introduced into (A. IX), $\Delta\chi$ is obtained in a closed form:

$$\Delta\chi = (3\gamma_0^2 - 5\gamma_0^4)\left(\frac{\partial\gamma}{\partial\alpha}\right)_{\alpha\,=\,0} / \ C\bar{V}X^2$$

It is seen that $\Delta\chi$ is related to the slope of $\gamma_{\exp}$ against α at the origin. Knowing $\Delta\chi$,

$\chi_{RA,RH}$ is computed according to Equation (A. IV). γ_{exp} was computed from the viscosity data.

References

1. Rieman III, W. and Walton, H. F.: *Ion Exchange in Analitical Chemistry*, Pergamon Press, 1970, p. 226.
2. Högfeldt, E.: *Ion Exchange*, vol. 1 (ed. by J. A. Marinsky), Marcel Dekker, Inc., N.Y., 1966, p. 139.
3. Sollner, K.: *Annals N.Y. Acad. Sci.* **148**, 154 (1968).
4. Eisenman, G.: *Anal. Chem.* **40**, 310 (1968).
5. Lauger, P.: *Ang. Chem.* **81**, 56 (1969).
6. Lauger, P.: *Science* **178**, 24 (1972).
7. Eigen, M. and Winkler, R.: *De la Physique théorique à la Biologie*, Editions du C.N.R.S., 1971, p. 17.
8. Mathis, A., Varoqui, R., Skoulios, A., and Schmitt, A.: *Europ. Polymer J.* **10**, 1011 (1974).
9. François, J., Varoqui, R., and Schmitt, A.: *C.R. Acad. Sci. Paris* **270**, 788 (1970).
10. Wojtczak, Z., Strazielle, C., Varoqui, R., and Benoît, H.: *Europ. Polym. J.* **6**, 247 (1970).
11. Leo, A., Hansch, C., and Elkins, D.: *Chem. Rev.* **71**, 525 (1971).
12. Singer, J. and Tasaki, S.: *Biological Membranes, Physical Fact and Function* (ed. by D. Chapman), Academic Press, N.Y., 1968, p. 347.
13. Varoqui, R. and Strauss U.P.: *J. Phys. Chem.* **72**, 2507 (1968).
14. Eisenberg, H. and Casassa, E. F.: *J. Polymer Sci.* **47**, 29 (1960).
15. Eisenberg, H.: *J. Chem. Phys.* **36**, 1837 (1962).
16. Stabinger, H., Leopold, H., and Kratky, O.: 'Digital Densimeter DMA 002', Institute for Physical Chemistry, University of Graz, Austria.
17. Gaines, G. L. and Thomas, H. C.: *J. Chem. Phys.* **21**, 714 (1953).
17bis. Sélégny, E., Metayer, M., and Merle, Y: *C. R. Acad. Sci. Paris* **266**, 157 (1968). Merle, Y.: *Europ. Polymer J.* **8**, 1265 (1972).
18. Holm, L. W.: *Ark. Kemi* **10**, 151 (1956).
19. Robinson, R. A. and Stokes, R. H.: *Electrolyte Solutions*, Butterworth and Co. Ltd, London (1955).
20. Rosseinsky, D. R.: *Chem. Rev.* **65**, 467 (1965).
21. Diamond, R. M. and Whitney, A. C.: *Ion Exchange*, vol. 1 (ed. by J. A. Marinsky), Marcel Dekker, Inc., N. Y., 1966 p. 277.
22. Fuoss, R. M. and Kraus, C. A.: *J. Am. Chem. Soc.* **55**, 1019 (1933).
23. Gilkerson, W. R.: *J. Chem. Phys.* **26**, 1199 (1956).
24. Fuoss, R. M. and Shedlovsky, T.: *J. Am. Chem. Soc.* **71**, 1496 (1949).
25. Marchall, E.: personal communication.
26. Flory, P. J.: *Principles of Polymer Chemistry*, Cornell University Press, Ithaca, N. Y., 1953, Chapter 14.
27. Haynes, D. H., Kowalsky, A., and Pressman, B. C.: *J. Biol. Chem.* **244**, 502 (1969).
28. Haynes, D. H., Pressman, B. C., and Kowalsky, A.: *Biochemistry* **10**, 852 (1971).
29. Eisenman, G., Ciani, S. M., and Szabo, G.: *Fed. Proc.* **27**, 1289 (1968).
30. Kirch, M.: Thesis, Strasbourg, 1974.
31. Froehlich, D.: Thesis, Strasbourg, 1966.
32. Tanford, C.: *Physical Chemistry of Macromolecules*, John Wiley and Sons, Inc., N. Y., 1966, p. 403.
33. Muller, P. and Rudin, D. O.: *Biochem. Biophys. Res. Comm.* **26**, 398 (1967).
34. Eisenman, G., Mc Laughin, S. G., and Szabo, C.: Symposium on Physical Chemical Bases of Ion Transport through Biological Membranes, Riga, U.S.S.R., 1970; and this volume, p. 107.
35. Tosteson, D. C.: *Fed. Proceedings* **27**, 1269 (1968).
36. Wipf, H. K., Pache, W., Jordan, P., Zähner, H., Keller-Schierlein, W., and Simon, W.: *Biochem. Biophys. Res. Comm.* **36**, 387 (1969).
37. Pedersen, C. J.: *J. Am. Chem. Soc.* **89**, 7017 (1967).

38. Mc Laughin, S. G., Szabo, G., Ciani, S., and Eisenman, G.: *J. Membrane Biol.* **9,** 3 (1972).
39. Sollner, K.: *Diffusion Processes*, Proceedings of the Thomas Graham Memorial Symposium, University of Strathclyde (2), Gordon and Breach, 1971.
40. Cussler, E. L., Fennell Evans, D., and Matesich, M. A.: *Science* **172,** 377 (1971).
41. Moore, J. H. and Schechter, R. S.: *Nature* **222,** 476 (1969).
42. Patterson, R.: *Nature* **217,** 545 (1968).
42bis Sélégny E.: in *Polyelectrolytes*, Reidel Publ. Dordrecht, Holland, 1974, p. 459.
43. Rosano, H. L., Duby, P., and Schulman, J. H.: *J. Phys. Chem.* **6,** 1704 (1961).
44. Meares, P. and Sutton, A. H.: *J. Colloid Interface Sci.* **28,** 117 (1968).
45. Caplan, S. R.,: *J. Theor. Biol.* **10,** 209 (1966); *ibid.* **11,** 346 (1966).
46. Sandblom, J., Eisenman, G., and Walker Jr., J. L.: *J. Phys. Chem.* **71,** 3862 (1967).
47. Sandblom, J.: *J. Phys. Chem.* **73,** 249 (1969).
48. Pefferkorn, E. and Varoqui, R.: *European Polym. J.* **6,** 663 (1970).
49. Gilbert, D. L. and Ehrenstein, G.: *Biophys. J.* **9,** 447 (1969).

V

OXIDATION–REDUCTIONS

SYNTHESIS, ELECTROCHEMISTRY AND APPLICATION OF SOME OXIDATION-REDUCTION POLYMERS

GEORG MANECKE

Institut für Organische Chemie
Freie Universität Berlin und Fritz-Haber-Institut der Max-Planck-Gesellschaft, Berlin

Abstract. Polymers containing redox systems which can be reversibly oxidized resp. reduced are called oxidation-reduction polymers or electron transfer polymers. The chemical structure of the incorporated redox systems can be varied widely. They may be parts of the polymer backbone or may be substituents on the polymer chain.

This review summarizes our work on the synthesis and redox properties of polymers with various quinone/hydroquinone systems. Information is given on our attempts to increase the chemical stability of monomers and the corresponding redox resins by suitable substitution and other methods. The redox properties were thoroughly investigated (e.g. redox potentials, redox capacity, rate of the redox reaction). A theoretical treatment is given for the redox behaviour of tetra- and hexavalent redox systems.

Redox resins can be used as insoluble, regenerable oxidants resp. reductants. By means of these resins some organic substrates could be dehydrogenated and several metal ions could be reduced or deposited. Redox resins in the form of films can be regarded as models of membranes allowing transport of protons coupled with the exchange of electrons.

Further applications seem to be promising.

1. Introduction

With the aim to synthesize membranes which contain reversible redox systems we intended to build up a system which allows a transport of protons through this membrane. This transport should occur by performing an oxidation on one side of the membrane and a reduction on the other side. Since several years we have synthesized polymer redox systems and have investigated their electrochemical properties. Polymer redox systems are known by the names oxidation-reduction polymers, redox polymers, redox resins and electron-transfer polymers [1, 2, 3].

2. Synthesis of Redox Resins

Redox resins are water-swellable polymers which can be reversibly oxidized or reduced. The redox systems can be links of the polymer backbone or they can be substituents of the chain. The chemical structure of the redox systems can be varied widely.

Redox polymers can be synthesized by all known classical methods of macromolecular chemistry (e.g. polymerization, condensation, polymer analogous reaction).

Examples of addition polymers are the redox resins on the basis of vinylhydroquinone [4]. As the hydroquinone/quinone system inhibits polymerization, protecting groups (e.g. acetate, benzoate, ether) are usually introduced. These groups can be quantitatively removed after the polymerization. To improve the properties of the

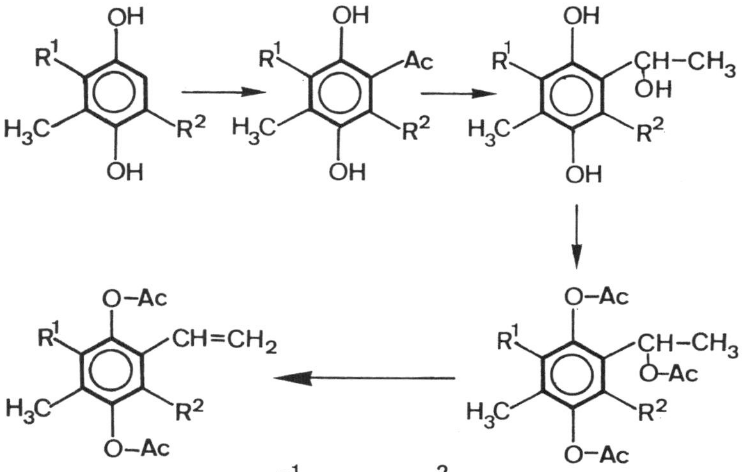

R^1	R^2
H	H
CH₃	H
H	CH₃
CH₃	CH₃

Fig. 1.

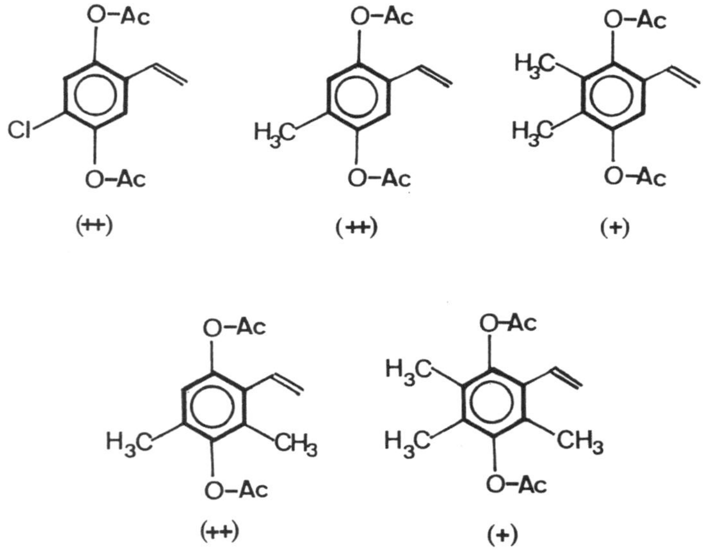

(++) = good polymerizability (80-90% yield)

(+) = bad polymerizability (approx. 20% yield)

Fig. 2.

redox resins vinylhydroquinone (its hydroxyl groups were blocked by acetylation) was copolymerized with styrene and divinylbenzene (DVB) as crosslinking agent. The copolymers were converted into waterswellable polymers by sulfonation. Their redox system was restored by splitting off the protecting groups.

The redox reaction of some redox resins can involve to a certain extent an irreversible oxidation of the polymer, that means the redox polymer is partially destroyed. This could be caused by radical attack on the hydrogen atoms of the unsubstituted ring of the hydroquinone/quinone system during oxidation.

Better stabilities were expected from methylsubstituted hydroquinone quinone systems [5–8]. So we synthesized several methylated vinylhydroquinone acetates and investigated their polymerizability.

We found out that the monomers with methyl groups at position 5 (meta position to the vinyl group) polymerize badly [9]. The polymerization was radically initiated by benzoyl peroxide or α, α'-azo-bis-isobutyronitrile in n-butanol.

Redox resins produced from 1-vinyl-2, 4-dimethyl-3, 6-diacetoxy-benzene by the usual procedure had a good chemical resistance [9]. Their redox capacity was approx. 2.1 meq/g, that means 1 g of dry resin reduces 2.1 mmol of Fe^{3+} to Fe^{2+}. As the resin contains SO_3H-groups, the redox resin showed an ion-exchange capacity (3.5 meq/g). By incorporating methyl substituents as electron-donating groups into the hydroquinone-quinone system the redox potential decreases. Therefore, depending on the substitution, redox resins can be produced with different redox potentials.

Another possibility of stabilizing the hydroquinone-quinone redox system can be achieved by annelation the quinone system with aromatic rings. 2-Methyl-3-vinylnaphthoquinone [10] was synthesized, which polymerized badly as hydroquinone diacetate. Therefore, the 2, 3-dimethyl-5-vinylnaphthoquinone [11] was prepared. Un-

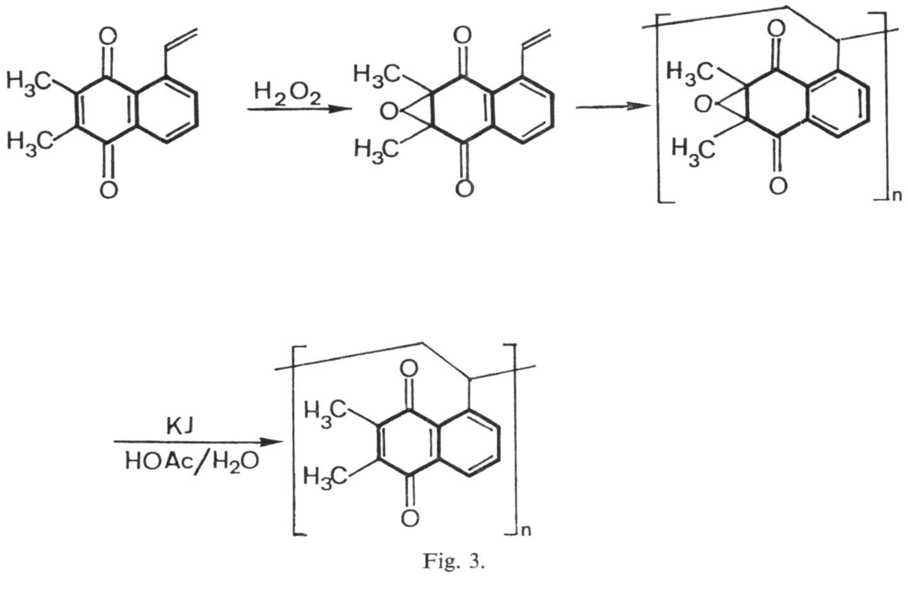

Fig. 3.

fortunately this monomer and its hydroquinone derivates were not polymerizable. It was found, however, that an epoxidation of the vinylquinone yielded a compound which could easily be copolymerized. The polymer quinone epoxide was converted into the corresponding poly(vinylnaphthoquinone) by using KJ in aqueous acetic acid.

The following polymerizable vinylnaphthohydroquinone diacetates were also synthesized [12, 13].

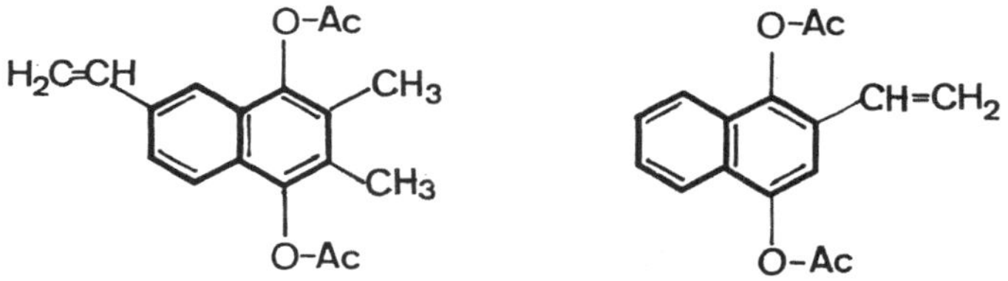

Fig. 4.

By annelating both sides of the benzoquinone the very stable anthraquinone system is obtained. Both vinyl-anthraquinone isomers were synthesized [14].

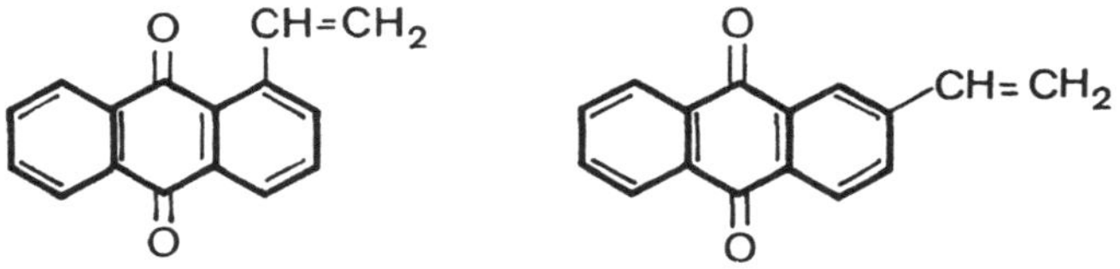

Fig. 5.

The 1-vinylanthraquinone polymerized very badly. The 2-vinylanthraquinone could be polymerized and copolymerized with styrene and DVB [15]. The copolymerization had to be performed in DMSO by thermal initiation, due to the bad solubility of this monomer. After sulfonation resins were obtained with redox capacities up to 4.76 meq/g and ion exchange capacities from 2.0 to 4.5 meq/g. The midpoint redox potential of this redox resin was apprcx. 127 mV (pH 1.09). Also macroreticular resins could be prepared with this monomer. Although the anthraquinone-redox resin had a relatively good chemical resistance, it was noticed that hydrogen peroxide attacks the resin and decreases the crosslinking.

To improve the oxidation resistance of the polymer chain we synthesized 2-isopropenylanthraquinone [16].

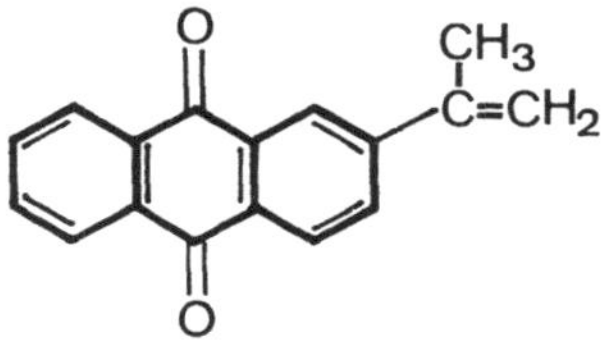

Fig. 6.

We obtained stable resins [17] by copolymerizing this monomer with styrene and DVB or acrylic or methycrylic acid. The resins could be prepared in beadform and also with a macroreticular structure because the monomer had a good solubility. These redox resins are now commercially available [18].

A large amount of model compounds and polymers in which hydroquinone groups and derivates are linked by the sulfone function have been prepared in our laboratory [19–24].

Also hydroquinonesulfonamides were included in our investigations. We prepared a lot of model substances and polymers which were potentiometrically measured [25–27].

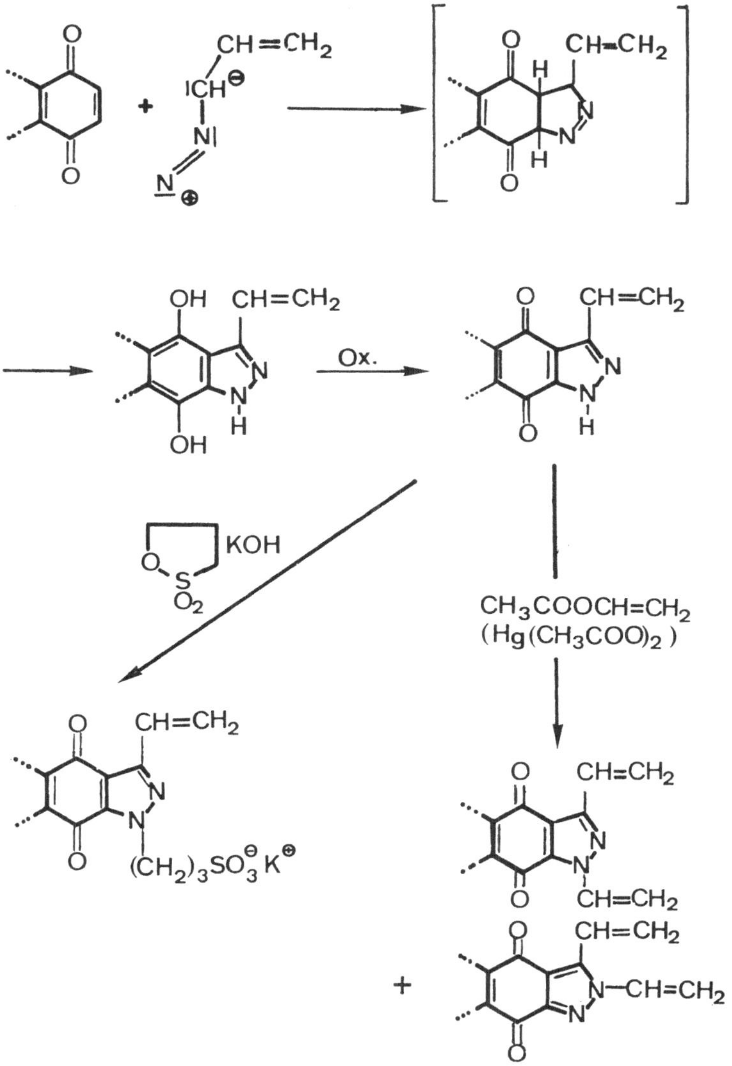

Fig. 7.

Most of the redox resins which contain sulfone and sulfoneamide groups had not perfect properties. Our investigations have shown that substituted pyrazoloquinones yield very stable redox polymers [28–31]. Vinylpyrazoloquinones can be prepared by 1, 3-dipolar cycloaddition of diazopropen-2 to suitable quinones. Because of their acid N-H pyrazole function these compounds can be sulfalkylated with excellent yields or N-vinylated with vinyl acetate.

Most of the N-vinylated pyrazoloquinones can not be polymerized by radicals unless an epoxy group is introduced into the quinone moiety. This epoxy group can be reduced later with KJ in acetic acid as it was mentioned of 2, 3-dimethyl-5-vinylnaphthoquinone.

3-Vinyl-1H-benzo [ƒ] indazole-4, 9-dione and the cross-linking agent [31] 3, 7-divinyl-1H, 5H-pyrazolo [5, 4-ƒ] indazole-4, 8-dione were copolymerized after being sulfalkylated. The swellability of the resulting product was too high, leading to bad mechanical stability. Resins with much better mechanical stabilities were obtained from these monomers performing the sulfalkylation *after* the copolymerization.

Most of the polymers of this type have low redox potentials. Therefore, we tried to synthesize polymers with higher potentials [28, 32] starting e.g. with 5, 6-dichloro-3-

TABLE I

Some N-vinylpyrazoloquinone derivates and their polymerizability[a]

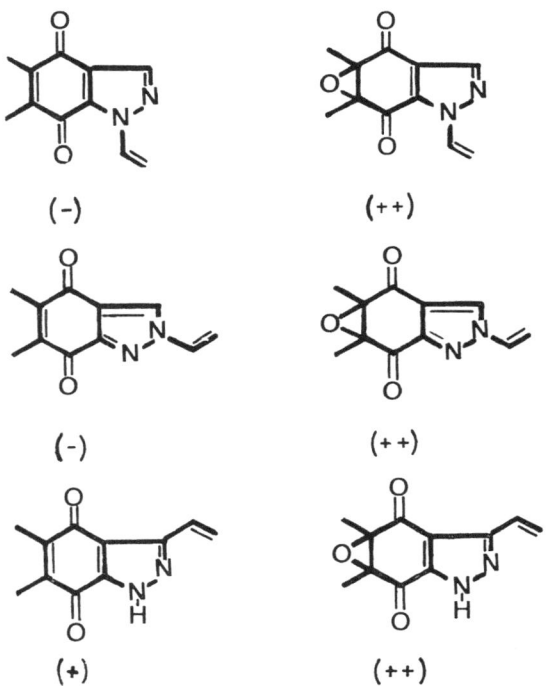

(-)	(++)
(-)	(++)
(+)	(++)

[a]as indicated in brackets.

vinyl-1H-indazole-4, 7-dione. The swelling volume of these redox resins changes only slightly within a broad pH range. This is very favorable for column application.

Good results are achieved with some new nitrogen-containing redox polymers [28, 33]. As the quinone system of some vinylquinones inhibits itself their radical poly-merization we tried to find another way to neutralize this inhibition. By introducing the aziridinyl group into certain vinylated quinones we obtained monomers which can easily be polymerized by cations. Uniform products were obtained by adding aziridine to the vinyl group of some vinylanthraquinones or vinylpyrazoloquinones. The reaction was performed in sealed tubes with a large excess of aziridine.

Vinylquinones not having a completely substituted quinone nucleus can also add aziridine to the quinone nucleus by 1, 4-addition [28, 34].

1-[2(1-aziridinyl) ethyl]-anthraquinone (I) obtained from 1-vinylanthraquinone was copolymerized with various amounts of the crosslinking agent 1, 4-bis [2-(1-aziridinyl)-ethyl]-benzene (II) but the products were too soft. In order to obtain polymers with better mechanical properties, large amounts of the cross-linking agent were used. But this was detrimental to the swelling properties of the redox resins. By synthesizing a terpolymer with [2-(1-aziridinyl)-ethyl]-benzene (III), a polymer with both good strength and good swelling characteristics could be obtained.

Because these polymers possess tertiary amino groups an additional introduction of hydrophilic groups into the polymer was unnecessary. For practical application it is important that these resins have good chemical stability and a high redox capacity in addition to good swelling and mechanical properties in acid solutions. The poly-condensation of formaldehyde with phenol and di- or trihydroxybenzenes is an easy practicable route to redox resins [35, 36]. We investigated with special accuracy the polycondensates of phenol-formaldehyde-hydroquinone which yielded rather stable

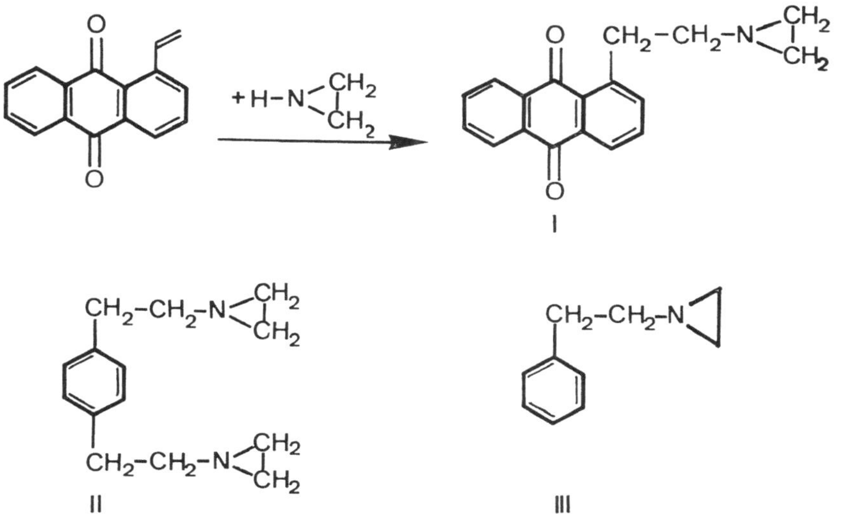

Fig. 8.

spongy redox resins with good redox characteristics. Also polycondensates of hy-
droxylated naphthoquinones and anthraquinones were prepared [37, 38].

Polymer analogous reactions are another possibility to synthesize redox resins, e.g.
poly (α-methylstyrene) was reacted with phthalic acid anhydride on Friedel-Crafts
conditions and cyclized with polyphosphoric acid. In this way we obtained a polymer
containing 8 mole% 2-iso-propenylanthraquinone units [39].

3. Electrochemical Properties

The main data for the characterization of redox resins are their redox capacities and
their redox potentials. Besides, the velocity of their redox reactions is an important
factor for their practical use. The redox capacities depend on the composition of the
polymers and are measured in meq/g resin. For estimates of their possible application
in oxidation-reduction reactions it is necessary to know their redox potentials. The
potentiometric titration of redox systems is a convenient method to determine their
redox potentials.

The redox process of the hydroquinone/quinone system can be described by the
following equation:

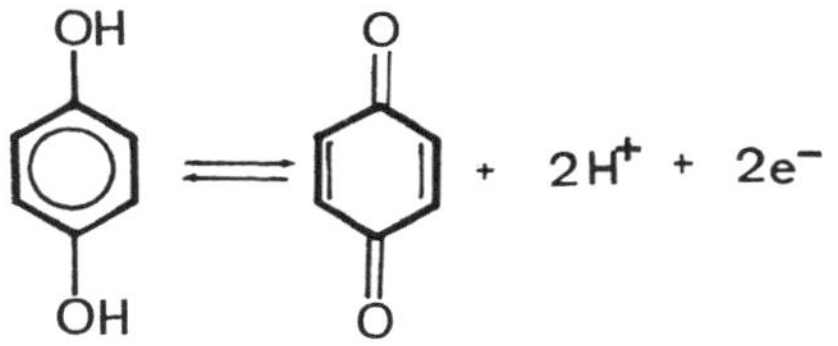

Fig. 9.

We shall use this equation in the following abbreviated form:

$$H \rightleftharpoons Q + 2e^-$$

The potentiometric titration curve of this system can be described by the simple
Nernst equation for processes consisting of a two-electron step.

$$E = E'_m + \frac{RT}{2F} \ln \frac{[Q]}{[H]}$$

E'_m is the pH-dependent midpoint potential which is measured when half the oxidant
necessary for complete oxidation has been used up (50% oxidation) ($E'_m = E_0 - 0.059 \, pH$).

The shape of the titration curve can be characterized by the index potential E_i, this
is the difference of the potentials at 50% and 25% oxidation or 50% and 75% oxidation [40]. A deviation of the limiting value of the index potential ($E_i = 14.1$ mV at
25 °C) indicates the presence of redox species other than H and Q e.g. of semiquinones.

The redox potential of a dissolved reversible redox system can be measured at an
inert electrode against the usual reference electrodes. For insoluble systems a direct
determination of the redox potential is impossible. Therefore, it is convenient in most

cases to add small amounts of a soluble redox system (called mediator) to the titra-tion mixture. The mediator must have a standard potential similar to the polymer and must reach equilibrium with the redox resin. In this case the determined redox potential of the dissolved system equals the redox potential of the polymer. The equilibration process within the redox resin usually takes a long time. Therefore, it is difficult to determine whether the equilibrium is quantitatively reached or not.

TABLE II

Potentiometric data of some tetravalent redox systems.
(acetic acid/water $=1:1$, v/v; 25°C; $K_2Cr_2O_7$) [41]

	E_{mh}' (mV)	E_i (mV)	$E_2'-E_1'$ (mV)	K
	587.2	21.3	39.6	21.9
	574.3	18.5	32.6	13.2[42]
	568.2	16.8	(17.8)	(4.0)
	575.3	15.2	21.9	5.5
	569.7	14.5	18.7	4.3
	621.3	14.2	–	–

As the redox process involves also protons, the redox potential is pH-dependent too. Because most of the redox resins contain ionogenic groups as hydrophilic parts ($-SO_3H$, $-COOH$; OH etc.) the distribution of the protons between the outer solution and the resin will be ruled by the Donnan equilibrium. This influences the potential of the resin in a complicated way, so that the E_m' of the redox resins are usually measured at definite conditions.

Polymer redox systems generally show higher midpoint potentials (E_m') than the corresponding monomer units. The slope of their potentiometric titration curves show a steeper increase and sometimes the curves are unsymmetrical. As there are many possible explanations for this behaviour, soluble model compounds of definite structure were synthesized and their redox behaviour was investigated.

Tetravalent redox systems with the following structure have been synthesized and measured potentiometrically (cf. Table II).

Tetravalent systems may be described by the following abbreviation

$$H \sim H \rightleftharpoons Q \sim Q + 4e^-$$

In the case of a four-electron step the slope of the titration curve would be very flat according to the Nernst equation with an E_i of 7.1 mV at 25 °C. Actually the measurements always result in an $E_i \geqslant 14.1$ mV (25°). So it can be assumed that the redox reaction consists of two bivalent steps, inserting an intermediate [43, 44].

$$\text{step 1: } H \sim H \rightleftharpoons I + 2e^-$$
$$\text{step 2: } \quad I \quad \rightleftharpoons Q \sim Q + 2e^-$$

From the Nernst equations at constant pH

$$E = E_1' - \frac{RT}{2F} \ln \frac{[I]}{[H \sim H]} \tag{1}$$

$$E = E_2' + \frac{RT}{2F} \ln \frac{[Q \sim Q]}{[I]} \tag{2}$$

$$E = \frac{E_1' + E_2'}{2} + \frac{RT}{4F} \ln \frac{[Q \sim Q]}{[H \sim H]} \tag{3}$$

and the redistribution reaction

$$Q \sim Q + H \sim H \rightleftharpoons 2I$$

$$\frac{[I]^2}{[H \sim H] [Q \sim Q]} = K \tag{4}$$

results equation (5)

$$E_2' - E_1' = \frac{RT}{2F} \ln K \tag{5}$$

Equation (5) connects the difference of the redox potentials of the individual bivalent steps with the equilibrium constant K.

E_1' und E_2' are the pH-dependent midpoint potentials of the both bivalent oxidation steps and

$$E_m' = \frac{E_1' + E_2'}{2}$$

is the pH-dependent measured midpoint potential of the total redox system.

As it is indicated in Equation (6) the constant K can be obtained from the titration curves by means of the index potential.

$$E_i = \frac{RT}{2F} \ln \left[1/2K^{1/2} + 1/2(K + 12)^{1/2} \right] \tag{6}$$

The limiting value for K is 4. This follows from considerations similar to those used for deriving the ratio of the first and second dissociation constants of dibasic acids which should be at least 4 [45]. This means that according to Equation (5) the difference $E_2' - E_1'$ should not be less than 17.8 mV (25°C). This limiting value will be obtained only if no interactions exist between the redox groups.

The maximal content of the intermediate I is given by

$$\left(\frac{[\text{I}]}{[\text{H} \sim \text{H}] + [\text{I}] + [\text{Q} \sim \text{Q}]} \right)_{max} = \frac{K^{1/2}}{2 + K^{1/2}} \tag{7}$$

Thus the species of the following tetravalent system were studied by NMR (CD_3—COOD, 100 °C) [41].

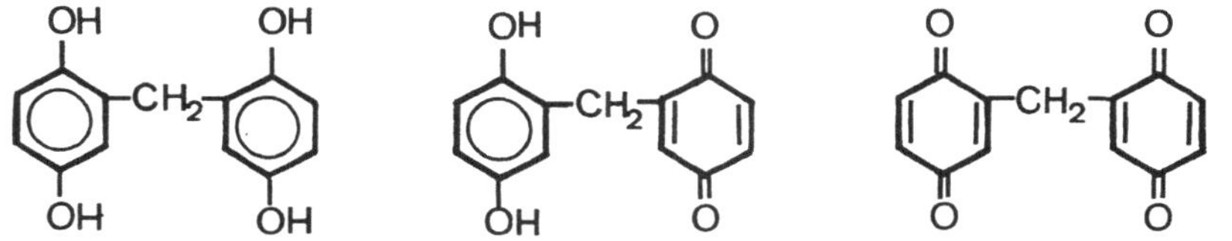

Fig. 10.

Table II shows the decrease of the index potential if the connecting chain between the redox systems is elongated. The limiting value of 14.1 mV is not completely reached because there are other factors influencing the index potential. Besides the inductive effect also a partial intramolecular complex formation (quinhydrones) can act on the index potential.

Table III illustrates the results of our calculations of existing complexes in tetravalent redox systems at 50% oxidation. The explanation of this procedure would be too extensive.

TABLE III

Species present at 50% oxidation in solutions of some tetravalent systems. (Acetic acid/water 1:1, v/v; 25°C; total concentration ca 5×10^{-4} mol./l) [41]

	E_i (mV)	H~H (%)	H~Q (%)	(H~Q)complex (%)	Q~Q (%)
	16.8	19.1	38.1	23.1	19.1
	15.0	23.4	46.7	6.6	23.4
	24.7	11.9	30.3	45.9	11.9

The longer the intramolecular distance the smaller is the complexing. Compounds with sterically favoured complex formation show also higher amounts of the quinhydrone.

Also steric effects arising from hindered free rotation of the connected redox systems have an obvious influence on the midpoint potential.

This is shown by some tetravalent naphthoquinone derivatives [46] which can be obtained in meso and racemic forms. A methyl group in the quinone nuclei changes distinctly the midpoint potential due to hindered rotation.

The index potential for the racemic forms are always higher because they can exist in a conformation in which the complex formation is favoured.

Hexavalent redox systems have been synthesized either with a symmetric structure [41] (See Table V) or a reduced symmetric structure [43, 47] (see Figure 11).

First let us treat the symmetric systems. During the oxidation the symmetric hexavalent systems produce in every bivalent redox step only one intermediate.

TABLE IV

Potentiometric data of tetravalent naphthoquinones. (acetic acid/water$=4:1$, v/v; 25°C; potassium dichromate)

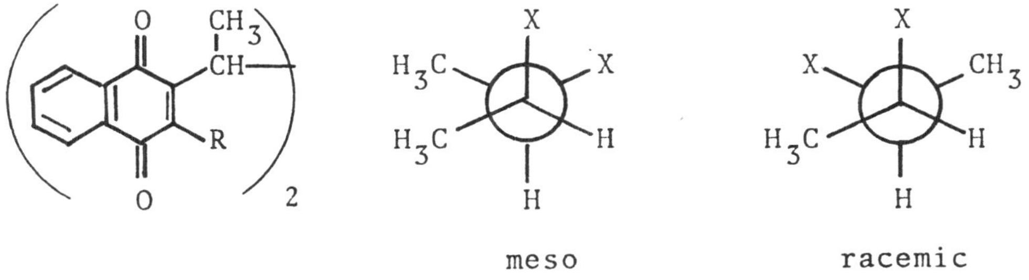

meso   racemic

Constitution	R	E_{mh}' (mV)	E_i (mV)	Colour at 50% oxidation
meso	H	383.4	17.9	brownish
racemic	H	383.3	21.0	reddish
meso	CH$_3$	299.6	17.5	brownish
racemic	CH$_3$	286.3	18.5	reddish
benzoquinone	–	670.2	14.4	–

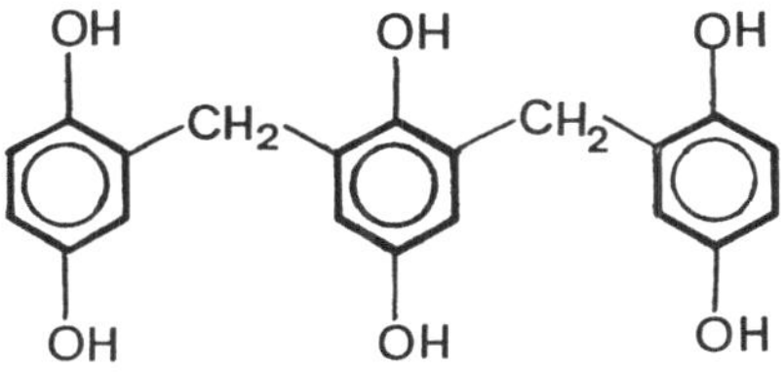

Fig. 11.

In abbreviated form:

$$\overline{H \sim H \sim H} \overset{E'_1}{\rightleftharpoons} \overline{H \sim Q \sim H} \overset{E'_2}{\rightleftharpoons} \overline{H \sim Q \sim Q} \overset{E'_3}{\rightleftharpoons} \overline{Q \sim Q \sim Q}$$

The analytical treatment of the potentiometric titration curves is similar to that of the symmetric tetravalent systems [41, 48].

$$K = \frac{[\overline{H \sim Q \sim H}]^2}{[\overline{H \sim H \sim H}][\overline{H \sim Q \sim Q}]} = \frac{[\overline{H \sim Q \sim Q}]^2}{[\overline{H \sim Q \sim H}][\overline{Q \sim Q \sim Q}]}$$

$$E = \frac{E'_1 + E'_2 + E'_3}{3} + \frac{RT}{6F} \ln \frac{[\overline{Q \sim Q \sim Q}]}{[\overline{H \sim H \sim H}]}$$

$$E'_1 = E'_2 - E'_1 = E'_3 - E'_2 = \frac{RT}{2F} \ln K$$

$$E_m = E'_2$$

TABLE V

Potentiometric data of symmetrical hexavalent systems. (acetic acid/water 1:1, v/v; 25°C; potassium dichromate) [41]

	E_{mh}' (mV)	E_i (mV)	K	E_1' (mV)	E_2' (mV)	E_3' (mV)
	584.4	25.7	13.1	551.3	584.4	617.5
n=1	575.8	16.1	4.03	557.9	575.8	593.7
n=2	568.6	15.2	3.54	552.4	568.6	584.8
Benzoquinone	621.3	14.2	–	–	–	–

Some of the results are:

(1) The titration curves are symmetrical to the midpoint.

(2) The formation constant K for the intermediates $\overparen{H \sim Q \sim H}$ and $\overparen{H \sim Q \sim Q}$ of the first and second redox step is equal. This is concluded from the symmetry of the molecule.

(3) K has a limiting value of 3. (cf. the similar case of the limiting ratio between successive ionization constants of tribasic acids). If the formation constant equals 3 the titration curve has an index potential of 14.1 mV (25 °C) showing that in this ideal case there are no interactions between the redox systems. The difference of the midpoint potentials $E_2' - E_1'$ and $E_3' - E_2'$ are equal and have the same limiting value of 14.1 mV (25 °C).

(4) The equilibrium constant K for the intermediates can be determined using the index potentials. The results of the potentiometric titration of some compounds are shown in Table V. They confirm our analytical approach.

The titration curve of hexavalent systems with reduced symmetry [47] can be analyzed only by making several assumptions. The redox curve can approximately be considered to result from two tetravalent steps overlapping each other.

$$H \sim H \sim H \overset{E'_1}{\rightleftharpoons} H \sim Q \sim H \overset{E'_2}{\rightleftharpoons} H \sim Q \sim Q$$

$$H \sim Q \sim H \overset{E'_2}{\rightleftharpoons} H \sim Q \sim Q \overset{E'_3}{\rightleftharpoons} Q \sim Q \sim Q$$

These compounds give two different intermediate compounds in each step.

Some problems may occur in potentiometric titrations of redox resins. Due to low diffusion of the reactants into the resin the establishment of the equilibrium redox potential can be very slow. If the penetration of the reactants should be facilitated by improving the swellability of the resin care must be taken to avoid an unwanted low mechanical stability. The redox potential of polymer redox systems depends on the microenvironment, i.e. resins with higher contents of the redox component show different midpoint and index potentials compared to those with a lower redox content [48]. The latter are similar to the corresponding monomer redox units [49–52].

4. Application

Redox resins are regenerable oxidating or reducing agents. After their use in redox reactions there is no contamination of the substrate solution because the resins can easily be filtrated.

There have been many proposals for the application of redox resins [53, 54]. Only some examples can be mentioned here.

Redox resins in their reduced form have been used for the reduction of ions and for depositing metals [53]. The redox resins can remove oxygen from solutions [55] and they can reduce oxygen to hydrogen peroxide in the presence of stabilizers [38, 53,

56–58]. In organic chemistry they are mainly used in their oxidized form as dehydrogenation agents [54, 57].

Cycloheptatrien can be dehydrogenated to the tropylium cation, 1, 4-dihydronaphthalene to naphthalene, 9, 10-dihydroanthracene to anthracene. 1, 4-Dihydrocarbazole and 1, 2, 3, 4-tetrahydrocarbazole could be dehydrogenated to carbazole, while the hexa- and dodeca-hydrocarbazole could not be dehydrogenated. Hydrazobenzene could be dehydrogenated to azobenzene, cysteine to cystine, ascorbic acid to dehydro-ascorbic acid.

The possible applications of the redox resins in biochemistry and pharmacy are of great interest. For instance we could dehydrogenate NADH to NAD$^+$ with a hydroquinone redox resin in the oxidized form. The redox resin in the quinone form performed the Strecker degradation of alanine to acetaldehyde and also of other α-amino acids [54]. Redox resins in the reduced form (hydroquinone form) were used successfully as antioxidants. Many other applications have been reported [3].

In this lecture some possibilities to synthesize redox polymers with various redox potentials are outlined. The author intended to excite the interest of biochemists and biophysicists in this field of macromolecular chemistry. The electrochemical investigations which have been carried out with these redox resins indicate that further applications should be promising.

In conclusion I would like to thank Dipl. Chem. E. Ehrenthal and Dr W. Storck for their helpful assistance in compiling and preparing this manuscript.

References

1. Cassidy, H. G. and Kun, K. A.: *Oxidation-Reduction Polymers (Redox Polymers)*, Interscience Publishers, New York, 1965.
2. Manecke, G.: *Angew. Makromol. Chem.* **4/5**, 26 (1968).
3. Cassidy, H. G.: *J. Polymer Sci., Pt. D* **6**, 1 (1972).
4. Updegraff, I. H. and Cassidy, H. G.: *J. Am. Chem. Soc.* **71**, 407 (1949).
5. Manecke, G. and Bourwieg, G.: *Chem. Ber.* **92**, 2958 (1959).
6. Manecke, G. and Bourwieg, G.: *Chem. Ber.* **95**, 1413 (1962).
7. Manecke, G. and Bourwieg, G.: *Chem. Ber.* **96**, 2013 (1963).
8. Kun, K. A. and Cassidy, H. G.: *J. Polymer Sci.* **56**, 83 (1962).
9. Manecke, G. and Bourwieg, G.: *Makromol. Chem.* **99**, 175 (1966).
10. Manecke, G. and Storck, W.: *Chem. Ber.* **94**, 300 (1961).
11. Manecke, G., Ramlow, G., and Storck, W.: *Chem. Ber.* **100**, 836 (1967).
12. Manecke, G. and Melzer, J.: unpublished.
13. Manecke, G. and Storck, W.: *Chem. Ber.* **102**, 3584 (1969).
14. Manecke, G. and Storck, W.: *Chem. Ber.* **94**, 3239 (1961).
15. Manecke, G. and Storck, W.: *Angew. Chem.* **74**, 903 (1962).
16. Manecke, G., Creutzburg, K., and Klawitter, J.: *Chem. Ber.* **99**, 2440 (1966).
17. Manecke, G. and Creutzburg, K.: *Makromol. Chem.* **93**, 271 (1966).
18. Merck, E.: Darmstadt.
19. Manecke, G. and Beyer, H. J.: *Makromol. Chem.* **123**, 223 (1969).
20. Manecke, G. and Wehr, G.: *Makromol. Chem.* **136**, 121 (1970).
21. Manecke, G., Beyer, H. J., Wehr, G., and Beier, J.: *J. Electroanal. Chem.* **28**, 139 (1970).
22. Manecke, G. and Beyer, H. J.: *Makromol. Chem.* **116**, 26 (1968).
23. Manecke, G., Beier, J., and Wehr, G.: *Makromol. Chem.* **134**, 231 (1970).

24. Manecke, G., Rühl, Chr.-St., and Wehr, G.: *Makromol. Chem.* **154**, 121 (1972).
25. Manecke, G., Beyer, H. J., and Mölleken, R.: *J. Electroanal. Chem.* **22**, 152 (1969).
26. Manecke, G. and Mölleken, R.: *Makromol. Chem.* **145**, 53 (1971).
27. Manecke, G., Wetzel, G., and Storck, W. *Makromol. Chem.* **176**, 3251 (1975).
28. Manecke, G., Kretzschmar, H.-J., and Hübner, W.: *J. Macromol. Sci.-Chem.* **A7**, 1181 (1973).
29. Manecke, G. and Ramlow, G.: *Chem. Ber.* **101**, 1987 (1968).
30. Manecke, G., Ramlow, G., Storck, W., and Hübner, W.: *Chem. Ber.* **100**, 3413 (1967).
31. Manecke, G. and Ramlow, G.: *J. Polymer. Sci. Pt. C* **22**, 957 (1969).
32. Acosta, F., Storck, W., and Manecke, G.: *Makromol. Chem.* **175**, 1813 (1974).
33. Manecke, G. and Kretzschmar, H. J.: *Makromol. Chem.* **169**, 15 (1973).
34. Manecke, G. and Kretzschmar, H. J.: *Chem. Ber.* **103**, 3862 (1970).
35. Manecke, G.: *Z. Elektrochem., Ber. Bunsenges. Physik. Chem.* **57**, 189 (1953).
36. Manecke, G.: *Z. Elektrochem., Ber. Bunsenges. Physik. Chem.* **58**, 369 (1954).
37. Manecke, G. and Bahr, Ch.: *Naturwissenschaften* **44**, 260 (1957).
38. Manecke, G. and Bahr, Ch.: *Z. Elektrochem., Ber. Bunsenges. Physik. Chem.* **62**, 311 (1958).
39. Manecke, G. and Koßmehl, G.: *Makromol. Chem.* **80**, 22 (1964).
40. Michaelis, L.: *J. Biol. Chem.* **96**, 703 (1932).
41. Storck, W. and Manecke, G.: *Makromol. Chem.* **176**, 97 (1975).
42. Manecke, G. and Zerpner, D.: *Makromol. Chem.* **108**, 198 (1967).
43. Manecke, G. and Förster, H. J.: *Makromol. Chem.* **52**, 147 (1962).
44. Manecke, G., Förster, H. J., and Panoch, H. J.: *Electrochim. Acta* **13**, 1101 (1968).
45. Adams, E. Q.: *J. Am. Chem. Soc.* **38**, 1503 (1916).
46. Manecke, G. and Storck, W.: *Chem. Ber.* **104**, 1207 (1971).
47. Manecke, G. and Panoch, H. J.: *Makromol. Chem.* **96**, 1 (1966).
48. Mills, J. E. C.: *Redox Polymers.* Dissertation, University of Toronto, 1971.
49. Manecke, G. and Storck, W.: *J. Polymer Sci., Pt. C* **4**, 1457 (1963).
50. Manecke, G. and Storck, W.: *Makromol. Chem.* **75**, 159 (1964).
51. Wegner, G., Nakabayashi, N., Duncan, S., and Cassidy, H. G.: *J. Polymer Sci.* [A–1] **6**, 3395 (1968).
52. Nakabayashi, N., Wegner, G., and Cassidy, H. G.: *J. Polymer Sci.* [A-1] **7**, 583 (1969).
53. Manecke, G., Koßmehl, G., Hartwich, G., and Gawlik, R.: *Angew. Makromol. Chem.* **2**, 86 (1968).
54. Manecke, G., Koßmehl, G., Gawlik, R., and Hartwich, G.: *Angew. Makromol. Chem.* **6**, 89 (1969).
55. Manecke, G.: *Angew. Chem.* **67**, 613 (1955).
56. Manecke, G.: *Angew. Chem.* **68**, 582 (1956).
57. Manecke, G., Bahr, Ch., and Reich, Ch.: *Angew. Chem.* **71**, 646 (1959).
58. Manecke, G. and Koßmehl, G.: *Makromol. Chem.* **70**, 112 (1964).

VI

FLUCTUATION, OSCILLATION, EXCITATION

FIELD FLUCTUATION IN IONIC SOLUTIONS AND MEMBRANES

FUMIO OOSAWA

Dept. of Biophysical Engineering, Osaka, University and Institute of Molecular Biology, Nagoya University

Abstract. The thermal fluctuation of the electric potential and the electric field of various modes is analysed in a homogeneous solution of simple electrolytes. The mean square of the fluctuating field at a point, averaged in a finite time, is calculated. The calculation is extended to the field fluctuation around a membrane with fixed charges. The state fluctuation of macromolecules in the solution or the membrane due to the field fluctuation is discussed. The problem is formulated as a reversible reaction with fluctuating rate constants. Apparent correlation between the states of independent macromolecules is produced by the spatial correlation of the fluctuating field around those molecules. The result is available for estimation of the probability of spontaneous excitation of the cell membrane without specific input. Special importance of the field fluctuation is expected in the biological system because of its capacity to amplify and digitalize the effect of fluctuation.

1. Introduction

The response of the membrane to the electric field is involved in the elementary process of excitation of living cells. The electric field across and along the membrane thermal lyfluctuates due to the Brownian movement of small ions inside and outside the cell. The membrane may respond to such a fluctuating field. Actually, most of the sensory cells show spontaneous firing without specific input. The mechanoreceptor of a muscle spindle near its resting length generates impulses with great irregularity. The rate of impulses is very low and successive intervals vary in a random manner. This phenomenon has been attributed to fluctuations of the membrane potential [1, 2, 3].

The contribution of random fluctuation of the electric field at the membrane is also suggested in the behavior of lower organisms. Swimming bacteria and protozoa occasionally change their direction. When a mechanical stimulus is given to a cell of paramecium, for example, it shows the avoiding reaction; it swims backward for a short while and then swims forward again. This direction change was found to be induced by generation of action potential at the cell membrane [4, 5]. Therefore, it is expected that the spontaneous direction change of swimming paramecium is also induced by the action potential generated as a result of thermal fluctuation of the membrane potential.

The frequency of the spontaneous direction change depends on the environmental condition. At low concentrations of calcium ions outside, for example, paramecium swims straightly and does not frequently change the swimming direction. However, with increasing calcium ion concentration, the frequency of direction change increases. At high concentrations, discontinuous direction change decreases again [6].

The average membrane potential of paramecium increases from negative toward zero with increasing calcium ion concentration [7]. The fluctuation of the membrane potential also increases and sharp action potential-like fluctuation often appears. The frequency of spontaneous direction change is related to the frequency of spontaneous generation of action potential.

At low temperatures, cells of E. coli swim straightly and with raising temperature, the frequency of discontinuous direction change increases. The frequency shows a sharp maximum near 35 °C. At higher temperatures, the trajectory becomes circular [8]. This may be due to the increase of thermal fluctuation of the membrane with raising temperature.

Such behaviours are commonly observed in different species of protozoa and bacteria, suggesting important effects of thermal fluctuation of the field (Figure 1).

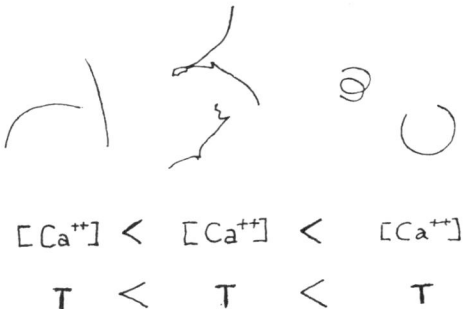

Fig. 1.   Illustration of trajectories of swimming paramecia or bacteria at different ionic conditions or different temperatures.

Fluctuation of the electric field and the current in electro-conductive media has been investigated for many years [9, 10]. However, most of the theories and experiments have been concerned with objects of the macroscopic scale. In the case of living cells, it is more important to know the magnitude of local and transient fluctuations of the field, which must have influences on molecules in the membrane. Therefore, calculations will be made first on the field fluctuation of various modes due to moving small ions, having different space and time correlations [11]. Then, the effect of such fluctuating field on the state of a macromolecular system will be discussed.

2. Fluctuations in Homogeneous Ionic Solutions

On the time average a small volume in a homogeneous ionic solution is neutralized by cations and anions, but the Brownian movement of those ions produces deviation from neutrality fluctuating with time. Let us express the electric potential ψ due to this deviation as a Fourier series of various modes; that is,

$$\psi = \sum_{\mathbf{k}} c_{\mathbf{k}} e^{i\mathbf{k}\mathbf{r}} \tag{1}$$

where $\mathbf{k}$ is the wave number vector, $\mathbf{r}$ is the spatial coordinate and $c_{\mathbf{k}}$ is the amplitude of the fluctuating potential of mode $\mathbf{k}$. Then, the electric field and the charge density associated with this potential are expressed as

$$\mathbf{E} = - \operatorname{grad} \psi = \sum_{\mathbf{k}} - ikc_{\mathbf{k}}e^{i\mathbf{k}\mathbf{r}} \tag{2}$$

$$e_0\delta\varrho = - (\varepsilon/4\pi) \nabla^2\psi = (\varepsilon/4\pi) \sum \mathbf{k}^2 c_{\mathbf{k}} e^{i\mathbf{k}\mathbf{r}} \tag{3}$$

where e_0 is the electronic charge and ε is the dielectric constant of the solvent. Such fluctuation has the excess internal energy given by

$$U = \int (\varepsilon/8\pi) (\operatorname{grad} \psi)^2 \, dv \tag{4}$$

The excess entropy S due to deviation from neutrality is given by

$$TS = (1/2) k_B T \int ((\delta\varrho)^2/\varrho_0) \, dv \tag{5}$$

where $\varrho_0 = \varrho_{+0} + \varrho_{-0}$; ϱ_{+0} and ϱ_{-0} are the average number densities of cations and anions. By the use of the Fourier series (1)–(3), it is found that the excess free energy is expressed as

$$F = U - TS = (\varepsilon/8\pi) \sum (\mathbf{k}^2 + \mathbf{k}^4/\kappa^2) c_{\mathbf{k}}^2 V \tag{6}$$

where

$$\kappa^2 = (4\pi e_0^2/\varepsilon k_B T) (n_0/V) \tag{7}$$

n_0 is the total number of cations or anions and V is the total volume of the solution. The constant κ is the same quantity as the Debye-Huckel parameter in the simple electrolyte theory. The fluctuation takes place in proportion to the factor $\exp(-F/k_B T)$. Therefore, the above Equation (7) means that fluctuations of different modes $\mathbf{k}$ are independent of each other.

The mean square of the amplitude of each mode $\langle c_{\mathbf{k}}^2 \rangle$ is given by

$$\langle c_{\mathbf{k}}^2 \rangle = (4\pi/\varepsilon) k_B T/((\mathbf{k}^2 + \mathbf{k}^4/\kappa^2) V) \tag{8}$$

Therefore, the mean square of the components of fluctuating field $\mathbf{E}_{\mathbf{k}}$ and density $\delta\varrho_{\mathbf{k}}$ are given by

$$\langle \mathbf{E}_{\mathbf{k}}^2 \rangle = (4\pi/\varepsilon) k_B T/((1 + \mathbf{k}^2/\kappa^2) V) \tag{9}$$

$$\langle \delta\varrho_{\mathbf{k}}^2 \rangle = (\varepsilon/4\pi e_{\mathbf{k}}^2) k_B T (\mathbf{k}^2/(1 + \mathbf{k}^2/\kappa^2) V) \tag{10}$$

The mean square amplitude $\langle c_{\mathbf{k}}^2 \rangle$ decreases with increasing k or decreasing wavelength of fluctuation. The amplitude tends to infinity when $k \to 0$; the uniform rise or drop of the potential has no excess energy. The amplitude of fluctuating field increases with increasing wavelength, tending to a finite value at $k \to 0$. On the other hand, the amplitude of fluctuating charge density decreases with increasing wavelength. The uniform accumulation of cations or anions produces high excess energy. Cations and anions are more alternatively distributed. These situations are illustrated in Figure 2.

The spatial correlation functions of fluctuating potential, field and charge density are obtained as

$$\langle \psi(r)\,\psi(0)\rangle = \sum \langle c_{\mathbf{k}}^2\rangle\, e^{i\mathbf{k}\mathbf{r}} = \sum (4\pi/\varepsilon)\, k_B T/((\mathbf{k}^2 + \mathbf{k}^4/\kappa^2)\, V)\cdot e^{i\mathbf{k}\mathbf{r}}$$

$$= (4\pi/\varepsilon)\, k_B T\,(1/2\pi)^3 \int\int\int (e^{i\mathbf{k}\mathbf{r}}/(\mathbf{k}^2 + \mathbf{k}^4/\kappa^2))\, d\mathbf{k}$$

$$= (k_B T/\varepsilon)\,(1 - e^{-\kappa r})/r \tag{11}$$

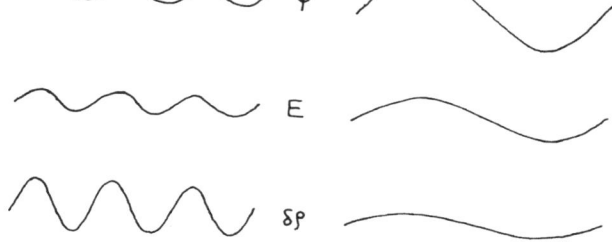

Fig. 2.   Fluctuation of electric potential, electric field, and charge density of different modes.

where the summation with respect to $\mathbf{k}$ (k_x, k_y, k_z) was replaced by the integration $(V/(2\pi)^3)\, dk_x dk_y dk_z$. Similarly,

$$\langle \mathbf{E}(r)\,\mathbf{E}(0)\rangle = \sum \langle \mathbf{E}_{\mathbf{k}}^2\rangle\, e^{i\mathbf{k}\mathbf{r}}$$

$$= (k_B T/\varepsilon)\,\kappa^2 e^{-\kappa r}/r \tag{12}$$

$$e_0^2\,\langle \delta\varrho(r)\,\delta\varrho(0)\rangle = e_0^2 \sum \langle \delta\varrho_{\mathbf{k}}^2\rangle\, e^{i\mathbf{k}\mathbf{r}}$$

$$= \delta^*(r) - (4e_0^4\varrho_0^2/\varepsilon k_B T)\, e^{-\kappa r}/r \tag{13}$$

where δ^* is a delta function. The last equation is the same as expected from the Debye-Huckel theory of electrolytes. In other words, the Deybe-Huckel approximation is equivalent to that only the second order term of fluctuation was taken into account.

The mean square of the potential difference between two points at distance r is given, from (11), by

$$\langle (\psi(r) - \psi(0))^2\rangle = (2k_B T/\varepsilon)\,\kappa\,(1 - (1/\kappa r)(1 - e^{-\kappa r}))$$

$$= ((k_B T/e_0)^2\,((2e_0^2/\varepsilon k_B T)\,\kappa)\,(1 - (1/\kappa r)(1 - e^{-\kappa r})) \tag{14}$$

The first factor in the right hand side $(k_B T/e_0)$ corresponds to about 25 mV at room temperature. The second factor, a non-dimensional quantity, is not very different from unity in ordinary conditions, and slowly increases with increasing concentration of small ions. The third factor tends to unity for large values of r, where the spatial correlation vanishes. For example, when the ionic strength $I=0.15$ and $r=100$ A, the root mean square of the potential difference becomes about 35 mV.

Thus, the magnitude of the fluctuating potential difference or field can be of the same order as the average potential difference or field at the cell membrane. It seems to be very probable that the state of the membrane is influenced by this fluctuation. However, the response of the molecule in the membrane requires a finite time. Very fast fluctuations are of little significance. Therefore, it is necessary to know the relaxation time of the fluctuation.

The decay of the fluctuating potential or charge to the statistical average in a homogeneous ionic solution can be assumed to be expressed by the following equations for cations and anions, separately

$$\zeta_+ (\partial \delta \varrho_+ / \partial t) = k_B T \nabla^2 \delta \varrho_+ + e_0 \text{div} (\varrho_{+0} \text{grad } \psi)$$
$$\zeta_- (\partial \delta \varrho_- / \partial t) = k_B T \nabla^2 \delta \varrho_- - e_0 \text{div} (\varrho_{-0} \text{grad } \psi) \tag{15}$$

in the first order of fluctuation. If, for the sake of simplicity, the frictional coefficients ζ_+ and ζ_- for cations and anions were assumed to be equal, we have, for $\delta \varrho = \delta \varrho_+ - \delta \varrho_-$

$$(\partial \delta \varrho / \partial t) = k_B T \nabla^2 \delta \varrho + e_0 \text{div} (\varrho_0 \text{ grad } \psi) \tag{16}$$

By the use of the Fourier series, the relaxation time of the mode $\mathbf{k}$, $\tau_{\mathbf{k}}$, is obtained as

$$-\zeta/\tau_{\mathbf{k}} = -k_B T \mathbf{k}^2 - 4\pi e_0^2 \varrho_0/\varepsilon$$
$$\tau_{\mathbf{k}} = 1/(D(\mathbf{k}^2 + \kappa^2)) \tag{17}$$

where $D(=k_B T/\zeta)$ is the diffusion constant of ions. The relaxation time increases with decreasing k or increasing wavelength, but it has an upper limit $1/D\kappa^2$. The fluctuation of long wavelengths has a strong restoring force due to coulomb interaction of the long range. In the case of small ions, the relaxation time is of the order of 10^{-7}–10^{-9} s, under the ordinary condition. This is probably very much smaller than the time necessary for the change of the state of molecules in the membrane.

3. Fluctuation in Heterogeneous Systems

When there are fixed charges in an ionic solution, moving ions make ionic atmospheres around the charge and the average potential and field are not zero. In this case fluctuations of various modes have correlations. For example, let us consider a small particle or a thin membrane having fixed charges of low density immersed in an electrolyte solution. The excess free energy due to deviation from the average of the potential or the charge density is approximately expressed as

$$F = (1/2)k_B T \sum \sum b_{ij} c_{\mathbf{k}_i} c_{\mathbf{k}_j} \tag{18}$$

where

$$b_{ij} = (\varepsilon/4\pi k_B T) \int\int\int (\mathbf{k}_i \mathbf{k}_j + (\mathbf{k}_i^2 \mathbf{k}_j^2/\kappa^*(\mathbf{r})^2)) \exp(i(\mathbf{k}_i - \mathbf{k}_j)\mathbf{r}) \, d\mathbf{r} \tag{19}$$

$$\kappa^{*2} = (4\pi e_0^2/\varepsilon k_B T) \varrho_0 \tag{20}$$

The parameter κ^* is a function of the spatial coordinate r because the number density of moving small ions in the equilibrium ϱ_0 depends on r. Then, b_{ij}'s are not zero even for $i \neq j$ and therefore the correlation of amplitudes of two different modes is given by

$$\langle c_{\mathbf{k}_i} c_{\mathbf{k}_j} \rangle = (b_{ij}^{-1}) \tag{21}$$

where b_{ij}^{-1} is the ij component of the inverse matrix of b_{ij}. If the density of ions ϱ_0 is constant $(=2n_0/V)$ except in and around a small volume occupied by the particle or the membrane and the deviation of ϱ_0 from the constant in this volume is small, the non-diagonal components of b_{ij} are much smaller than diagonal's. Then, the approximate solution is given by

$$\langle c_{\mathbf{k}_i} c_{\mathbf{k}_j} \rangle = - b_{ij}/b_{ii} b_{jj} \qquad (i \neq j) \tag{22}$$

The correlation of the fluctuating potential is given by

$$\langle \delta\psi(\mathbf{r}_1)\,\delta\psi(\mathbf{r}_2)\rangle = \sum\sum \langle c_{\mathbf{k}_i} c_{\mathbf{k}_j}\rangle \exp\left(-(\mathbf{k}_i\mathbf{r}_1 - \mathbf{k}_j\mathbf{r}_2)\right) \tag{23}$$

$\delta\psi$ is the fluctuating component of the potential at r. Summation or integration with respect to k_i and k_j can be performed if (22) with (19) is applicable. Only the final result is described here. The mean square of the fluctuation of the potential difference between outside and inside the particle having uniform fixed charges is approximately given by

$$\langle (\delta\psi(a) - \delta\psi(\infty))^2 \rangle = (2k_B T/\varepsilon)\kappa$$
$$+ (2k_B T/\varepsilon)\kappa(1/2)\left((\varrho_{0a} - \varrho_{00})/\varrho_{0a}\right)(1 - e^{-\kappa a}) \tag{24}$$

where a is the radius of the particle, ϱ_{0a} is the average density of moving small ions inside the particle and ϱ_{00} is that outside. κ is given by (20) with $\varrho_0 = \varrho_{00}$. The second term in the right hand side indicates that the fluctuating potential difference is greater in a particle having larger fixed charges and it is smaller around a particle repelling small ions.

The above result was obtained under the condition that the difference between ϱ_{00} and ϱ_{0a} is not very much greater than ϱ_{00}. The same method of calculation is not easily extended to obtain the general solution of the fluctuating potential in the case of particles or membranes which have high charge or exclude small ions completely.

In connection with this problem it must be noticed that the electrolyte solution containing highly charged colloidal particles or macromolecules has a large dielectric constant. The large dielectric increment means the presence of a large fluctuating dipole in the absence of the external field [12]. In the case of a highly charged spherical particle, the observed increment is understandable if the particle has the mean square dipole moment thermally fluctuating of the order of

$$\langle \delta\mu^2 \rangle = (n^{*1/2} e_0 a)^2 \tag{25}$$

where a is the radius of the particle and n^* is the number of charges (or counter-ions) on the surface of the particle. $n^{*1/2}$ gives the average deviation of counter-ions from the uniform distribution. Then, the fluctuating potential difference between inside and outside the particle is given

$$\langle(\delta\psi\,(a) - \delta\psi\,(\infty))^2\rangle = (k_B T/e_0)^2\,(n^{*1/2}e_0{}^2/\varepsilon k_B Ta)^2 \tag{26}$$

This is analogous to the previous expression (14). For a particle of $a = 30$ A and $n^* = 100$, the potential difference is about 75 mV. The relaxation time of this fluctuating field becomes of the order of

$$\tau = a^2/D \tag{27}$$

where D is the diffusion constant of counter-ions along the surface [13].

For a highly charged rod-like particle also, many theories have been proposed [14, 12]. Most of them have shown that the mean square dipole moment along the rod and its relaxation time are approximately given by the formulae similar to (25) and (27), respectively, where a must be the length of the rod. With decreasing wavelength of fluctuation on the rod, the field becomes stronger but the relaxation time becomes shorter [12].

It must be emphasized here that the relaxation time given by (27) can be very much longer than that expected in homogeneous ionic solution. In fact, the lowest dispersion frequency of dielectric constant of polyelectrolyte solutions has been found in 10^3–10^4 Hz. The restoring force of fluctuations of long wave lengths due to conlomb repulsion between counter-ions is not very large [12, 14, 15, 16].

4. Reaction in Fluctuating Environment

Let us consider macromolecules in an ionic solution, which can be in one of the two states A and B. The rate constants of reactions from A to B and from B to A are denoted by k_A and k_B. Then, the change with time of the number of molecules in A, n_A, is given by

$$dn_A/dt = -k_A n_A + k_B(N - n_A) \tag{28}$$

where N is the total number of molecules. The rate constants k's are determined by the environmental condition. At a given macroscopic environmental condition, k's are usually considered to be 'constant'. Then, n_A in equilibrium is given by

$$n_A = k_B N/(k_A + k_B) \tag{29}$$

Actually, however, the environmental condition fluctuates, so that k's are not necessarily constant but change with time. n_A in (29) must be regarded as the average of n_A determined by the average of k's.

Concerning the probability that a specified molecule is in A, we have a similar equation to (28):

$$dp_A/dt = -k_A p_A + k_B(1 - p_A) \tag{30}$$

The environmental condition surrounding this molecule fluctuates thermally; k's can be written as

$$k_A = \langle k_A \rangle + \delta k_A \tag{31}$$

$$k_B = \langle k_B \rangle + \delta k_B \tag{31}$$

where $\langle k_A \rangle$ and $\langle k_B \rangle$ are the average values of k's. Then,

$$dp_A/dt = \langle -k_A \rangle p_A + \langle k_B \rangle (1 - p_A) + F(t) \tag{32}$$

$$F(t) = -\delta k_A \cdot p_A + \delta k_B \cdot (1 - p_A) \tag{33}$$

The last term $F(t)$ is the fluctuating force for the reaction, the average of which must be zero. The above equation is similar to that for the Brownian motion of a particle in a solution; where the fluctuating force is due to the collision of solvent molecules with the particle. In the present case, however, the force $F(t)$ does not directly come from such a microscopic interaction. It comes from the submacroscopic environmental fluctuation. Its relaxation time must be very much longer than that of the microscopic collision.

As the first approximation $F(t)$ can be given by

$$F(t) = -\delta k_A \cdot \langle p_A \rangle + \delta k_B \cdot (1 - \langle p_A \rangle) \tag{34}$$

$$\langle p_A \rangle = \langle k_B \rangle / (\langle k_A \rangle + \langle k_B \rangle) \tag{35}$$

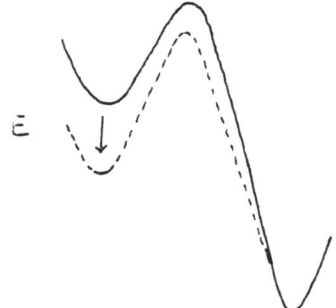

Fig. 3.   Potential profile of two states of a macromolecule in fluctuating field.

For example, let us assume that the free energy of the state A or B depends on the electric field on the molecule, as illustrated in Figure 3. The rate constants are written as

$$k_A = k_{A0} \exp(+\mathbf{u}\mathbf{E}/k_B T)$$
$$k_B = k_{B0} \exp(-\mathbf{u}\mathbf{E}/k_B T) \tag{36}$$

If the field fluctuates as shown in the previous section, the rate constants fluctuate. Their average is given by

$$\langle k_A \rangle = k_{A0} \langle \exp(+\gamma \mathbf{E}) \rangle = k_{A0} \exp\left(\tfrac{1}{6}\gamma^2 \langle \mathbf{E}^2 \rangle\right)$$
$$\langle k_B \rangle = k_{B0} \langle \exp(-\gamma \mathbf{E}) \rangle = k_{B0} \exp\left(\tfrac{1}{6}\gamma^2 \langle \mathbf{E}^2 \rangle\right) \tag{37}$$

where $\gamma = u/k_B T$, and the fluctuation was assumed to be isotropic. Thus, even if the average of the fluctuating field is zero, the average of rate constants is larger than that in the complete absence of the field. The deviation of the rate constant from the average is simply obtained from the difference between (36) and (37).

The formal solution of the Equation (32) is written as

$$p_A(t) = \langle p_A \rangle + \int F(t') \exp\left(- (\langle k_A \rangle + \langle k_B \rangle)(t - t')\right) dt' \tag{38}$$

for $t \gg 1/(k_A + k_B)$. By the use of (34) with (35) and (36) we can follow the change of p_A with time in the fluctuating field. The above treatment is similar to the case of magnetic spins in the time-dependent magnetic field [17].

5. Correlating Transition in the Fluctuating Field

Let us suppose two macromolecules at a short distance in an ionic solution. They undergo reversible transition between two states. If the rate constants of the transition depend on the electric field, they show response to the fluctuating field. When the fluctuation of the field has correlation in a range longer than the distance between molecules, the fluctuation of the state of these molecules may take place in the same direction. Even when they have no direct energetic interaction, correlation appears between their states. Similar idea of the apparent correlation due to the correlating fluctuation of the environmental condition can be applied to the case of the transition of two different parts of a single macromolecule or to the case of the transition of the assembly of molecules.

The probability that a single molecule is in the state A, p_A, changes with time in accordance with (30). The probability that two molecules 1 and 2 are in the state A at the same time is given by the product $p_{1A}(t)p_{2A}(t)$, if there is no direct correlation between them. In the fluctuating environment the average of p_A, $\langle p_A \rangle$, is given by (35). However, the average of the product $p_{1A}(t) \cdot p_{2A}(t)$ is not always given by the product of the averages $\langle p_{1A} \rangle \langle p_{2A} \rangle$, if the fluctuation of the environment has correlation between two points 1 and 2. From (38)

$$\langle p_{1A}(t) p_{2A}(t) \rangle = \langle p_A \rangle^2 + \left\langle \int^t \int^t F_1(t') \exp\left(- (\langle k_A \rangle + \langle k_B \rangle) \times \right.\right.$$
$$\left.\left. \times (t - t')\right) dt' F_2(t'') \exp\left(- (\langle k_A \rangle + \langle k_B \rangle)(t - t'')\right) dt'' \right\rangle \tag{39}$$

where F_1 and F_2 are the fluctuating forces acting on the molecule 1 and 2. The second term in the right hand side is easily transformed in the following way;

$$\int \int \langle F_1(t') F_2(t'') \rangle \exp\left(- (\langle k_A \rangle + \langle k_B \rangle)((t - t') + (t - t''))\right) dt' \, dt'' =$$
$$= (1/(\langle k_A \rangle + \langle k_B \rangle)) \int \langle F_1(0) F_2(\tau) \rangle \exp\left(- (\langle k_A \rangle + \langle k_B \rangle) \tau\right) d\tau \tag{40}$$

The correlation function of the fluctuating force is written as

$$\langle F_1(0)F_2(\tau)\rangle = \langle \delta k_{1A}(0)\cdot \delta k_{2A}(\tau)\rangle \langle p_A\rangle^2$$
$$- 2\langle \delta k_{1A}(0)\cdot \delta k_{2B}(\tau)\rangle \langle p_A\rangle (1-\langle p_A\rangle)$$
$$+ \langle \delta k_{1B}(0)\cdot \delta k_{2B}(\tau)\rangle (1-\langle p_A\rangle)^2 \tag{41}$$

If the rate constants are given by (36) as functions of the electric field, the correlation function of k's are expressed by the correlation function of the fluctuating field.

$$\langle \delta k_{1A}(0)\cdot \delta k_{2A}(\tau)\rangle = k_{A0}^2 \langle (\exp(+\gamma E_1(0)) - \langle \exp(+\gamma E_1)\rangle)$$
$$\cdot (\exp(+\gamma E_2(\tau)) - \langle \exp(+\gamma E_2)\rangle)\rangle$$
$$= \langle k_A\rangle^2 (\exp((\gamma^2/3)\langle E_1(0)E_2(\tau)\rangle) - 1) \tag{42}$$

Thus, by using this equation and others for $\langle \delta k_{1A}(0)\cdot \delta k_{2B}(\tau)\rangle$ and $\langle \delta k_{1B}(0)\cdot \delta k_{2B}(\tau)\rangle$, the final expression for the correlation function of the fluctuating force is given by

$$\langle F_1(0)F_2(\tau)\rangle = 4\langle p_A\rangle^2 (1-\langle p_A\rangle)^2 (\langle k_A\rangle + \langle k_B\rangle)^2$$
$$\sinh((\gamma^2/3)\langle E_1(0)E_2(\tau)\rangle) \tag{43}$$

From (39), (40) and (43), we have

$$\langle p_{1A}(t) p_{2A}(t)\rangle = \langle p_A\rangle^2 + \langle p_A\rangle^2 (1-\langle p_A\rangle)^2 (\langle k_A\rangle + \langle k_B\rangle) \times$$
$$\int 4\sinh((\gamma^2/3)\langle E_1(0)E_2(\tau)\rangle) \exp(-(\langle k_A\rangle + \langle k_B\rangle)\tau)\cdot d\tau \tag{44}$$

The second term in the right hand side gives the contribution of correlation in the environmental fluctuation. The probability that the two molecules are in the state A simultaneously was increased by this correlation. The relaxation time of the reaction $A \rightleftarrows B$, τ_D, is given by

$$\tau_D = 1/(\langle k_A\rangle + \langle k_B\rangle) \tag{45}$$

This relaxation time is usually very much longer than the relaxation time of the fluctuating field τ_E. Then, (44) can be rewritten as

$$\langle p_{1A}(t)p_{2A}(t)\rangle = \langle p_A\rangle^2 + \langle p_A\rangle^2 (1-\langle p_A\rangle))^2$$
$$(\tau_E/\tau_D)\cdot 4\sinh((\gamma^2/3)\langle E_1(0)E_2(0)\rangle) \tag{46}$$

The probability of simultaneous transition is determined by the following factors; the probability of the singel transition p_A, the ratio of two relaxation times τ_E/τ_D, the correlation of the fluctuating field $E_1 E_2$ and the factor on the dependence of the free energy of two states on the electric field. If uE is larger than $k_B T$, the sinh term becomes very much larger than unity.

6. Discussion

It is very probable that the excitation of the nerve membrane is due to the transition of (a) macromolecule(s) involved in specific channels for permeation of small ions [18, 19, 20]. Each channel may be composed of a few macromolecules which undergo reversible transitions, the rate of which depends on the electric field. In the Hodgkin-Huxley theory [18] it was assumed that the channel becomes permeable only when these molecules are in the second state simultaneously; however, it was not necessary to introduce direct correlation between transitions of these molecules. Direct cooperation between different channels was not assumed either. Their states were correlated each other only through the electric field which is determined as a result of the average permeability for small ions in a certain area of the membrane. Therefore, the above calculation is applicable for estimation of the probability of spontaneous transition of single channels. The voltage clamp experiment on the nerve membrane gave the relation between the rate constant and the electric field, according to which the movement of a few electronic charges in the membrane seemed to participate in the transition of each macromolecule [18]. Therefore, it is very probable that the ratio $\mathbf{u}E/k_BT$ becomes larger than unity due to the thermal fluctuation. Then, the second factor in (46) makes important contribution to the spontaneous opening of the channel.

Electric noise has been measured on various cell membranes and charged membranes immersed in ionic solutions. The amplitude of fluctuating field is very small when macroscopic electrodes were used; however, it is very remarkable that in the membrane, fluctuations of long relaxation times appear in addition to the Nyquist noise [21, 22]. This suggests the importance of the effect of correlating fluctuation of the environmental field on the spontaneous and transient change of the membrane permeability. If a certain number of channels are simultaneously opened by chance, the fluctuation of the electric field is amplified and all channels are opened. Then, action potential is generated. A small change in the probability of opening of single channels may result in a large change in the probability of generation of action potential.

The above theory treated the effect of the environmental fluctuation on the state of molecules in an ionic solution. The free energy of the molecule was assumed to depend on the electric field. This means that the molecule has an induced or permanent dipole or non-uniform charge distribution. If the molecule has different dipoles in two states, the transition between these states must have some influence on the distribution of small ions in the environment. Such a reaction from the molecule to the environment, however, was not taken into consideration in the above theory. In this sense, the theory was not consistent. If the relaxation time of the environmental fluctuation and that of the transition of the molecule became of the same order of magnitude, this reaction can not be neglected. A self-consistent treatment is necessary on the whole system of the molecule and the environment.

It must be remarked also that in the case of living membranes both sides of the membrane have different ionic compositions, and the system is not in thermodynamic equilibrium. It is desirable to extend the theory to the fluctuation around a non-equilibrium state.

References

1. Fatt, P. and Katz, B.: *Nature* **166**, 597 (1950).
2. Buller, A., Nicholls, J., and Strom, L.: *J. Physiol.* **122**, 409 (1953).

3. Buller, A.: *J. Physiol.* **179**, 402 (1965).
4. Naitoh, Y.: *J. gen. Physiol.* **51**, 85 (1968).
5. Naitoh, Y. and Eckert, R.: *Science* **164**, 963 (1969).
6. Nakaoka, Y.: unpublished.
7. Eckert, R., Naitoh, Y., and Friedman, K.: *J. Exp. Biol.* **56**, 683 (1972).
8. Maeda, K.: unpublished.
9. Nyquist, H.: *Phys. Rev.* **32**, 110 (1928).
10. Johnson, J.: *Phys. Rev.* **32**, 97 (1928).
11. Oosawa, F.: *J. Ther. Biol.* **39**, 373 (1973).
12. Oosawa, F.: *Biopolymers* **9**, 677 (1970); Oosawa, F.: *Polyelectrolytes*, Marcel Dekker, New York, 1971, Chap. V.
13. Schwarz, G.: *J. Phys. Chem.* **66**, 2636 (1962).
14. Mandel, M.: *Mol. Phys.* **4**, 489 (1961).
15. Mandel, M.: in E. Sélégny, M. Mandel, and V. P. Strauss (ed.), *Polyelectrolytes*, D. Reidel Publishing Co., Dordrecht, Holland, 1974.
16. Takashima, S.: *J. Phys. Chem.* **70**, 1372 (1966).
17. Glauber, R.: *J. Math. Phys.* **4**, 294 (1963).
18. Huxley, A. and Hodgkin, A.: *J. Physiol.* **116**, 500 (1952).
19. Tasaki, I.: *Nerve Excitation*, C. C. Thomas, 1968.
20. Hille, B.: *Prog. Biophys. and Mol. Biol.* **21**, 1 (1970).
21. Verveen, A., Derksen, H., and Schick, K.: *Nature* **216**, 588 (1967).
22. Poussart, D.: *Proc. Natl. Acad. Sci.* **64**, 95 (1969).

NON-LINEAR TRANSPORT AND OSCILLATIONS IN FIXED CHARGE MEMBRANES: SOME POSSIBLE BIOLOGICAL IMPLICATIONS

TORSTEN TEORELL

Institute of Physiology and Medical Biophysics, Biomedical Center,
University of Uppsala, Uppsala, Sweden

Abstract. (1) We have tried to obtain a formal 'analog' able to reproduce and describe in both time and intensity the relations observed in smooth muscle and in heart tissue. The present article deals only with the effects of external mechanical stimuli. Internal 'myogenic' phenomena can also be translated in this way.

(2) The above formalism derives essentially from the previously described physical and mathematical properties of the 'membrane oscillator' which is an excitable system; it is composed of compartments and a membrane in which the electric potential is coupled to the fluxes of ions and water across the membrane, thus creating modifications in pressure and volume. It is supposed that oscillations of the potential in the system can simulate biological action potentials. Moreover the simultaneous variations in the transmembrane pressure are considered to be representative of the force of 'active' muscular contractions in function of time.

(3) This excitation-contraction analog is combined with another formal analog; the latter describes the typical passive and non-linear visco-elastic properties ascribed to muscular function.

(4) The 'electro-contraction' analog is computed with recent experiments on periodic, mechanical stretching of snail heart (*Helix pomatia*). There is good agreement between experimental results and theoretically computed responses of the analog.

(5) We emphasize the highly formal character of the 'electro-contraction' analog. We also give some preliminary discussion about our as yet incomplete knowledge of the real relationship that exists between the contraction of muscle and its stimulus.

Résumé. (1) Nous avons cherché à obtenir un 'analogue' formel capable de reproduire et de décrire les relations, en durée et en intensité, qui sont observées sur les muscles lisses et sur le tissu cardiaque. La présente étude est limitée aux effets des stimuli mécaniques externes. Elle peut aussi traduire des phénomènes 'myogéniques' internes.

(2) L'essentiel du formalisme ci-dessus repose sur les propriétés physiques et mathématiques précédemment décrites de l''oscillateur à membrane'. Celui-ci est un système excitable, compartiment-membrane, où le potentiel électrique est fortement couplé aux flux d'eau et d'ions à travers la membrane, ce qui entraîne des variations de pression et de volume. On suppose que les oscillations de potentiels, amorties ou non amorties, de ce système peuvent simuler les potentiels d'action biologiques. De plus, les fluctuations concomitantes de la pression transmembrannaire sont tacitement considérées comme représentant le décours en fonction du temps et la force des contractions musculaires 'actives'.

(3) L'analogue excitabilité-contraction ci-dessus est combiné avec un autre analogue formel, lequel décrit les propriétés visco-élastiques typiques, passives et non linéaires, invoquées dans la fonction musculaire.

(4) L'analogue 'électro-contraction' a été comparé avec des expériences récentes sur la distension mécanique, périodique, du coeur de l'Escargot (*Helix pomatia*). L'accord est très sa istaisant entre les résultats expérimentaux et les résultats donnés par l'analogue après traitement de ceux-ci sur ordinateur.

(5) Nous insistons sur le caractère très formel de l'analogue électrocontraction. Nous présentons aussi des discussions préliminaires portant sur notre connaissance encore incomplète, des véritables liaisons entre un stimulus et la réponse contractile d'un tissu musculaire.

Eric Sélégny (ed.), Charged Gels and Membranes II, 205–211. All rights reserved.
Copyright © 1976 by D. Reidel Publishing Company, Dordrecht-Holland.

Many biological transport phenomena, especially the ionic exchanges in the nerves and the heart tissue are considered as oscillatory. According to the well known 'ionic theory' of Hodgkin and Huxley the nerve action potentials are due to an alternating change in the membranes of the K and the Na permeability. The amplitude of the potential 'spikes' is of the order of 100 mV. The formalism of the ionic theory works with certain assumptions leading to 4th order non-linear differential equations describing the potentials as relaxation oscillations of relatively constant frequency. Howsever, the biological 'signal' frequency has a variable span, roughly in the range 1–100 Hz. Furthermore, the frequency is easily 'modulated' by electrical current stimuli and in many tissues, as in the so-called presso-receptors, a mechanical (pressure) external forcing acts as stimulus also influencing the frequency. In fact, the mechanism of 'frequency modulation' of an oscillatory membrane poses an intriguing problem for theoretical membranologists.

In the early discussions of the possible nature of the biological oscillations one has considered known chemical systems which exhibited rhythmicity (the iron wire of Ostwald-Lillie, the oil film model by A. Monnier and recently the Müller-Rudin bilayer membranes). The knowledge of rhythmical-chemical reactions has been greatly advanced in the recent years. At present the formal treatment and the biological application of these types of reactions are difficult.

Another type of model for excitable oscillatory systems was developed in the late fifties by the author [1]–[3]. It has been named the 'membrane oscillator'. It is a

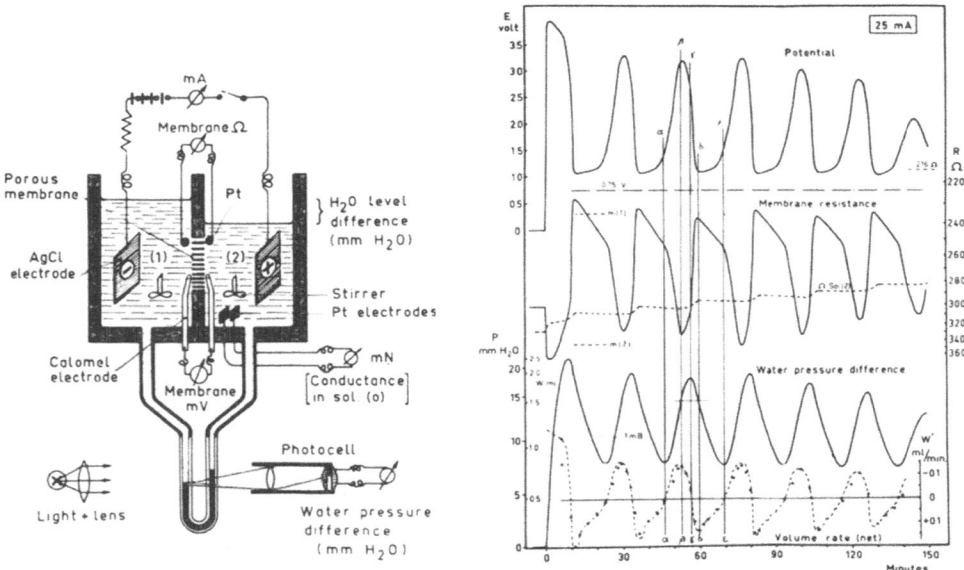

Fig. 1. The 'membrane oscillator' and the oscillatory changes of potential, resistance, hydrostatic pressure and water flow. (Reproduced from *Exp. Cell Research* (Suppl. 3), 1955 (left) – *J. Gen. Physiol.* **42**, 1959 (right)).

relatively simple system: The basic conditions are the presence of a porous fixed charge membrane separating two electrolyte compartments and a steady galvanic current flow across the membrane. Under these conditions transport of ions as well as of water (i.e. electro-osmosis) takes place which easily becomes oscillatory (Figure 1). The theoretical explanation lies in the complicated interplay between the driving forces within the membrane (chemical, electrical and hydrostatic gradients). The salient feature of the membrane oscillator is that it is 'excitable' by electrical as well as by hydrostatic pressure stimuli. In this respect the membrane oscillator is a rather unique model of a 'mechano-receptor'.

The original membrane oscillator theory has been extended to come closer to the actual biological membrane situation [4]–[6]. The new model, called the 'electro-hydraulic excitability analog', is a closed system, a 'membrane balloon'. Here the transmembrane pressure (electro-osmotic) difference appears as an interior cell 'turgor' pressure, balanced by *visco-elastic* structures in the membrane (Figure 2). This excitability analog also exhibits a pronounced frequency modulation.

The oscillation equations are described in the earlier papers, particularly in [5] and [6]. There are essentially six basic expressions (cf. Figure 3) derived from diffusion equations and the laws for electro-osmosis, arising from membrane channels lined with negative *fixed charges*. The electro-osmosis in the membrane is the central feature. The detailed formulations are given in the references. Here it may suffice to point out that the 'basic equations' can be condensed into a non-linear 2nd order differential equation (where also the coefficients are non-linear) of the type

$$a_0 \cdot \frac{d^2 x}{dt^2} + a_1 \cdot \frac{dx}{dt} + a_2 \cdot x = f(t)$$

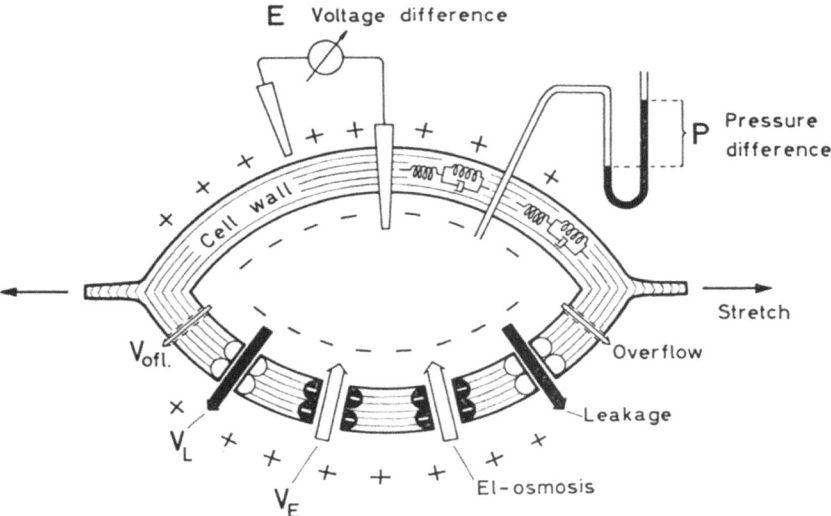

Fig. 2.   A scheme of an excitable cell (the 'electrohydraulic excitability analog').
(Reproduced from [6]).

The $f(t)$ represents the externally applied forcing, or 'stimulus', which can be electrical currents or mechanical displacements, affecting the 'turgor' pressure. The equation represents in a general form a variety of types of oscillators well-known within the realm of non-linear mechanics.

1. Some Biological Applications

1.1. An excitation model

The electrohydraulic excitability analog (=a specialized fixed charge membrane device) has been compared with actual biological mechano-receptors, which are sensitive to touch, stretching or pressure (examples are the touch receptors in the skin, the muscle spindles or the 'baroreceptors', which gauge the arterial blood pressure) [4]–[6].

It is well known that smooth muscles are particularly sensitive to stretching. The heart tissue has many functional similarities with the smooth muscle. Like a smooth muscle the heart tissue can respond to a mechanical distension with a twitch or contraction. This is particularly well documented for the hearts of lower animals (for example the snail). In these more primitive hearts and in the smooth muscle the sequence of events is a mechanical stimulus (distension) → action potentials → developments of contractile force (contraction). This is called 'excitation-contraction coupling'.

1.2. A muscle elasticity model

In our laboratory in Uppsala a great deal of experimental and theoretical work has been devoted to normal (and abnormal) heart rhythmicity. In the attempts to simulate theoretically the heart actions, it was soon found that the electrohydraulic model had to be supplemented with some equation which formalized the *elastic* properties of the muscle tissue. It is well known from muscle physiology, that the active contractile force development is subject also to passive processes, which cause a decay of single contractions, or hysteresis phenomena in repeated contractions. Usually these time dependent phenomena are ascribed to the interference of 'visco-elasticity'. There exist many formal descriptions in terms of elastic springs, dashpots etc., called the Maxwell, the Voigt models etc. However, another simpler, non-linear 'visco-elasticity analog' expression was found satisfactory. It relates the length displacement (ΔL) (which was varied also in the actual heart experiments) with the ensuing tension (T). The equation to be used is written in the scheme, Figure 3.

1.3. A formalism for muscle contractions

The general observation in contractile tissues (striated and smooth muscles, heart tissue) is that the active contraction force is triggered by a preceding action potential (AP), which, in turn, can either be spontaneously rhythmical, or be a consequence of a stimulus (electrical, mechanical, osmotic, chemical). The detailed links between the electrical action potentials and the ensuing tissue contraction are not well known at

1. The "Electrohydraulic" equations : $\boxed{\text{Eqs. 1–6}}$

2. The Volume–Pressure relation (M–P) of the excitable unit ("cell") :
$$\boxed{M = f(P)}$$

3. A hydraulic "viscoelasticy analog" (non-linear) :
$$\boxed{\frac{dT}{dt} = K\left[f'(t) - \bar{s}\cdot T\right]}$$

T   tension

f'(t)   the derivative of a sinusoidal, step or ramp function, applied as a length displacement (ΔL)

K   a function f(T) and of geometry

s   a "flow" constant

t   time

4. Total contractile force CF
$$\boxed{CF = \text{Passive tension } T + K(P_{rest} - P)}$$

GRAPHS

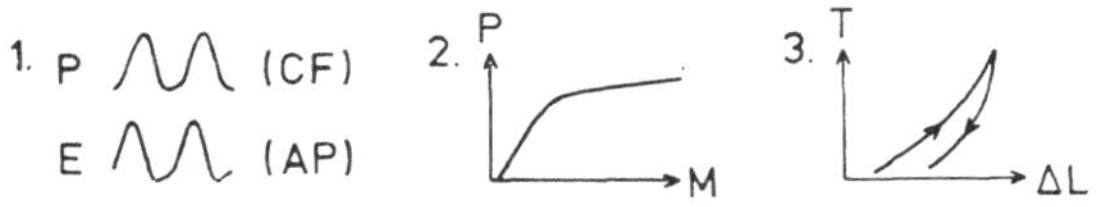

1. P $\wedge\!\wedge$ (CF)
   E $\wedge\!\wedge$ (AP)

2. (graph: P vs M)

3. (graph: T vs ΔL)

Fig. 3.   The basic set up of the 'excitation-contraction' formalism.

the present time. However, there exist a number of hypotheses, which suggest that the electrical potential somehow drives ions (Ca?) and that the ions, in turn, interact with biochemical systems of proteins and energy rich compounds leading to the shortening of the muscle cell.

There is at the present time no knowledge of any single biochemical or biophysical factor that can be used to describe the time and strength relations of the contractions apart from the 'total contraction force' itself (as available in the experimental force transducers). Nevertheless, from a purely biophysical descriptive point of view, it might be argued that there could exist an event, or a chain of events, which acts as a 'precursor', which represents or reflects the measured contractile force. Such a postulated event, or a 'factor', although enclosed in a 'black box', may be used as an essential part in a mathematical model. This mode of analysis of complicated events in purely descriptive and mathematical terms is not uncommon in pure biophysics. It may be justifiable to operate with symbolic parameters, which may be 'stand-ins' for actual unknown phenomena. The means might justify the end, in the sense that if the model reasoning is not only capable of describing but also of predicting events, it may be useful.

Following this philosophy it is now proposed that the *transmembrane pressure difference represents that part of the total contractile forces* which in muscle physiology is called '*active state*'. This assumption implies that the *contractile twitches are re-presented by cell pressure decreases. In summary*, the active part of the contraction is postulated to occur parallel and proportional to the change of the cell membrane pressure difference referred to some constant resting bias of state (in our notation $K(P_{rest}-P)$). The *total* contractile force (CF) is assumed to be the sum of this 'active' contribution and the passive tension, T.

1.4. THE OVERALL MATHEMATICAL FORMALISM

is summarized in Figure 3, which represents the basis for the excitation-contraction coupling formalism. The equations are essentially a set of simultaneous differential equations and can be solved by analog or digital computer techniques. The details of this procedure must be omitted in this paper.

1.5. A COMPARISON BETWEEN THE PROPOSED OSCILLATION FORMALISM AND SOME EXPERIMENTAL OBSERVATIONS ON THE HEART

For the purpose of this presentation we will limit the comparison between the theoretical formalism and actual physiological experiments to a few cases produced in this laboratory by Dr M. Almquist and obtained under well controlled and reproducible conditions [7]. The material was excised heart strips from the snail, *Helix pomatia*. The technique allowed rhythmical, or step wise, *stretch* stimulation to the *Helix* heart. This is very sensitive to changes in tension and responds with good regularity to the various types of mechanical stimulation. The preparation can exhibit numerous configurations both with regard to the contraction pattern and with regard to the action potential behaviour (the stretches were recorded through suitable presso-recording devices, the potentials were measured by means of suction electrodes applied

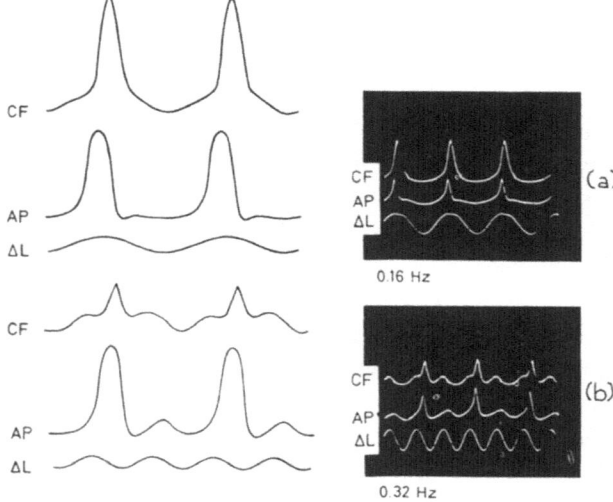

Fig. 4.   A comparison between the *computed* formalism (*left*) and *experiments* on heart strips from *Helix pomatia* (*right*, these oscillograms are reproduced from Almquist (7)). – *ΔL*, length of dis-placement; *AP*, action potentials; *CF*, total contractile force.

to the preparation). Especially interesting patterns were obtained by Almquist when applying sinusoidal mechanical stretches of frequencies of the same magnitude as the frequency of the spontaneous beating of the preparation (pseudo-sinusoidal). In such cases quite conspicuous interference may arise from the applied stimulus and the inherent rhythmicity. In Figures 4a and 4b the *insets* are reproductions of oscillo-grams from Almquist's work [3]. The experimental results in the insets should be compared with the theoretically computed responses in the *left-hand part* of the figures. Disregarding differences in scaling, the overall characteristics of the experiments are also found as a consequence of the applied mathematical formalism. One could especially notice the phase-relations between the wave of stimulus (bottom curve), the electrical response (middle) and the ensuing contractile force curve (top). The agree-ment between the experiments and the theory seems to be good and not simply explainable by a trivial addition of elementary components. Especially noteworthy from this point of view is the sometimes intricate shape of the action potentials, which may indicate the presence also of a 'contraction-excitation' coupling. It should also be mentioned that the several other experimental findings in Almquist's thesis were subject to a comparison with the computerized formalism, usually with amazingly good general agreement.

1.6. CONCLUSION

The application of proper kinetics to a fixed charge membrane may help to describe at least the formalism of some oscillatory and very non-linear biological systems, including the beating of the heart.

Acknowledgements

This work has been supported by grants from the Swedish Medical Research Council (14X-629-10B), the National Institutes of Health (3 R01-HL-12960), the M. Bergvall Foundation and the Jeansson Foundation, which are gratefully acknowledged. The author also owes his sincere thanks to Monica Ohlson, laboratory engineer, for con-ducting the biological experiments, Helena Billander and Ulla-Britta Berg, laboratory assistants, for assistance with the computation work and to M. Almquist, M. D., for the courtesy to use some of his material on the *Helix* heart and also for stimulating discussions. Mrs Ebon Arnelund has also given valuable assistance to our work.

References

1. Teorell, T.: *J. gen. Physiol.* **42**, 831 (1959).
2. Teorell, T.: *J. gen. Physiol.* **42**, 847 (1959).
3. Teorell, T.: *Biophys J.* **2**, No. 2, Part 2 (Suppl.), 27 (1962).
4. Teorell, T.: *Arch. int. Pharmacodyn.* **140**, 563 (1962).
5. Teorell, T.: *Ann. N.Y. Acad. Sci.* **137**, 950 (1966).
6. Teorell, T.: in *Handbook of Sensory Physiology* (Vol. I) (ed. by W. Loewenstein), Springer-Verlag, Berlin, Heidelberg, pp. 291–339, 1971.
7. Almquist, M.: 'Mechano-Electrical Transduction in Isolated Myocardium of *Helix pomatia*,' *Acta Univ. Uppsal.* Dissertation, 161, 1973.

PHYSICO-CHEMICAL PROPERTIES OF THE NERVE MEMBRANE

I. TASAKI

Laboratory of Neurobiology, National Institute of Mental Health, Bethesda, Md., U.S.A.

Abstract. (1) The membrane of a squid giant axon was examined from a physicochemical point of view. The Donnan phase-boundary potential at the outer membrane surface is shown to play an important role in determining the membrane potential.

(2) Under internal perfusion with a dilute salt solution, the excitability of a squid axon membrane can be maintained in an external medium containing a divalent-cation salt but no univalent-cation salt. The effects of adding univalent-cation salts to the external $CaCl_2$ solution are described.

(3) Under internal perfusion with a dilute NaF solution, the axon membrane can be excited by non-electrical means, such as a rise in the univalent-divalent cation-concentration ratio or lowering of the temperature of the external medium. The phenomenon of excitation of the squid axon membrane is explained in terms of a cooperative cation-exchange process involving divalent and univalent cations.

(4) The existence of large fluctuation in electrochemical properties near the critical points of the membrane is described. Small domains of the membrane in the excited state are shown to exist surrounded by the resting membrane. The importance of electric interaction between different parts of the membrane is emphasized.

(5) The effects of external Na-ions and internal K-ions on the action potential are discussed on the basis of the macromolecular concept of nerve excitation.

(6) The significance of the voltage-clamp procedure and of the equivalent circuit of the membrane is described.

(7) The results of recent experiments obtained by various optical techniques are briefly reviewed.

1. Introduction

From an electrochemical point of view, a live nerve fiber (axon) immersed in normal tissue fluid is an extremely complex system. In such natural environment, the external fluid medium of the axon contains many univalent and divalent ions. The ionic composition in the axon interior is very different from that in the external medium. There are continuous chemical reactions taking place in- and outside the axon. Furthermore, there exist in the axon interior polyelectrolytes (mainly proteins) which often make measurements of the membrane potential quite unreliable.

This membrane system can be greatly simplified by the *technique of internal perfusion* applicable to squid giant axons. By this technique, the polyelectrolytes inside the axon can be completely removed and a perfusion solution of a well-defined chemical composition can be introduced into the axon. The ability of the axon to develop 'action potentials' can be maintained for many hours when the electrolyte compositions in- and outside the axon are properly chosen. Since all the water soluble substrates are continuously washed away by the perfusion fluid, it is most unlikely that there are normal metabolic processes going on in the axon under these conditions.

Although an axon membrane under these experimental conditions can be treated like an inanimate membrane, this membrane possesses peculiar properties which are not generally encountered in inanimate membranes. The axon membrane is extremely

Eric Sélégny (ed.), Charged Gels and Membranes II, 213–238. All rights reserved.
Copyright © 1976 by D. Reidel Publishing Company, Dordrecht-Holland.

labile. Its ability to develop action potentials (i.e., excitability) is suppressed irreversibly by a large number of physical factors. Or rather, the axon excitability is maintained only in a limited range of pH, temperature, osmolarity, redox potential, electrolyte compositions in- and outside the axon.

It is important to note in this connection that this lability of the axon membrane is directly related to the process of action potential production itself. We believe that the process of action potential production is an electrochemical manifestation of a drastic, but reversible, physicochemical change in the macromolecular complex of which the axon membrane is composed [1, 2]. Under usual experimental conditions, this drastic physicochemical change is brought about by a very weak electrical stimulus, namely by a change in the transmembrane potential difference of about 25 mV which corresponds to the familiar thermodynamic quantity RT/F at room temperature. This fact indicates that the membrane system is unstable against perturbations which is comparable to the thermal agitation of the ions in the system.

In this presentation, various properties of this labile membrane system of an excitable axon are discussed. Here, the axon membrane is treated mainly from a macromolecular point of view, rather than on the basis of the equivalent circuit concept of Hodgkin and Huxley [3]. Because most macromolecules of biological origin are negatively charged in neutral and alkaline ranges of pH, it is expected that the axon membrane behaves like a cation-exchanger membrane. The properties of a negatively charged membrane are known to be strongly affected by multivalent cations. It would not be surprising, therefore, to find that the axon membrane is strongly influenced by divalent cations. Thus, analyses of the process of nerve excitation involve investigation of the modes of interaction between the divalent cations and the membrane macmolerocules.

2. Does the Axon Membrane at Rest Behave Like a Potassium Electrode?

In most elementary textbooks of physiology, the axon membrane is regarded as being permeable specifically to K-ions in its resting state. However, many investigators in the field are not as confident as these textbook writers on this point. In his monograph, Hodgkin makes the following cautious remark: "One of the main pieces of evidence for regarding the resting potential *at least partly* due to the potassium concentration cell is that *at high potassium concentrations* the membrane behaves like a potassium electrode" (p. 31 in ref. [3]). Arhem and Frankenhaeuser [4] say on this point: "The behavior like a potassium electrode might therefore apply to a fiber with low membrane potential *only*." Stämpfli [5] raises the same question: Does the axon membrane in the resting state behave like an ideal potassium concentration cell? He concludes: "Our experimental results clearly indicate that the answer must be a *negative* one." [Emphasis by italicizing words in quotation was made by the present author.]

In electrophysiology, an axon with a low membrane potential is said to be in a 'depolarized' state. Nobody questions the experimental finding that the membrane potential is dominated by K-ions when an axon (containing a high concentration of K-ion inside) is depolarized by external application of KCl. However, it is important to note that the membrane potential varies with the external potassium concentration logarithmically *only* when the axon is rendered *inexcitable** by the increased potassium concentration (p. 63 in ref. [1]). In the range of K-ion concentration which does not suppress the axon excitability, a variation in the external K-ion concentration *does not* alter the membrane potential significantly.

By the technique of intracellular perfusion, the effect of varying the internal K-ion concentration on the membrane potential was examined. Using a mixture of NaF and KF solutions for perfusion, the potassium-sodium ratio in the axon interior was varied in a wide range, keeping the total concentration at a constant level [8].

In the entire range of internal K-ion concentration examined, the axon immersed in normal sea water remained excitable. It was found by this technique that the resting membrane potential was hardly affected by a large variation in the internal K-ion concentration. It is therefore quite clear that the resting membrane potential of an excitable squid giant axon is not determined by the K-ion concentration ratio across the membrane.

According to the theory of charged membrane (Teorell [9]; Meyer and Sievers [10]), the membrane potential is given by the sum of two phase-boundary potentials nad the intramembrane diffusion potential. In the presence of the salts of divalent actions in the external medium, the phase-boundary potential at the outer membrane surface is expected to depend strongly upon the divalent cation concentration. The experiment to be described below indicates that this is actually the case.

3. Negative Fixed Charges of the Axon Membrane

The axon was examined using internal perfusion with a potassium salt solution (Figure 1). [A very similar result could be obtained using intact, i.e., internally unperfused, axons.] The external surface of the axon membrane was exposed to a circulating solution of $MgCl_2$ (or $CaCl_2$). The concentration of this salt was varied by mixing the 0.4 M salt solution with an iso-osmotic sugar solution, the pH of the mixture being adjusted to 8.0 with a trace of tris-buffer. No univalent-cation salt was added to the external medium.

As can be seen in the figure, the membrane potential measured with calomel electrodes in- and outside the axon varied with the external $MgCl_2$ concentration in a perfectly reversible manner. For a four-fold change in the concentration $[Mg]_e$, the change was approximately 15 mV. The membrane potential under these conditions, E_r, is very close to the value given by the following equation.

$$E_r = \frac{RT}{2F} \ln [Mg]_e + \text{constant} \qquad [(1)$$

*Axons immersed in a high potassium-salt solution are inexcitable when tested with an ordinary stimulus, i.e., with an outwardly directed transmembrane current; however when tested with an inwardly directed transmembrane current, these axons show a distinct sign of excitability indicating a transition from the depolarized to the resting state in a cooperative fashion (see Segal [6]; Stämpfli [5]; Tasaki [7]).

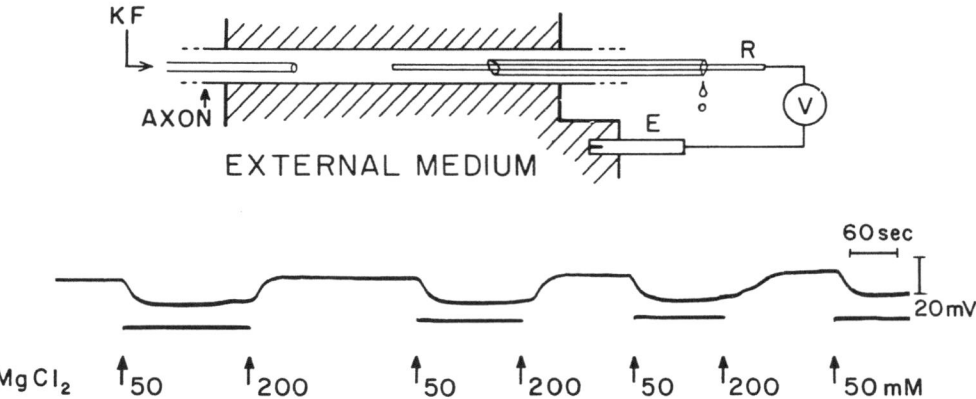

Fig. 1. *Top:* Schematic diagram showing the experimental setup used to determine the effect of changing the external divalent cation salt concentration on the membrane potential of a squid giant axon. *E* represents a large calomel electrode and *R* a small glass capillary electrode filled with 0.6 M KCl. *Bottom:* Oscillograph record taken from an axon internally perfused with 0.3 M KF solution. 21 °C. The concentrations of MgCl₂ solutions used are indicated below. (From Tasaki *et al.: Am. J. Physiol.* **213**, 1467 (1967).)

expressing the dependence of the Donnan phase-boundary potential on the divalent cation. Judging from the magnitude and the sign of these potential changes, it is evident that the axon membrane behaves, in the range of concentrations examined, like an ideally permselective cation-exchanger membrane. Since the external anions at the concentration of 0.4 equiv/l are excluded under these conditions, the effective fixed charge density in the external membrane layer is considered to be high enough to effectively exclude anions of this concentration in the external medium.

It is to be noted in this connection that $CaCl_2$ can be used in place of $MgCl_2$ in this experiment, yielding almost identical results. It is well known that the mobility of Ca-ion in the membrane is far smaller than that of K-ions [11]. The phase-boundary potential can vary in accordance with the Nernst equation even when the 'permeability' of the membrane to the ions under study is very low.

[In the range of $CaCl_2$ concentration lower than about 30 mM, the observed changes in the membrane potential was smaller than the value expected from an ideally permselective membrane. This result is not surprising in view of the fact that immersion of an axon in a salt-free sugar solution for a short period of time exerts a serious detrimental effect upon the axon excitability. Probably, an electrostatic force across the phase-boundary, required to maintain the Donnan distribution of ions, brings about an undesirable side-effect on the labile membrane macromolecules.]

4. Ions Required to Maintain Excitability

It is well-known that, under ordinary experimental conditions, the action potential amplitude varies directly with the logarithm of the external Na-ion concentration

(Hodgkin and Katz [12]). However, the presence of Na-ion in the external medium is not essential for the maintenance of excitability. Hydrazinium-ion, guanidinium-ion, etc.* can be used in place of Na-ion to sustain excitability in frog nerve fibers [13] as well as in squid giant axons [14]. It is also known that, under internal perfusion of squid giant axons with favorable salt solutions (see below), large action potentials* can be observed in axons immersed in a Ca-salt solution to which no salt of univalent cations is added [15].

Most electrochemists find it very difficult to examine physicochemical properties of an ion-exchanger membrane system involving many counter-ions of different valences. In squid giant axons, it is possible to reduce the number of counter-ions considerably by the method of intracellular perfusion. However, it is not possible to eliminate divalent cation, Ca-ion in particular, because the axons lose their ability to develop action potentials when immersed in divalent-cation free media. It is important to note that, when applied intracellularly, divalent cations exert a strong detrimental effect on the axonal excitability. At concentrations lower than 1 mM, Ca-ion in the internal perfusion fluid destroys the axon membrane irreversibly in a relatively short period of time [11]. In light of these fact , the simplest environmental conditions under which the axon excitability can be maintained is to introduce a univalent cation salt internally and a divalent-cation salt externally.** For example, the following system is capable of developing all-or-none action potentials:

$$electrode \,|\, CaCl_2 \text{ solution} \,|\, axon \text{ membrane} \,|\, NaF \text{ solution} \,|\, electrode$$

[Note that the osmolarity of both internal and external media has to be maintained by addition of glycerol or sucrose.]

Figure 2 shows an example of the experimental results obtained recently by Inoue et al. [20], demonstrating that the behavior of the axons taken from squid available in Japan (Doryteuthis bleekeri) is very similar to that of the squid caught near Woods Hole (Loligo pealii). The resting membrane potential was about 30 mV (inside negative) under these conditions. It is seen in Record B that a pulse of outwardly directed transmembrane current produced an action potential of about 60 mV in amplitude in an all-or-none manner. Weak current pulses applied to the membrane in the resting (Record A) and excited state (Record C) of the membrane indicate that there is an approximately 10-fold increase in the membrane conductance associated with the process of nerve excitation.

---

*It is frequently stated in biological literature that a puffer-fish poison, tetrodotoxin, blocks specifically the inward movement of Na-ion across the nerve membrane. The action potentials of axons immersed in these Na-free media are suppressed and the inward current under voltage-clamp (see below) is eliminated by this poison. Tetrodotoxin cannot be regarded, therefore, as an agent which blocks the flux of Na-ions specifically (cf. Narahashi et al. [16]; Teorell [17] and Moore et al. [18] on this point).
**Replacement of one internal anion species with another without changing the cation species does not usually bring about any immediate change in axon excitability. However, to maintain excitability for a long period of time under continuous intracellular perfusion, a proper choice of anions is very important. Anions with high lyotropic numbers (SCN⁻, I⁻, Br⁻, etc.) are known to be unfavorable and those with low numbers (F⁻, phosphate, etc.) are favorable for the maintenance of the ability of the axon to produce large action potentials [19].

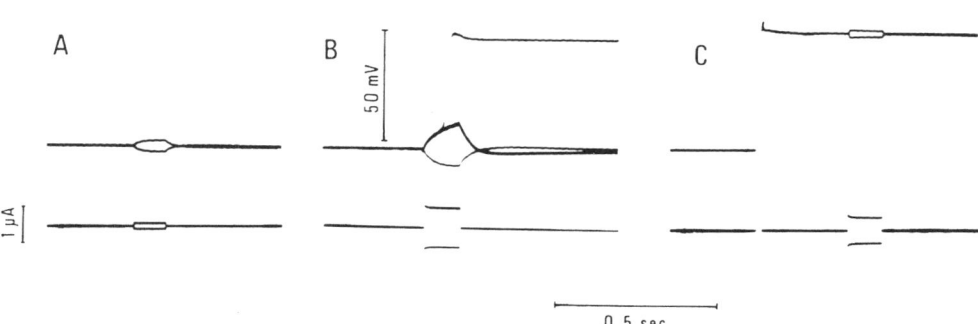

Fig. 2. Demonstration of electric excitability of a squid giant axon immersed in a CaCl₂ solution. The axon under study was internally perfused with a dilute NaF solution. The concentration of the external CaCl₂ was 0.1 M. Record A shows the effect of a subthreshold current pulse (of which the duration and intensity are indicated by the lower oscillograph trace) on the membrane potential (upper trace). Record B was taken at threshold intensity of stimulation; out of three trials at this intensity, an action potential (of about 50 mV in amplitude) was evoked only once. The action potential was a few sec in duration and, therefore, only its inital part is seen in the record. The action potential in Record C was evoked by a short, strong current pulse. In each record, the effect of an electric pulse of the same intensity but of the opposite sign on the membrane potential is shown. (From Inoue *et al.*, 1972, unpublished.)

It is interesting to compare the ionic compositions of these media with those of the normal environment of intact axons. The major components in the normal external medium are NaCl and divalent-cation salts; the normal protoplasm contains a high concentration of K-ion and a low concentration of Na-ion. Therefore, the ionic environment of the axon under the condition of the experiment shown in Figure 2 is very close to that of a normal axon from which *the external Na-ion and internal K-ion have been removed.* On this ground one might say that both the external Na- and internal K-ion are simply 'modifiers' of the action potential and that the fluxes of these ions during the process of nerve excitation are merely 'symptoms' of the conformation changes of the membrane macromolecules (see Section 9).

The amplitude and configuration of the action potential were found to be affected to a considerable extent when the internal Na-ion was replaced with other alkali metal ions or polyatomic univalent cations [15]. With 30 mM tetramethylammonium-ion internally and 100 mM CaCl₂ externally, action potentials of about 150 mV in amplitude were observed. Very large univalent cations, such as tetrabutylammonium-ion or benzylamine, were incapable of sustaining action potentials. Small polyatomic cations with hydrophilic side-groups, such as hydrazinium or aminoguanidinium, gave rise to relatively small, but all-or-none, action potentials.

The action potentials observed under these conditions are invariably very long in duration and look almost rectangular on the screen of a recording oscilloscope. It is possible, therefore, to treat the excited state of the membrane as a quasi-stationary state and to speak of the mobilities and selectivities of ions in this state. Undoubtedly, the process of excitation brings about a sudden change in the mobilities and selectivities of the cations in the system. Judging from the difference in behavior among different internal cations, it is concluded that the selectivity for strongly hydrated ions (Na^+, $NH_2NH_3^+$, etc.) increase more markedly than that for cations with hydrophobic side-groups (Me_4N^+, Et_4N^+, etc.) when the axon is excited by electric stimuli [1, 15]. This conclusion was borne out by the finding that the amplitude of the observed action potential is enhanced by addition of strongly hydrated univalent cations to the external medium, but not by addition of those with hydrophobic side-chains [15].

5. Non-Electric Means of Exciting the Axonal Membrane

In the experiments illustrated in Figure 2, pulses of electric current were used to induce transitions of the axonal membrane from the resting to the excited state. In electrophysiology, application of an electric current is a favorite means of exciting the nerve membrane. In the case of electric pulses, the intensity is readily adjustable, the delivery can be accurately timed and no undesirable after-effects are encountered. Because of this manipulative convenience, electric pulses are used exclusively in almost every quantitative study of the axonal membrane.

It is important to note, however, that the axonal membrane can be *excited by non-electric means* without being troubled by irreversible after-effects. Records A and B in Figure 3 show that a transition can be induced by addition of a univalent-cation salt to the external medium. The axon under study was internally perfused with a dilute NaF solution and was immersed initially in a $CaCl_2$ solution. As indicated by the lower oscillograph trace in Figure 3A, NaCl was added to the external medium, keeping the $CaCl_2$ concentration at a constant level. The membrane potential, shown by the upper oscillograph trace, did not change appreciably when the NaCl concen-

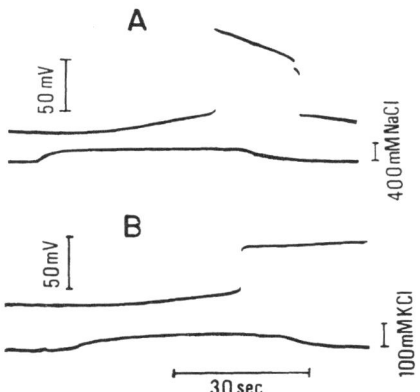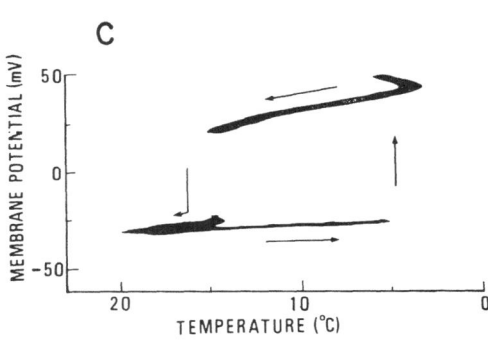

Fig. 3. Demonstration of 'excitation' of the squid axon membrane by non-electric means. The axons were under continuous internal perfusion with a dilute NaF solution. Record A shows excitation by addition of NaCl to the external $CaCl_2$ solution. Record B demonstrates excitation by addition of KCl. Record C shows excitation by cooling; the hysteresis loop in the recorded membrane potential was obtained by a cyclic change in the ambient temperature. The external medium contained both $CaCl_2$ and NaCl in this case. (From Inoue *et al.*: *Biochem. Biophys. Acta* **307**, 473 (1973).)

tration was relatively low. As the external NaCl concentration was increased gradually, a critical point was finally reached at which a large potential change appeared suddenly.

In the experiment illustrated in the lower part of Figure 3 (Record B), an abrupt change in the axonal membrane was induced by addition of KCl to the external medium. In this case, the observed abrupt change in the membrane potential was definitely smaller than that produced by addition of NaCl (Figure 3A). However, the critical concentration of KCl required was far smaller than that for NaCl (approximately 1/10). Very similar results were obtained using axons internally perfused with a dilute solution of CsF.

The ability of alkali metal ions to induce this type of transition follows the following sequence:

$$Na^+ < Cs^+ < Rb^+ < K^+$$

(p. 145 in ref. [1]). The critical concentration of these univalent cations varies to a considerable extent with the divalent-cation concentration in the medium. A four-fold increase in the CaCl$_2$ concentration tends to double the critical concentration of KCl; however, the reproducibility of these measurements is limited [21].

Figure 3C shows abrupt changes in the state of the axon membrane induced by changes in the temperature. The external medium of the axon under study contained both CaCl$_2$ and NaCl. The axon interior was perfused with a dilute NaF solution. The temperature of the external medium was lowered uniformly along the axon. When the temperature reached about 5 °C in this case, there was a sudden rise in the intracellular potential (depolarization). A further lowering of the temperature brought about only a small and smooth change in the membrane potential. When the temperature was raised following the appearance of an abrupt change, there was an abrupt change in the membrane potential in the reversed direction (repolarization). However, the temperature at which the abrupt repolarization was observed was much higher than the critical temperature for abrupt depolarization. In other words, a distinct hysteresis was observed in response to a slow cyclic change in the temperature.

It may be pointed out in this connection that these abrupt potential jumps can be observed only in excitable cells which produce prolonged action potentials (compare ref. [22] and [23]). This difference in behavior is explained in the following manner. In cells which produce very brief action potentials, it is not possible to alter the temperature or the chemical composition of the external medium uniformly over the whole surface of the cell during the time comparable to the duration of the action potential; this reults in asynchronous transitions at different parts of the membrane. The abruptness can not be recognized when the membrane is spatially non-uniform and transitions at different parts are asynchronous (see Section 8).

6. Cooperative Cation-Exchange Process in Axonal Membrane

In inanimate ion-exchanger membranes, the mobilities and selectivities of counter-ions are strongly affected by the composition of the electrolytes in the surrounding fluid media. A change in the ionic composition of the medium alters the concentrations and the ratios of various ions within the membrane, thereby affecting the ion-mobilities and selectivities. The experimental results shown in Figures 2 and 3 indicate that, at critical points created by various means, the ion mobilities and selectivities change abruptly from one set of values to a different set. This fact may be interpreted as an indication of a cooperative ion-exchange process which alters the population of intra-membrane ions abruptly and profoundly.

The ion-exchange isotherm shown in Figure 4 is constructed with a view toward explaining this proposed ion-exchange process [24]. Here, the membrane is assumed to be composed of a cation-exchanger in which *univalent* and *divalent* cations are competing for the ionized sites. To obtain this isotherm, it is assumed that occupancy of two neighboring sites by cations of different valencies is energetically unfavorable. This was done by describing the activity-coefficients of the ions in the membrane in terms of Kieland's equation [25]. [The statistical mechanical significance of Kieland's equation is discussed by Barrer and Falconner [26]].

By the aid of the isotherm shown in Figure 4, it is possible to offer a reasonable interpretation of the cooperative cation-exchange process in the axonal membrane.

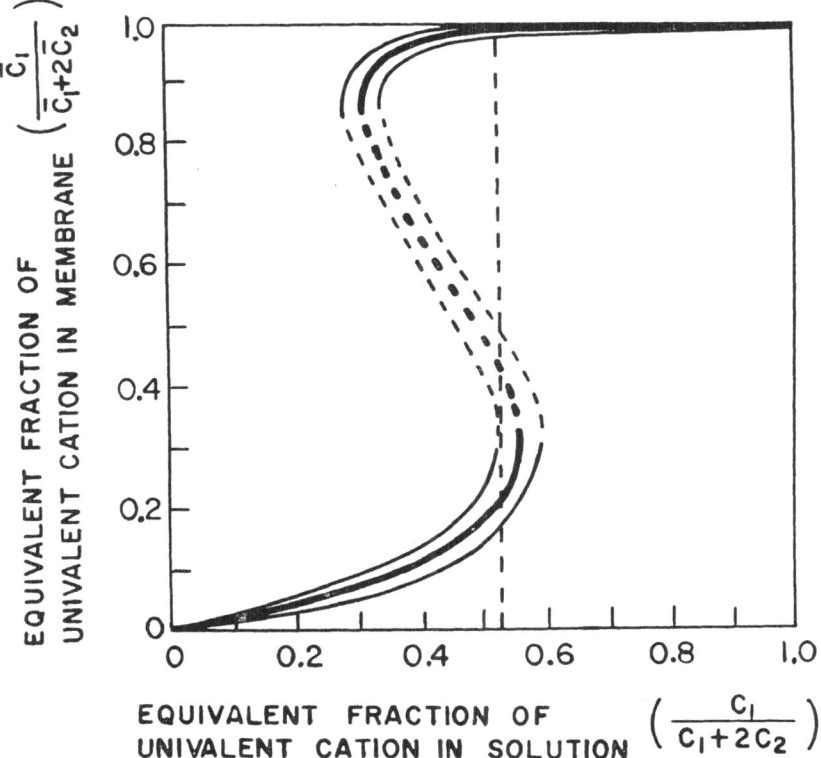

Fig. 4. A hypothetical ion-exchange isotherm introduced to explain the process of nerve excitation by electrical, thermal and chemical means. (From Tasaki: *J. Gen. Physiol.* **46**, 755 (1965).)

As the fraction of the external univalent cation is raised from a small value, the corresponding fraction within the membrane rises smoothly (up to 0.2 in the figure). When the equivalent fraction within the membrane is raised to the critical point (0.3 in the figure), there is a sudden rise in the fraction in the membrane (indicated by the vertical line). A further increase in the equivalent fraction in the medium raises the corresponding fraction in the membrane smoothly. When the equivalent fraction in the medium is lowered from a high level, the corresponding fraction in the membrane changes in a cyclic manner. The transition represented by the vertical line C–E must be accompanied by a sudden change in the potential difference and resistance across the membrane. Figure 4 also indicates that a cyclic change in the external equivalent fraction may produce a hysteresis loop in the observed membrane potential. In fact, it is quite easy to demonstrate such a hysteresis in the membrane potential and resistance under these conditions [20, 21].

The mechanism of thermally induced abrupt change in membrane properties can be explained in a similar manner in terms of ion-exchange isotherm shown in Figure 4. In inanimate cation-exchangers, substitution of univalent cations for divalent cations is known to be exothermic, involving an enthalpy change in the range between 1.5 and 3 kcal/equivalent [27, 28, 29, 30]. Replacement of a univalent cation species with another is associated with a very small enthalpy change. These facts suggest that the critical point for transition is displaced to the left in the figure when the temperature is lowered. Therefore, when the equivalent fraction of the external medium is not far from the critical value at room temperature, cooling is expected to induce a transition of the membrane from a divalent-cation rich state to a univalent-cation rich state. Similarly, raising the temperature is expected to shift the critical point for abrupt repolarization to the right in the figure.

It is also possible to interpret the process of electrically induced abrupt depolarization (see Figure 2) on the basis of the diagram in Figure 4. An outwardly directed current through the membrane in the resting state tends to drive the intracellular univalent cations into the membrane. Furthermore, an electric field produced by the current may alter the density of the electric charge near the external surface of the membrane, producing a change in the ion-distribution across the surface. Near the critical point, the ion-exchange process is assumed to be cooperative. Therefore, a relatively small amount of electric energy (comparable to the energy thermal motion of ions) is sufficient to induce a transition. Similarly, a pulse of inwardly directed current applied to an axonal membrane in the excited state is expected to induce a transition of the membrane to the resting state. – Note that the transference number of Ca-ion in the membrane is very small and consequently that an inwardly directed current tends to accumulate divalent cations in the outer layer of the axon membrane. – This transition of the membrane from the excited to the resting state is wellknown and is called 'abolition of an action potential' (Tasaki [1, 31]).

In the diagram shown in Figure 4, the lower solid line corresponds to the resting state and the upper solid line represents the excited (i.e., depolarized) state of the axon membrane. A jump from one state to another is regarded as a kind of phase-transition occurring in the membrane macromolecules. The two thin lines drawn near the heavy line in the figure represent a variation in the isotherm at different parts of the axonal membrane and their significance will be discussed later.

7. Critical Fluctuations in the Axonal Membrane

Quite recently, large fluctuations in the membrane potential were observed in the axonal membrane brought close to a critical point for transition from one state to the other (Inoue et al. [20]). Figure 5 shows an example of the results obtained. Here, the axon was internally perfused with a dilute NaF solution and was immersed in a $CaCl_2$ solution initially. As NaCl was added to the external medium, the amplitude of the fluctuating membrane potential increased gradually. Immediately before the onset of abrupt depolarization, the amplitude reached an extremely high level (Record C). It is seen in the figure that a large potential jump took off from one of the peaks of the fluctuating membrane potential. Similarly, large fluctuations were observed when the axonal membrane was brought close to the critical point for abrupt repolarization.

This increase in the fluctuation of the membrane potential is very similar to 'giant fluctuations' observed in systems in thermodynamic equilibrium [32]. However, it is important to realize that an axon membrane separating the external and the internal perfusion solutions is an open system with a continuous dissipation of free energy. Transitions between two stable dissipative membrane structures were analyzed by Kobatake [33]. Glansdorff and Prigogine [34] discussed the general properties of instability and fluctuations in open systems.

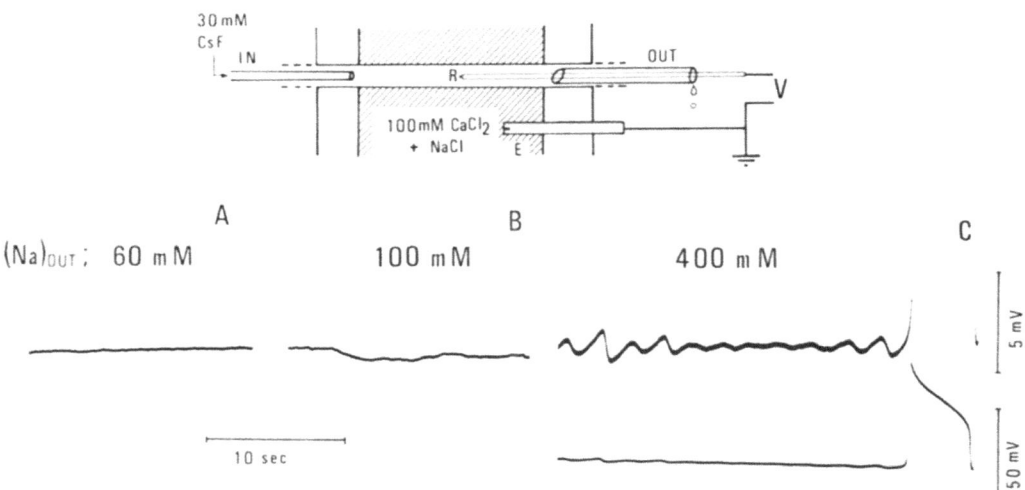

Fig. 5. Oscillograph records showing large fluctuations in the membrane potential observed near the critical point for transition. The axon under study was brought close to the critical point by addition of NaCl to the external $CaCl_2$ solution. The lower oscillograph trace in Record C was taken simultaneously at a low oscilloscope sensitivity. (From Inoue et al.: *Biochim. Biophys. Acta* **307**, 475 (1973).)

8. Spatial Non-Uniformity of the Axonal Membrane

In the preceding sections we have described the experimental bases for regarding the axon membrane as consisting of macromolecules which have two distinct conformational states. We consider these two states to be stable against small perturbations. Changes in the electrolyte compostion of the media, temperature changes or application of electric current pulses through the membrane can be used to perturb these macromolecules. These two stable states are separated by an unstable state. When the perturbation is intensified to a critical point, the macromolecules may undergo a transition from one state to the other.

According to this picture in mind, the axonal membrane can possess in principle a 'mixed' or 'mosaic' structure in which a macroscopically uniform membrane is composed of zones in two different states. Such a mixed state has been observed on the surface of an Ostwald-Lillie model of nerve, namely, an iron wire immersed in concentrated nitric acid (see e.g. Franck [35]). The surface of the iron wire under these conditions can be in either one of the two states, oxidized or reduced. Under favorable conditions, a pattern indicating the presence of a mixture of the two states on the surface can be seen with an optical microscope.

Last spring, Inoue *et al.* [36] obtained direct experimental evidence for the presence in the axonal membrane of patches in the excited state surrounded by resting areas. The experimental arrangement used for this experiment is illustrated schematically in Figure 6, top. The axon under study was internally perfused with a dilute NaF solution and was immersed in a medium containing both NaCl and $CaCl_2$. The potential difference across the membrane was recorded with a small calomel electrode in the axon (referred to a large electrode outside). A small galss-pipette electrode, with a diameter less than one micron at the tip, was used to record small potential variations produced by localized membrane currents. The tip of the micro-electrode was pressing the membrane surface without penetration. A negative potential recorded with this micro-electrode (referred to the large external electrode) indicates the existence of an inwardly directed current through the membrane in the vicinity of the tip of the micro-electrode. Conversely, a positive potential recorded with the micro-electrode is a sign of an outwardly directed membrane current at the position of the micro-electrode. When there are patches of the membrane in the excited state surrounded by a resting membrane area, the patches in the excited state are traversed by inwardly directed currents and the resting area by outwardly directed currents.

The oscillograph trace marked 'M' in the figure shows the potential variations recorded with the micro-electrode near the surface of the axonal membrane. The trace marked 'R' indicates the 'average potential' inside the axon recorded with the electrode at the center of the axon. The axon was electrically excited by a pulse of outwardly directed current through an internal stimulating electrode. As can be seen in the figure, small potential changes with a rectangular configuration were observed with the micro-electrode during the prolonged action potential. The small rectangular potential variations were *either positive or negative* (relative to the large electrode in the external

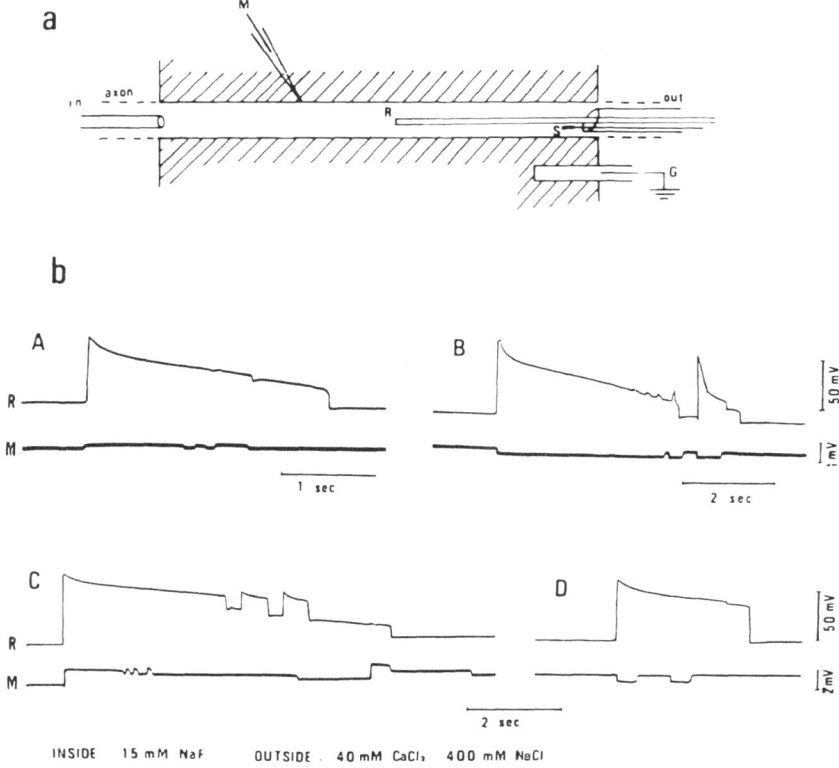

Fig. 6. Demonstration of 'domains' of the axon membrane in the excited state surrounded by resting membrane areas. The diagram on the top shows the experimental arrangement used. M, a glass-pipette microelectrode; R, an intracellular glass-tubing electrode; S, a metal electrode used for stimulation; G, a large Ag-AgCl electrode used for grounding the external medium. The oscillograph traces marked R and M indicate the potential variations recorded with electrodes R and M, respectively. (From Inoue et al., *Biophys. Chem.* **2**, 116 (1974).

medium). At a given position of the micro-electrode on the surface of the membrane, the sign and amplitude of the potential variation was approximately constant.

These records may be regarded as the direct evidence for the co-existence of 'domains' in the excited state together with resting areas of the membrane. We made an attempt to determine the size of the domains of the membrane in the excited state. However, the present method did not yield reliable results. From the distance between sites giving rise to potential variations with different signs, we roughly estimate the area of individual domains in the excited state to be of the order of several microns in diameter under the conditions of this experiment.

On the basis of these observations, it is evident that the membrane potential recorded with a large electrode at the center of an axon represents the sum of the contributions from all the domains located in the vicinity of the internal electrode. When the axon as a whole is at rest, a large fraction of the axonal membrane must be in its resting state. When the axon membrane is brought close to its critical point by various forms of perturbation (see Figures 2 and 3), the fraction of the membrane in the excited state increases [cf. Changeux *et al.* [37] for a similar view].

There is good reason to believe that the axonal membrane is essentially non-uniform in the sense that the critical points (represented by the position of the vertical line in Figure 4) are slightly different at different parts of the membrane. [Note that there are many Schwann cells immediately outside the axonal membrane, and these cells are expected to produce a slight variation in the current density along the surface of the membrane.] Under such circumstances, the ion-exchange isotherm for the axon membrane should be represented, not by a single sigmoid curve, but by a large number of curves with slight variations in the critical points (see the three curves in Figure 4). At the end of a stimulating pulse, a limited number of domains with a low threshold undergo a transition from the resting state to the excited. Since there is a difference in the potential between these domains and the remaining resting areas, there is a flow of electricity between different parts of the membrane. Undoubtedly, the dissipation of free energy associated with these currents, which we may call 'ring-currents', play an essential role in the creation of instability of the axonal membrane at the end of a stimulating current pulse of the threshold strength.

It is important to realize that a domain in the excited state (surrounded by the areas at rest) is traversed by an inwardly directed current. This current tends to bring about a transition of this domain to its resting state. When, however, the fraction of the membrane in the excited state is increased to a certain (i.e., critical) level, the outwardly directed currents through the resting areas become strong enough to bring about depolarization of the entire (or almost entire) membrane. The critical fraction of the axonal membrane is known to be a few percent or less of the entire functional surface of the membrane (p. 133 in ref. [1])

Physicochemical properties of mosaic membranes were discussed by Kedem and Katchalski [38]. In this conference, Caplan [39] expanded on his investigation of mosaic membranes. Most of the arguments developed by Caplan is expected to be applicable to the nerve membrane in the mixed state.

9. Relationship between a Prolonged and a Normal Action Potential

The action potentials recorded in the experiments described up to this point have an abnormally long duration. Under the conditions of the experiments shown in Figures 2, and 3, the cations in- and outside the axons studied were those which exist in the normal environment of the axon. In these experiments, however, the external Na-ions and the internal K-ions which, are abundant in the normal ionic environment, have been either completely or partially removed. We now examine the effects of putting back these missing univalent cations into the media on the electrochemical properties of the axons.

The oscillograph records presented in Figure 7 were obtained by Inoue *et al.* [36] to illustrate the effect of Na-ions added to the external medium. Action potentials of the

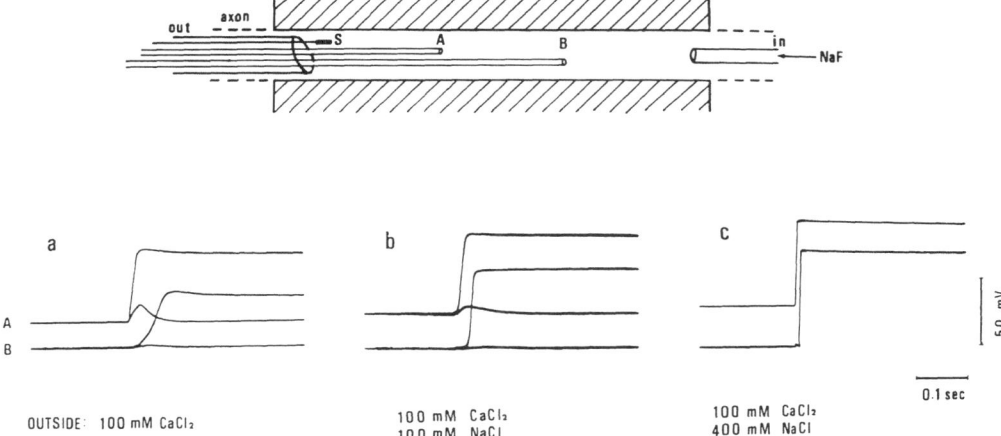

Fig. 7. Oscillograph records showing the effect of addition of NaCl to the external CaCl₂ solution on the amplitude and the conduction velocity of the action potential. The axon under study was internally perfused with a dilute NaF solution. (Inoue *et al.*, *Biophys. Chem.* **2**, 116 (1974).

axon were recorded with two intracellular electrodes of which the recording tips were separated by a distance of about 6.5 mm. Pulses of stimulating current were delivered between an intracellular metal electrode and a large external electrode. The temporal separation between the rising phases of the two action potentials recorded simultaneously was taken as a measure of the conduction velocity of the action potential.

When the external medium contained $CaCl_2$ and glycerol, but not Na-salt, both the amplitude and the conduction velocity of the action potential were small, being about 60 mV in amplitude and about 100 cm/s or less in velocity. It is seen in the figure that addition of NaCl to the external medium markedly enhanced both the amplitude and the velocity. In previous studies under similar experimental conditions [15] the action potential amplitude, E_a, was shown to increase roughly in accordance with the following rule described by Hodgkin and Katz [12]

$$E_a = \frac{RT}{F} \ln [Na]_e + \text{constant} \tag{2}$$

in the range of external Na-ion concentration, $[Na]_e$, is much higher than $[Ca]_e$. It was argued on preceding pages (p. 219) that there is, in the excited state of the membrane, a large increase in the selectivity for univalent cations of hydrophilic nature. Under these circumstances, the Donnan phase-boundary potential at the external surface of the membrane is expected to vary in accordance with the equation cited above under the condition that $[Ca]_e \ll [Na]_e$.

The membrane resistance is affected strongly by addition of Na-ions to the external medium. In the medium containing only $CaCl_2$, the membrane resistance of the axon is roughly 5-times (or slightly more) as high as that of the normal axon in sea water. When 400 mM NaCl is added to the 100 mM $CaCl_2$ solution, the membrane resistance becomes close to the normal level of the membrane resistance $(1–2\ K\Omega \cdot cm^2)$. This fact is consistent with the view that the membrane resistance at rest is determined by the divalent-univalent cation-concentration ratio af the negatively charged sites within the membrane. At the peak of nerve excitation, the membrane resistance of these axons falls to a low level. The extent of this fall approaches that of a normal axon when NaCl is added to the external medium. The observed increase in the conduction velocity with $[Na]_e$ is caused by the enhancement of both the action potential amplitude and the membrane conductance at the peak of excitation.

The effect of K-ion introduced into the axon interior is shown in Figure 8. The external medium of the axon studied was a mixture of $CaCl_2$ and NaCl solutions. The

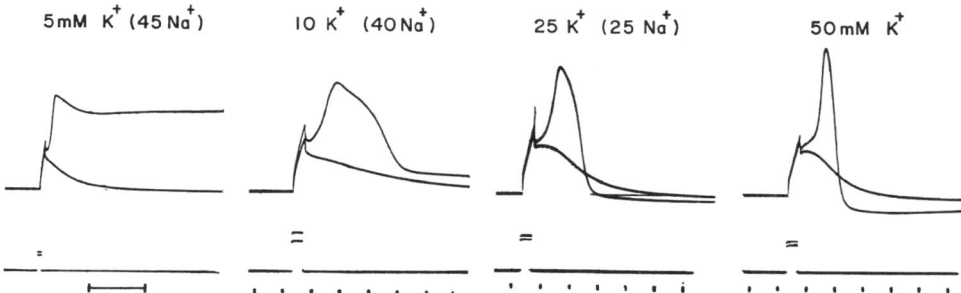

Fig. 8. Oscillograph records showing the effect of replacing the internal Na-ion with K-ion on the duration of the action potential. The sum of Na- and K-ion concentration was kept at 50 mM. The external medium contained both CaCl₂ and NaCl. The bar represents 10 ms; the dots indicate intervals of 1 ms. (From Tasaki: *Amer. J. Physiol.* **216**, 135 (1969).)

internal K-ion concentration was varied in a wide range, keeping the sum of Na- and K-ion concentrations at a constant level. It is seen in the figure that the action potential duration was markedly reduced by addition of K-ion to the fluid medium inside the axon. There was a slight increase in the action potential amplitude when the internal Na-ion was replaced with K-ion; however, this effect on the action potential amplitude was far smaller than what is expected from the Nernst equation applied to the internal Na-ions. As has been stated earlier (p. 215), the resting potential of the axon under these conditions remained practically *unaffected* by a large change in the Na-K ion-concentration ratio in the axon interior.

The effect of the internal K-ion on the action potential duration may be interpreted in the following manner. At the peak of excitation, the membrane conductance increases enormously (see Cole and Curtis [40]). This increase in the membrane conductance is accompanied by an increase in the interdiffusion fluxes of cations across the membrane. – Note that the ion fluxes across a membrane in the absence of current is related to the membrane conductance by virtue of the Nernst-Einstein relationship (see Gottlieb and Sollner [41], Kobatake and Tasaki [42]). – The interdiffusion involving Na- and K-ion is far more intense than that involving Ca- and Na-ion because of high intramembrane mobilities of these univalent cations. This enhanced cation-interdiffusion rapidly raises the Ca- and Na-ion concentration near the inner surface of the membrane and the K-ion concentration at the outer surface. [Note that the external surface of an axon membrane cannot be stirred because of the presence of Schwann cells.] Due to the strong Ca-displacing power of the accumulating K-ions (see p. 220) and probably also due to some relaxation process in the macromolecules which may occur following a sudden conformational change, there is a progressive shift in the membrane potential. Measurements of the membrane impedance during these prolonged action potentials indicate that a large impedance-loss at the onset is followed by a gradual impedance-restoration. From this fact, it is expected that the interdiffusion flux gradually diminishes during the action potential. Finally, the membrane returns to the initial Ca-ion rich state.

From an electrophysiological point of view, the properties of an axon producing prolonged action potentials may appear quite 'abnormal' and have nothing to do with the properties of a 'normal' axon. It is true that the time-courses of the membrane currents observed by the voltage-clamp technique (see below) are so different from those obtained from intact axons that the mathematical formulae describing the currents in intact axons may not be applicable to the currents observed under these conditions. However, the effect of varying the internal Na-K ratio on the action potential is *reversible*. There is no sign of an irreversible change in the membrane macromolecules during these experiments. From the standpoint of an investigator interested in physicochemical properties of the membrane macromolecules, therefore, there is no reason to regard the behavior of the axons described above as being abnormal.

From an electrochemical point of view, analyses of transient, i.e., highly time-dependent, phenomena in membrane systems involving a large number of uni- and divalent ions are extremely difficult. During the excited (or depolarized) state of an axon which is internally perfused with a dilute Na-salt solution, the membrane system may be regarded as being in a quasi-stationary state. The approach adopted in the investigation described above is (1) to study the process of transition of the axon membrane from its resting (stationary) state to the quasi-stationary excited state, and later (2) to analyze the highly time-dependent processes. This approach may be considered to be less prone to misinterpretation than the one adopted by many electrophysiologists who use the voltage-clamp technique (see below) to analyze highly time-dependent membrane phenomena.

10. The Voltage-Clamp Technique

The voltage-clamp technique developed by Cole [43] and Hodgkin *et al.* [44] is extremely popular among electrophysiologists. This technique is to apply, by the use of an electronic feed-back device, a rectangular voltage-pulse across an excitable membrane and to determine the membrane current produced by the pulse as a function of time. A detailed mathematical analysis of the rapidly changing membrane current observed under these conditions is regarded by many electrophysiologists as the ultimate goal of their research. This popularity seems to derive mainly from the situation

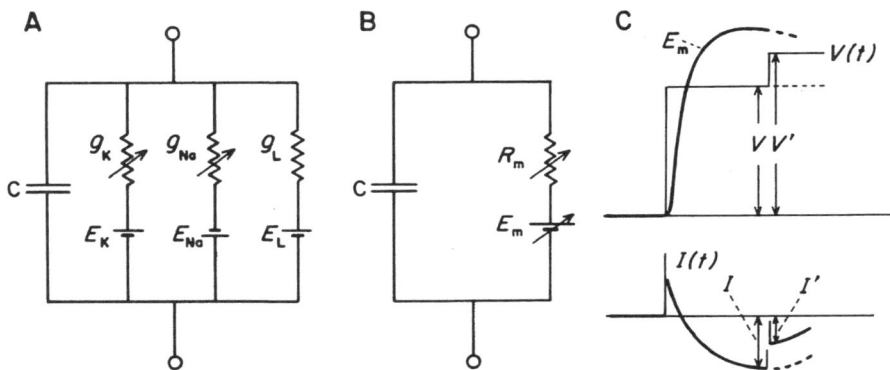

Fig. 9. Equivalent circuits for the axon membrane and the variation of the membrane emf, E_m, during voltage clamp. For further discussion, see text.

that many investigators in the field of electrophysiology regard biological membranes to be represented by an electric circuit shown in Figure 9A introduced by Hodgkin and Huxley [45].

In this electric circuit, the emf's of the batteries are given by the Nernst equations:

$$E_{Na} = \frac{RT}{F} \ln \frac{a''_{Na}}{a'_{Na}} \tag{3}$$

and

$$E_K = \frac{RT}{F} \ln \frac{a''_K}{a'_K} \tag{4}$$

where a'_j and a''_j are the activities of the ion-species j (either Na or K) in- and outside the axon and RT/F signifies the familiar electrochemical quantity (25.2 mV at 20 °C). The emf and the conductance of the 'leakage' pathway, E_L and g_L, are not well-defined and are introduced to account for the small current through a pathway with time-independent conductance. In this equivalent circuit, the flow of ions through each pathway of the circuit is independent of the processes taking place in other pathways, and the total current is obtained as the sum of the currents through individual pathways. All of these emf's are considered to be time-independent. The currents carrying Na-, K- and leakage-ions in the circuit are described by Ohm's law:

$$I_{Na} = g_{Na}(V - E_{Na}), \tag{5}$$

$$I_K = g_K(V - E_K) \tag{6}$$

and

$$I_L = g_L(V - E_L) \tag{7}$$

where V represents the potential difference between the two terminals of the circuit, and g's are the electric conductances of individual pathways. When the potential difference between the two terminals is shifted suddenly from the level of the resting potential to a different level, dV/dt is zero except for an extremely short period of time. It is assumed that both g_{Na} and g_K change continuously and relatively slowly under these conditions. To separate the total current into three components, Hodgkin and Huxley used the dependence of the peak value of the action potential on $[Na]_e$. They employed the following formulae to describe the observed conductances:

$$g_{Na} = \bar{g}_{Na}m^3h, \tag{8}$$

$$dm/dt = \alpha(1 - m) - \beta m, \tag{9}$$

$$dh/dt = \alpha'(1 - h) - \beta'h, \tag{10}$$

and a similar set of equations for g_K. In these empirical formulae, the effects of varying the levels of V on g's are included in the parameters $\bar{g}$, α's and β's. We see that many parameters were required to describe the observed membrane currents in terms of these equations. Finally, using these experimentally determined values of α's and β's, the action potential of an 'unclamped' axon was calculated. This calculation yielded a result which was in excellent agreement with the observed action potential.

The reason for this excellent agreement between the observed and calculated action potentials is now considered. From a purely electrical point of view, the circuit shown by diagram A is equivalent to that in diagram B, where E_m and R_m are given by

$$E_m = (g_{Na}E_{Na} + g_K E_K + g_K E_L)/(g_{Na} + g_K + g_L) \tag{11}$$

and

$$R_m = 1/(g_{Na} + g_K + g_L). \tag{12}$$

This can be shown easily by adding the three Equations (5), (6) and (7), which yields

$$I = (g_{Na} + g_K + g_L)\left[V - \frac{g_{Na}E_{Na} + g_K E_K + g_L E_L}{g_{Na} + g_K + g_L}\right] \tag{13}$$

where I is the sum of the currents in the three pathways, namely,

$$I = I_{Na} + I_K + I_L \tag{14}$$

Under voltage clamp, both the membrane emf, E_m, and the membrane resistance, R_m, change relatively slowly with time (diagram C). The membrane current observed under voltage-clamp is described by Ohm's law applied to this circuit, namely, by Equation (13) which can be rewritten as

$$I = (V - E_m)/R_m \tag{15}$$

When the voltage is shifted suddenly from one value, V, to another value V', the current changes from one level to another. This change can be treated approximately by regarding both E_m and R_m as remaining unaltered during the short period of time involved; $I' = (V' - E_m)/R_m$. Therefore, by measuring I' following a sudden shift from V to V', it is possible to determine the value of E_m at any time during voltage clamp; $E_m = (IV' - I'V)/(I - I')$.

When E_m was determined by this two-step procedure, it was found that this emf reproduces the early part of the action potential fairly accurately. The process of reconstruction of the action potential using the (one-step) voltage-clamp procedure is not fundamentally different from the process of determining E_m by the two-step procedure. Therefore, it is not surprising to find good agreement between the observed and the calculated action potentials.

The time-course of E_m can also be determined by a slightly different technique. When a weak sinusoidally varying voltage is superposed on a rectangular voltage pulse, V, the observed membrane current is found to consist of two parts, (i) a slowly changing part which is essentially the same as I observed in the absence of the A.C., and (ii) a sinusoidal part which can be taken as the measure of the conductance $1/R_m$ of the membrane. By introducing R_m measured by this A.C. method into Equation (15), the entire time-course of E_m can be determined. Again, the result of such a determination indicated that E_m reproduces the early part of the action potential accurately. The discrepancy between E_m and the late part of the action potential is attributed to the effect of the membrane current observed under voltage-clamp (see ref. [1] and [46]).

From these arguments, it follows that the success in reconstructing the action potential by the (one-step) voltage-clamp technique derives from the equivalence between electric circuits A and B in Figure 9. This success does not by itself attest to the soundness of the physico-chemical basis of the electric circuit shown in diagram A. Neither does it show which one of the equivalent circuits, A or B, represents the properties of the real membrane more adequately. It simply tells us that both E's and g's change slowly with time and that there is an approximately linear relationship between I and V (Ohm's law).

11. Physico-Chemical Significance of the Equivalent Circuit

The equivalent circuit under consideration is the one illustrated in diagram A of Figure 9. The emf's of the batteries in this circuit are given by Equations (3) and (4). The conductance g_{Na} is considered to be determined solely by Na-ions; similarly, g_K is assumed to be dependent only on K-ions and not on other ion species in the system. [This assumption is known as the independence principle of Hodgkin and Huxley (see ref. [47] and [48]).] The contribution of the 'leakage channel' is considered to be very small.

The properties of the equivalent circutit mentioned above are possessed by a mosaic membrane consisting of (i) a zone which is ideally selective for Na-ion, (ii) a zone which is ideally selective for K-ions, and (iii) a zone with a high electric resistance $g_L \ll (g_{Na} + g_K)$. In the sodium theory of nerve excitation [45], the process of action potential production is explained by assuming that $g_K \gg g_{Na}$ in the resting state of the nerve membrane and $g_K \ll g_{Na}$ at the peak of excitation. The falling phase of the action potential is attributed to a relatively slow rise in g_K. The conductances are considered to be 'voltage-dependent'.

Now a question arises: Does the squid axon membrane actually have all these properties of the equivalent circuit? At present we have abundant experimental evidence indicating that the electrochemical properties of an internally perfused squid giant axon are far more complex than those of a mosaic ion-exchanger membrane. The major experimental findings showing this aspect of an axon membrane are as follows:

(i) The membrane potential of a resting, excitable axon is very insensitive to a large variation in $[K]_e$ or $[K]_i$ (p. 215).

(ii) The action potential varies directly with $[Na]_e$ when sodium is the only major cation species outside the axon; however, many other cations can substitute Na-ion externally. The dependence of the action potential amplitude on $[Na]_i$ is far smaller than the level expected from equations (3) (see p. 134 ref. [15]).

(iii) The voltage-dependence of g_K is affected by both $[Ca]_e$ and $[Na]_e$. Similarly, g_{Na} is influenced by the level of $[Ca]_e$, probably by $[K]_e$ and by intracellular ionic composition [49, 50, 51].

(iv) The transport of Ca-ion is increased tremendously at the onset as well as during the action potential [11, 52, 53].

(v) It has been argued that E_K (and E_{Na}) can not remain constant during nerve excitation because of an increased flux of cations (e.g., ref. [54] and [1]).

As a consequence of these and many other studies, the equivalent circuit in its original form is no longer accepted by most investigators in the field. However, it is assumed in almost every voltage-clamp study that the axon membrane can be represented by a somewhat analogous equivalent circuit. No one believes at present that each channel is specific for one cation species. Nevertheless, it is frequently assumed that two channels 'open' and 'close' in succession in the manner described by the mathematical expressions explained in the preceding section.

As to the question of whether or not these two channels occupy spatially distinct parts of the membrane, investigators seem to favor the view that this problem has not yet been solved. Moore *et al.* [18] use the term 'sodium channel' to signify the 'early channel' and 'potassium channel' to indicate the 'late channel'. They conclude that these two channels are *operationally distinct*. On this point, Frankenhaeuser [49] says (p. 103): "Some of the very basic properties of the specific permeabilities are still unclear although indirect evidence accumulates on these points. No finding decides whether sodium and potassium are moving through the same or through different sites. Calcium clearly affects both sodium and potassium systems."

It should be pointed out in this connection that there are investigators in the field who assume the existence of two kinds of *ion-pores* in the nerve membrane, one kind for sodium and the other for potassium (see e.g., Mullins [55]; Hille [56]; Armstrong [57]). In a recent article [56], Hille says: "The pore has a rectangular hole 3.1×5.1 A formed by a ring of oxygen atoms. This hole is both the pathway for ion flow and the selectivity filter which determines which ions can flow. The narrow part of the pore is further supposed to be very short and to bear a single negative charge. The charge attracts cations and repels anions." It seems quite seldom that a physical chemist working on well-defined ion-exchanger material can arrive at this sort of detailed description of the ion pathway. Nevertheless, because of this impressive description of the ion pathway, many electrophysiologists seem to prefer the concept of ion-pores to the macromolecular concept.

From an electrophysiologist's point of view, the macromolecular concept has a serious limitation in the sense that it does not give a simple interpretation of voltage-clamp data obtained under multi-ionic conditions. At the peak of an action potential where $I=0$ and $dV/dt=0$, the efflux of potassium (J_K) must be roughly equal to the influx of sodium (J_{Na}), regardless of the difference in mobilities (or 'permeabilities') of the two ions involved. [Note that the fluxes of divalent cations and of anions are very small.] When the membrane potential is slightly shifted from this level by the voltage-clamp technique, J_K is no longer equal to the magnitude of J_{Na} and a finite membrane current I given by

$$I = \{|J_{Na}| - |J_K|\} F$$

is observed. Under these conditions, it is difficult to determine the mobility-ratio (or 'permeability-ratio') of the two interdiffusing cations by measuring the voltage-current relationship by the clamping technique. Probably this difficulty is one of the main reasons which makes the macromolecular approach unpopular among voltage-clamp investigators. Most of the experimental findings presented in this article in support of the macromolecular concept are regarded by these investigators as too complex to be explained on the basis of the equivalent-circuit concept.

We may conclude our discussion on the equivalent circuit in the following manner. Mainly due to the lack of precise knowledge about the molecular organization of the nerve membrane, there is at present no widely accepted interpretation of the physico-chemical basis of the equivalent circuit. Electrophysiologists seem to favor the interpretations based on the assumption of the presence in the nerve membrane of ion-pores of a few Angstroms in diameter. Physico-chemically or bio-chemically oriented investigators prefer macromolecular interpretations* (see e.g., Kabotake et al. [58]; Changeux et al. [37]; Lehninger [59]; Neumann et al. [2]). In view of the fact that a variety of new techniques are now being applied to studies of the nerve membrane (see below), it seems possible that a better understanding of the molecular architecture of the nerve membrane will be reached in the near future.

12. Optical Studies of the Axonal Membrane

Since the discovery of the 'action current' in 1843 by Du Bois-Reymond [60], the advancement of neurophysiology has been closely linked with the development of sensitive galvanometers. Soon after the end of World War I, vacuum tubes and catho-de-ray oscillographs were introduced into the field of neurophysiology. Following the

---

*It may be pointed out in this connection that a mixed state of the axon membrane (see p. 224) can be represented by an equivalent circuit similar to that shown in diagram A. A variation in the fraction of the membrane in the excited state can be described by a parallel connection of two batteries with variable series resistances (see p. 135 in ref. [1]).

end of World War II, there was an increasing tendency among neurophysiologists to rely heavily on electronic equipment. One might say that neurophysiology has grown, in a sense, as a branch of applied electronics.

In the field outside the main stream of neurophysiology, however, there have been many attempts to find non-electrical signs of nerve excitation. The oldest and most successful experiment along this line is the demonstration of heat evolution accompanied by nerve stimulation [61]. More recently, Howart et al. [62] have shown that there is heat evolution at the onset and heat absorption at the end of an action potential. [Concerning enthalpy changes associated with nerve excitation, cf. p. 222.] However, the time resolution of a sensitive thermopile is not high enough to permit a close examination of the temporal relationship between the temperature changes and the action potential.

Quite recently, various optical methods have been applied to studies of the process of nerve excitation, yielding many interesting results. Particularly, the devices used to detect changes in light scattering [63, 64, 66], in birefringence [63, 65] and in extrinsic fluorescence [66, 67] have a high time resolution (better than about 0.1 m s) and therefore the time-courses of these optical signals (i.e., small changes associated with nerve excitation) can be compared directly with that of the action potential. The method of spin label EPR [68, 69] and infrared spectroscopy [70, 71, 72] seem to be promising. In addition, there are several unconfirmed experiments suggesting the applicability of other optical techniques [73, 74].

It is important to note that the results obtained by these optical techniques can not easily be assimilated at present to the framework of concepts established by traditional, electrical methods of investigation. As compared with electric methods, optical methods are not very sensitive. Furthermore, the information obtained by an optical method places special emphasis on a particular aspect of the membrane macromolecules, and this aspect may make little or no contribution to production of an electric potential difference across the membrane. This point may be illustrated by the following example.

ANS and TNS (anilino- and toluidinyl-naphthalene sulfonate) come under the category of fluorescent probes which are often called 'hydrophobic' probes. A change in the fluorescence intensity of these molecules may be brought about by (i) a change in the number of bound probe molecules, (ii) a change in the polarity of the microenvironment of the bound probe molecules, or (iii) a change in the number of colliding molecules or side-groups of macromolecules which act as quenchers. In an ionic membrane, a change in the membrane potential derives from an alteration of the phase-boundary potential and of the intramembrane diffusion potential. It is evident therefore that these two methods of recording signals of nerve excitation place emphasis upon very different features of the nerve membrane and that their sensitivities are very different in terms of the extent of molecular changes detectable.

In the case of 2,6-TNS, it was found possible to discriminate between several alternative processes contributing to observed fluorescence signals. When the incident light is polarized, a group of probe molecules with their absorption oscillator aligned in the

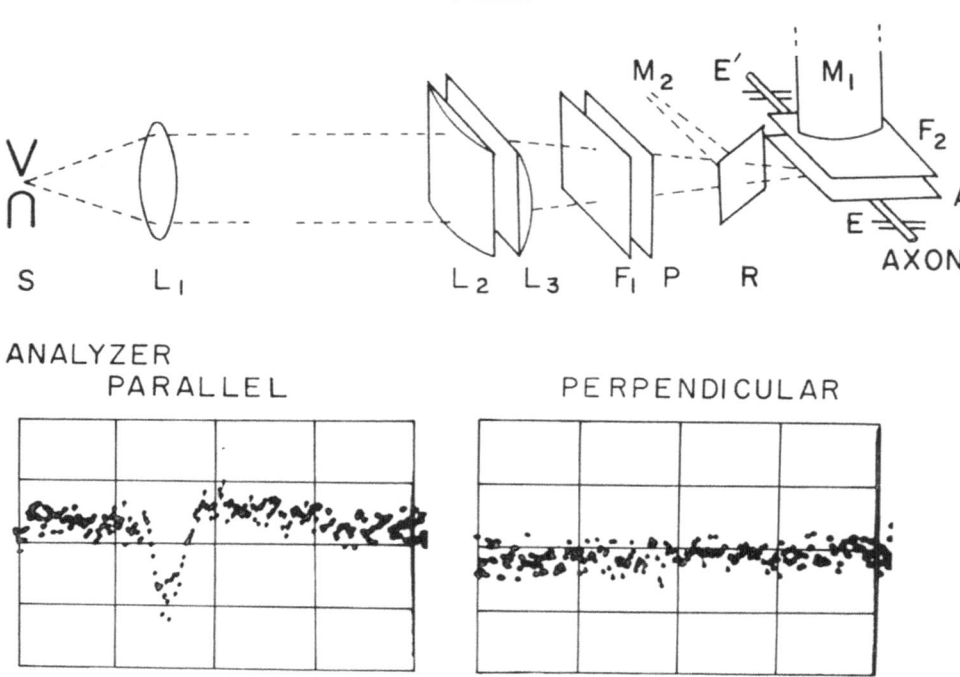

Fig. 10. *Top:* Experimental arrangement used to study fluorescence properties of a squid giant axon stained internally with 2, 6-TNS. *S*, a xenon-mercury lamp; L_1, a spherical quartz lens; L_2 and L_3, cylindrical lenses; F_1, primary filter for 365 nm; *P*, polarizer; *R*, quartz cover slip; *A*, Analyzer; F_2 secondary filter; *E*'s, electrodes; *M*'s photomultipliers. *Bottom:* Computer records showing the presence (left) and the absence (right) of a transient change in the fluorescent light intensity associated with nerve excitation. The directions of the polarizing axis of the analyzer (relative to the longitudinal axis of the axon) are indicated. (From Tasaki *et al.*: *Proc. Natl. Acad. Sci.* **68**, 939 (1971).)

direction of polarization are preferentially brought to the excited singlet state. The records furnished in Figure 10 were taken from a squid giant axon into which 2,6-TNS had been injected. The incident light was polarized in the direction of the long axis of the axon. As can be seen in the figure, the size of the fluorescence signal is very sensitive to the direction of the analyzer under these conditions. Only those probe molecules with their transition moment aligned roughly along the long axis of the axon contribute to production of the signal. A thorough examination of the spectrum of the light contributing to this fluorescence signal strongly suggests that the observed change in fluorescent light intensity is produced by a transient increase of the polarity of the micro-environment of the probe molecules [75].

From a biochemical point of view, a squid axon membrane is undoubtedly a heterogeneous system. In order to gain detailed information about this extremely thin and labile structure of the membrane, much more extensive studies are required using many different optical techniques.

References

1. Tasaki, I.: *Nerve Excitation: A Macromolecular Approach*, Charles C. Thomas, Springfield, Ill., 1968.
2. Neumann, E., Nachmansohn, D., and Katchalsky, A.: *Proc. Nat. Acad. Sci., U.S.A.*, **70** 727–731. (1973).
3. Hodgkin, A. L.: *The Conduction of the Nervous Impulse*, Liverpool University Press, Liverpool, 1964.
4. Arhem, P. and Frankenhaeuser, B.: *Acta Physiol. Scan.* **87**, 7A–8A (1973).
5. Stämpfli, R.: *Ann. New York Acad. Sci.*, **81**, 265 (1959).
6. Segal, J. R.: *Nature* **182**, 1370 (1958).
7. Tasaki, I.: *J. Physiol. (London)*, **148**, 306 (1959).
8. Tasaki, I. and Takenaka, T.: *Proc. Nat. Acad. Sci, U.S.A.*, **50**, 619 (1963).
9. Teorell, T.: *Proc. Soc. Exp. Biol. Med.* **33**, 283 (1935).
10. Meyer, H. and Sievers, M.: *Helv. Chim. Acta* **19**, 649 (1936).
11. Tasaki, I., Watanabe, A., and Lerman, L.: *Amer. J. Physiol.* **213**, 1465 (1967).
12. Hodgkin, A. L. and Katz, B.: *J. Physiol. (London)* **108**, 37 (1949).
13. Lorente de No, R., Vidal, F., and Larramendi, L. M. H.: *Nature (London)* **179**, 737 (1957).
14. Tasaki, I., Singer, I., and Watanabe, A.: *Amer. J. Physiol.* **211**, 746 (1966).
15. Tasaki, I., Lerman, L. and Watanabe, A.: *Amer. J. Physiol.* **216**, 130–138 (1969).
16. Narahashi, T., Moore, J. W., and Scott, W. R.: *J. Gen. Physiol.* **47**, 965 (1964).
17. Teorell, T.: in *Handbook of Sensory Physiology* (ed. by W. R. Loewenstein), Springer-Verlag, Berlin, Heidelberg, New York, 1972, p. 332.
18. Moore, J. W., Narahashi, T., and Anderson, N. C.: *Science* **157**, 220 (1967).
19. Tasaki, I., Singer, I., and Takenaka, T.: *J. Gen. Physiol.* **48**, 1095 (1965).
20. Inoue, I., Kobatake, Y., and Tasaki, I.: *Biochim. Biophys. Acta* **307**, 471 (1973).
21. Tasaki, I., Takenaka, T., and Yamagishi, S.: *Amer. J. Physiol.* **215**, 152 (1968).
22. Hill, S. E.: *J. Gen. Physiol.* **18**, 357 (1934).
23. Thorhaug, A.: *Biochim. Biophys. Acta* **225**, 151 (1971).
24. Tasaki, I.: *J. Gen. Physiol.* **46**, 755 (1963).
25. Kielland, J.: *J. Soc. Chem. Ind.* (London) **54**, 232T (1935).
26. Barrer, R. M. and Falconer, J. D.: *Proc. Roy. Soc. (London)* **A236**, 227 (1956).
27. Coleman, N. T.: *Soil Sci.* **74**, 115 (1952).
28. Flett, D. S. and Meares, P.: *Trans. Faraday Soc.* **62**, 1469 (1966).
29. Boyd, G. E.: in *Ion Exchange in the Process Industries*, Proc. Internat. Conf. Soc. Chem. Industry, 1970, p. 261.
30. Sherry, H. A.: *J. Phys. Chem.* **72**, 4086 (1968).
31. Tasaki, I.: *J. Gen. Physiol.* **39**, 377 (1956).
32. Callen, H.: in Donnelly, R. J., Herman, R. and Prigogine, I. (eds.), *Thermodynamic Fluctuations in Non-Equilibrium Thermodynamics*. Univ. of Chicago Press, Chicago, Illinois, 1966.
33. Kobatake, Y.: *Physica* **48**, 301 (1970).
34. Glansdorff, P. and Prigogine, I.: *Thermodynamic Theory of Structure, Stability and Fluctuations*, Wiley-Interscience, London, 1971.
35. Franck, U. F.: *Progr. Biophys.* **6**, 171 (1955).
36. Inoue, I., Tasaki, I., and Kobatake, Y.: *Biophys. Chem.* **2**, 116 (1974).
37. Changeux, J. P., Thiery, J., Tung, Y., and Kittel, C.: *Proc. Natl. Acad. Sci. U.S.A.* **57**, 335 (1967).
38. Kedem, O. and Katchalsky, A.: *Trans. Farady Soc.* **59**, 1931 (1963).
39. Caplan, R.: this volume, p. 89.
40. Cole, K. S. and Curtis, H. J.: *J. Gen. Physicol.* **22**, 649 (1939).
41. Gottlieb, M. H. and Sollner, K.: *Biophys. J.* **8**, 515 (1968).
42. Kobatake, Y. and Tasaki, I.: in *Nerve Excitation* Charles C. Thomas, Springfield, Ill., 1968, p. 181.
43. Cole, K. S.: *Arch. Sci. Physiol. (Paris)* **3**, 253 (1949).
44. Hodgkin, A. L., Huxley, A. F., and Katz, B.: *J. Physiol. (London)* **116**, 424 (1952).
45. Hodgkin, A. L. and Huxley, A. F.: *J. Physiol. (London)* **117**, 500 (1952).
46. Carnay, L. D. and Tasaki, I.: in W. J. Adelman, Jr. (ed.), *Biophysics and Physiology of Excitable Membranes*, Van Nostrand Reinhold Co., New York, 1971.
47. Hodgkin, A. L. and Huxley, A. F.: *J. Physiol. (London)* **116**, 449 (1952).

48. Frankenhaeuser, B.: *J. Physiol.* (*London*) **137**, 245 (1957).
49. Frankenhaeuser, B.: *Progr. Biophys. Molec. Biol.* **3**, (ed. by J. A. V. Buler and D. Noble) **99**, Pergamon Press, New York, 1968, p. 99.
50. Brismar, T.: *Acta Physiol. Scand.* **87**, 474 (1973).
51. Moore, J. W., Narahashi, T., and Ulbricht, W.: *Fed. Proc.* **22**, 174 (1963).
52. Baker, P. F., Hodgkin, A. L., and Ridgway, E. B.: *J. Physiol.* (London) **218**, 709 (1971).
53. Hallett, M. and Carbone, E.: *J. Cell. Physiol.* **80**, 219 (1972).
54. Adelman, W. J. Jr. and Palti, Y.: in *Current Topics in Membranes and Transport* **3**, (ed. by F. Bronner and A. Kleinzeller), Academic Press, New York, 1972, p. 199.
55. Mullins, L. J.: *J. Gen. Physiol.* **42**, 1013 (1959).
56. Hille, B.: *J. Gen. Physiol.* **59**, 637 (1972).
57. Armstrong, C. M.: *J. Gen. Physiol.* **54**, 553 (1969).
58. Kobatake, Y., Tasaki, I., and Watanabe, A.: *Adv. in Biophys.* **2**, 1 (1971).
59. Lehninger, A. L.: *Proc. Natl. Acad. Sci. U.S.A.* **60**, 1069 (1969).
60. Du Bois-Reymond, E.: *Ann. Physik. Chem.* **58**, 1 (1843).
61. Downing, A. C., Gerard, R. W., and Hill, A. V.: *Proc. Roy. Soc.* [*B*] **100**, 223 (1926).
62. Howarth, J. V., Keynes, R. D., and Ritchie, J. M.: *J. Physiol.* (*London*) **194**, 745 (1968).
63. Cohen, L. B., Keynes, R. D., and Hille, B.: *Nature* (*London*) **218**, 438 (1968).
64. Cohen, L. B., Keynes, R. D., and Landowne, D.: *J. Physiol.* (*London*) **224**, 727 (1972).
65. Cohen, L. B., Hille, B., Keynes, R. D., Landowne, D., and Rojas, E.: *J. Physiol.* (*London*) **218**, 205 (1971).
66. Tasaki, I., Watanabe, A., Sandlin, R., and Carnay, L.: *Proc. Natl. Acad. Sci. U.S.A.* **61**, 883 (1968).
67. Tasaki, I., Watanabe, A., and Hallett, M.: *Proc. Natl. Acad. Sci. U.S.A.* **68**, 938 (1971).
68. Hubbell, W. L. and McConnell, H. M.: *Proc. Natl. Acad. Sci. U.S.A.* **61**, 12 (1968).
69. Calvin, M., Wang, H. H., Entine, G., Gill, D., Ferrut, P., Harpold, M. A., and Klein, M. P.: *Proc. Natl. Acad. Sci. U.S.A.* **63**, 1 (1969).
70. Fraser, A. and Frey, A. H.: *Biophys. J.* **8**, 731 (1968).
71. Sherebrin, M. H., McClement, B. A. E., and Franko, A. J.: *Biophys. J.* **12**, 977 (1972).
72. Papakostidis, G., Zundel, G., and Mehl, E.: *Biochim. Biophys. Acta* **288**, 277. (1972).
73. Ungar, G., Aschheim, E., Psychoyos, S., and Romano, D. V.: *J. Gen. Physiol.* **40**, 635 (1957).
74. Makarov, P. O. and Krasovitskaya, M. V.: *Biophys. J.* **15**, 515 (1970).
75. Tasaki, I., Carbone, E., Sisco, K., and Singer, I.: *Biochim. Biophys. Acta.* **323**, 220 (1973).

INDEX

Acetylcholine 121
Action potential 193, 208, 213, 238
Active transport X, 68, 70
Active uptake 68
Activity of salt 89, 98
Affinities 49
Ag^+ 117–19
Alkali cation selectivity 109
Amide
  carbonyls 112
  solvents 112
Amphiphilic polyelectrolytes 138
Amplitude
  of the fluctuating potential 195
  of each mode 195
ANS 197, 235
Asymmetries X
  physical origin of 111
Asymmetrical interaction energies 111
ATP 70

Backbone RNA 23
  conformations 30
  hydrogen bonding in 24, 28, 31
  2′ OH group 29
Bacteria 193
Best fit 112
Bilayers 109
Bimolecular equilibrium reaction 146
Binding energy 109
Bioanalog compounds IX
Biological membranes 39, 63, 107, 210, 213, 238
  phenomena 120
Biomimetic phenomena IX
Biophysics IX
Brownian movement of small ions 193

Caplan, S. R. 89
Carbonyl 114
  ligands 113
Cardiac glucosides 70
Carrier 108, 109
  mobile 130
  transport 160
Cation
  and backbone conformation 29
  cross linking of fixed ions by 38
  exchange process 220
  exchangers 4, 5, 8, 9, 10, 216

influence on RNA structure 30
poly (A) Ca^{2+} salt 35
poly (U) Mg^{2+} salt 33
radius 110
specific effects of 33
23 S r. RNA structure due to 29
Cavity size 121
Cell membrane 193
Cellular death 81
Centre of mass velocity 54, 57
Channels 107
  early 233
  leakage 232
  potassium 233
  sodium 233
  specific for small ions 202
Charge density 61
Charged
  capillary 51
  colloidal particles 198
  slit 51, 59
Chemical potential 46
Complexes
  absorption spectra 5, 7, 11, 12, 13, 19
  analysis 6
  cobalt 3
  bromo 13
  chloro 3
  configuration of 6, 7
  copper 5
  donor-acceptor 183
  infrared spectra 11, 12
  preparation of 7
  nitrate 7
  non-aqueous solutions 8, 10
  sulfate 13
  thiocyanate 13, 18
  uranyl 13
Complexers and selectivity 109
Configuration (complexes) 6
Conductance 57, 58, 68
Contractions (of heart, smooth muscles) 208
Convective velocity 55
Coordination number
  cobalt 7
  uranyl 13
Copolymers
  of aziridinylethylanthraquinone 179

of 2-isopropenylanthraquinone 176
of maleic acid and cetylvinylether 181
of styrene and vinylbenzenesulfonate 50
of vinylanthraquinones 176
of vinylpyrazoloquinones 178
Correlating transition 201
Correlation function
    of the fluctuating force 202
    of the fluctuating field 202
Counterions 199
    conductance 57, 58
    flux 57
    mobility 54
    monovalent 52, 56
Coupled flows 162
Coupling effect Na^+-Ca^{2+} 164
    of excitation contraction 208
    of vibrations 21
Cross linking
    by cations 38
    degree of 61
    in biological membranes 39
Cu^+ 117
Cyclic polyethers 109

Debye-Hückel
    approximation 196
    parameter 195
Dielectric constant 58, 195
Diffusion constants of ions 197
Dimethyl formamide 112
Dinactin 113
Dipole moments 113
Discontinuity 78
Discrete charge arrays 117
Dissipation function 45, 54, 92
Dissociation constants 132
Divinyl benzene 51
DNA
    base pairing 21
    secondary structure 21
    structure band 23
Domain model 55–57
Donnan
    equilibria 129
    phase-boundary potential 216
Double layer (electrical) 51, 60
Double helical structures 22
    (G+C) 33
    poly (A) 36

E. Coli 194
Eigen, M. 111
Eisenman, G. 111
Electric
    conductance 54, 57, 59
    current 46, 57

noise 203
steady state current 89
Electrical double layer 51, 58, 59, 60
    potential 194
Electrodes
    reversible 46
    silver-silver chloride 47
Electro-osmosis 207
Electron transfer polymers 173
Energy
    dependences 110
    equation 54
    for the active transport X, 70
Entropy
    equation 54
    of mixing 154
    production rate of 45
Enzymes 107
    reactions X
    systems 70
Equivalent circuit (nerve) 232
Ester carbonyls 112
Ether oxygens 112
Excess free energy 195
Excitability, electro hydraulic analog 208
Excitable membranes 208
Excitation-contraction coupling 208

Facilitated diffusion of ions 159
Field strength 120, 121
    theory 113
Fixed charges 51, 125, 207
    cross linking of by cations 38
Flow
    bulk 48
    charge 48
    convective 51, 54
    counterion 57
    coupled 48, 102
    diffusive 48
    electric current 49, 56, 92
    electro-osmotic 49
    heat 46, 48, 49
    local 54
    primary 49, 51
    reduced heat 46, 54
    reversal 52, 53
    salt 46, 49, 92
    sign convention for 46, 51, 54
    secondary 49, 51
    stationary 54
    thermo-osmotic volume 51
    total heat 46
    viscous 54
    volume 46, 49, 51, 56, 57, 92
Fluctuations
    field 193, 201

of discontinuous direction change 194
of environment 199
of force of the reaction 200
of potential differences 197
thermal 193
Formal 'analog' heart 205
nerve 232
Formamide 112
methyl 112
Fourier's series 194
Free energy
excess 195
of mixing 154
Frequency
of discontinuous direction change 194
modulation 206

(G+C) pair 33
Gibbs-Duhem equation 47

Heart
contraction 208
rhythmicity 208
Heat capacity for transport 59, 60
Heat of transport
ions 49, 54, 58, 59
solvent 49, 55, 57, 58
Heitner-Wirguin, C. כ
Helical structures
DNA 21
(G+C) 33
poly (A) with Ca²⁺ 36
poly (U) with Mg²⁺ 33
RNA mono helices 31
Helix pomatia (snail, heart) 210
Henderson-Hasselbach equation 141
Hexadecavalinomycin 108, 112, 117, 119,
121
Highly charged rod-like particle 199
Hodgkin-Huxley theory 203
Homogeneous ionic solution 194
Hydration 31, 39
number 147
Hydrogen bonds 24, 28
of 2′ OH groups with RNA 22
RNA backbone structure due to 22, 31

Ice, thermal conductivity 61
Individual ligands' 112
Ion
association 149
cross linking by 38
diffusion 83
exchangers 3, 4, 6, 7, 9, 10, 16–19
exchange equilibria 138
heat of transport 49, 54, 58, 59
hydration energy 109

ligand interaction energies 117
pair 125
selective 'sites' 109
selectivity 107, 109, 149
shape 119
solvation 111, 112
specificity 125
transport 158
Ion exchangers 4
dowex A-1, chelating 4, 7, 9
dowex 1, anion-exchange 3, 6, 10, 12, 19
size of functional group 18
solvation shell 19
dowex 50, cation-exchange 4, 5, 8, 10
IRC-50, weak 4, 5, 9
liquid anion exchanger 3, 158
selective sorption by 4
Ionic strength 197
Incongruent transport 163
Irreversible thermodynamics X, 68, 89
IR spectra 3, 11, 12, 15, 17, 21
structure investigations 21

Kedem, O. 125
Krasne, S. 107

Ligands 3–12, 109
Ling, G. N. 111
Lipid 126
Liquid membranes 3, 158
Lorimer, J. W. 45

Macrocyclic carriers 112
compounds 127
Macromolecules, structure of 21
Macroreticular resins 177
Macroscopic environmental condition 199
Manecke, G. 173
Mean-square
amplitude 195
dipole moment 198
Mechanoreceptor 193
Mechanical permeabilities 58
Membranes 63
arrays (parallel or series) 89
biological 39, 107
capacity 51
cation exchanger 216
cell 193
cellophane 50,61
cellulose 50
clay 51
collodion-based 50
copper (II) ferrocyanide 51, 61
current rectifier 103
excitable 40
isotropic 54

liquid 31, 126, 158
lipid bilayer 133
Millipore 132
mosaic 89, 224
nerve 213, 238
osmotic pressure difference 93
oscillator 206
phenolsulfonate 51
plasma 70
plant cell 65
plasticized 128
porous glass 50
potential 129, 193
salt-free 57
solubility 61
stacks 89
styrene-vinylbenzene-sulfonate copolymer
    50–52
transport 63
Mg⁺ ions
    influence on backbone structure 31
    influence on RNA structure 29
    poly (U) 33
Mobile carriers 130
Mode 195
Monactin 112
Mono helices with RNA 31
Multidentate ligands 112

Navier-Stokes equation 54, 55
Nernst equation 1X 70, 180
Nerve
    fiber 213
    membrane 213, 238
NH₄⁺ transport 117
Nonactin 108, 110, 113, 119
NMR investigations 28
Nyquist noise 203

2′ OH group of RNA
    and RNA backbone 31
    hydrogen bonding 23
Onsager relations 48, 57
Oosawa, F. 193
Optical studies (axone) 234
Oscillations X
    iron wire model 206
    membrane oscillator 206
    oil film model 206
Osmosis (electro) 207
Osmotic pressure 47, 92
Ouabain 71
Oxidation-reduction polymers 173

Paramecium 193
Pattern of selectivity 111
Pefferkorn, E. 137

Peltier,-effect,-heat 50
Permeability 125
    hydraulic (mechanical) 50, 54, 57, 58
    ratios 109, 111
    thermo-osmotic 53, 57, 59
Perry, M. 125
Phase transition 81
Phenomenological
    coefficients 48
    equations 48, 92
    higher-order coefficients 51
Phosphate groups 23, 30, 32, 36
Physical origin of asymmetries 111
Plant cell 64
    membrane 65
Plasmalemma 64
Plasma-membrane 70
Poisson-Boltzmann equation 57, 59
Poisson equation 55
Polarization of solvent 58, 60, 61
Poly (A)
    backbone structure 28
    double helix with Ca²⁺ 36
    double helix with protons 36
Polyatomic cations 121
    ions 117
Poly (C) 23, 28, 33
    backbone structure 28
Polyelectrolyte
    amphiphilic 138
    rods 55, 59
    solutions 199
Polyether 111, 117, 119
Poly (U) 28, 29
    backbone structure 28
    Mg²⁺ induced structure 33
    osmotic 92
Polystyrene sulfonic acid salts 39, 51
Pressure difference, osmotic 93
Propylene carbonate 119
Protozoa 193
Polymer
    conformation 151
    electron-transfer 173
    redox (electrochemical properties) 180
    oxidation-reduction 173
    of styrene 60
    of vinylanthraquinones 176
    of vinylhydroquinones 174
    of vinylhydroquinone sulfonamides 177
    of vinylnaphthoquinones 175
    of vinylpyrazoloquinones 177
Polysoas 139
Polystyrene 60
Potassium 68
    specific 125
Potential

action – 193, 208, 213, 238
chemical 46
difference 68
electrical 194
electrochemical, index 180
    mid-point 180
    redox 180
electrostatic 58, 59
Donan phase-boundary 216
membrane – 129, 193
resting 80
Potentiometric data (redox) 181, 184–86
slope 127
titration 139

Redox
    polymers 180
    properties 180
    resins 173
    systems: bivalent 180
        tetravalent 182
        hexavalent 185, 187
Reduction 187
Relaxation time 199
    of the fluctuating field 202
    of the mode 197
    of reaction 202
Resistance coefficients 92
Resting potential 80
Rhythmicity (heart, smooth muscles) 208
RNA
    backbone 23
    influence of cations on 31
    hydration 29, 31
    secondary structure of 21
    structure band 23

Secondary structure 21
    Ca^{2+} poly (A) 36
    cross linking of fixed ions 38
    DNA 21
    Mg^{2+} poly (U) 33
    RNA backbone 23
Selectivity 109, 119
    fingerprints 118, 119, 121
    of nonactin 110
    sequences 107, 111
    Sub-Ia, Supra Ia, symmetrical 110
Sélégny, E. XI
Short circuit current 68, 71
Simon and Morf 'sites' 111
SO_3-ions 39
Solubility, X, 126
Solvation 112
Soret effect 49
Spectra
    electron spin resonance 3, 7, 9, 17

infrared 3, 11, 12, 15, 17, 21
NMR 28
Raman 3
reflectance 3
ultraviolet and visible 5, 7, 9, 19
Spectroscopic parameters 4, 6, 7, 9–11
Spontaneous direction change 194
Stability constants 4
Stationary state 4
Stepwise formation 1
Steric
    constraints 112
    factors 120
Submicroscopic environmental fluctuation 202
Symmetry
    of nitrate, sulfate, uranyl 11, 13, 15

Tasaki, I. 213
Tautomeric amino-imine equilibrium 36
Temperature
    dependency 68
    difference 45
    effect 63
    gradients X
Teorell, T. 205
Theories of ion-selectivity 120
Thermal
    conductivity 48, 60
    diffusion 49
    fluctuation 193
    of composite media 61
    potential 49
Thermocell, 'class 1' 48
Thermodynamic
    equilibrium constant 145
    non-equilibrium, X 68, 89
Thermo-osmosis 49, 59, 60
    osmotic permeability 53, 57, 59
    osmotic pressure (coefficient) 50
    osmotic volume flow 51, 52
Thorhaug, A. 63
TNS 235
Tonoplast 64
Trajectory of E. Coli 194
Transition zones 80
Transport number 125
    of cations 100
    of water 81
Trinactin 113
Triple helix Mg^{2+}-poly (U) 36
'Turgor' pressure 169

Valinomycin 108, 112, 119, 121, 125
Valonia 61, 68
    macrophysa, utricularis, ventricosa 68
Varoqui, R. 137
Vinylquinones 173

Visco-elasticity 208
'Vital force' (receding) IX

Water transport 77, 81

X-ray crystal data 119

Zundel, G. 21